China's Outbound Tourism

The People's Republic of China has changed from a country that discouraged tourism as a useless bourgeois activity into one of the major source markets for international tourism. In excess of 30 million Chinese travelled across the border in 2005, yet they are only the tip of the iceberg of the Chinese who have acquired the taste, money and freedom for international travel.

China's Outbound Tourism is the first book written about this major development using a multitude of sources from China and around the world. The topic is approached from many different angles, using methods from the fields of economics, political sciences, sociology, semiotics and cross-cultural studies. The book explains the economic and social background of the surge in tourism in China and the changes in tourism policy in China since 1949, which moved from prevention through controlled development to possibly encouragement of outbound travels of Chinese citizens.

Throughout the book facts and figures are given for the global development as well as in-depth information about major destinations of China's outbound tourism on all continents. The growing importance of tourists from China is however not just a question of quantity: the text explains the features that distinguish the travel motivations and behaviours of Chinese tourists from 'western' as well as Japanese tourists and the consequences for product adaptation and marketing methods for destinations interested to attract and satisfy Chinese visitors.

This ground-breaking and valuable book offers fresh insight into the fact that already one-quarter of all international tourists are coming from non-western countries and mainly from Asia. *China's Outbound Tourism* indicates some of the future lines of development in this area before concluding with a look into the future of tourism.

Wolfgang Georg Arlt has more than 25 years experience in Chinese–European business and tourism relations. As a trained sinologue, he was the co-owner of an Incoming Agency for Chinese experts and tourists visiting Europe with offices in Beijing and Berlin during the 1990s.

He is professor of Leisure and Tourism Economy at the University of Applied Sciences in Stralsund, Germany and visiting professor of several universities in China. He is also the president of the China Outbound Tourism Research Association, organizer of the UNWTO-sponsored European Chinese Tourists' Welcoming Award, which is awarded every year to European destinations and companies, and a board member of the European Leisure and Recreation Association.

Contemporary Geographies of Leisure, Tourism and Mobility
Series Editor: C. Michael Hall
Professor at the Department of Tourism, University of Otago, New Zealand

The aim of this series is to explore and communicate the intersections and relationships between leisure, tourism and human mobility within the social sciences.

It will incorporate both traditional and new perspectives on leisure and tourism from contemporary geography, e.g. notions of identity, representation and culture, while also providing for perspectives from cognate areas such as anthropology, cultural studies, gastronomy and food studies, marketing, policy studies and political economy, regional and urban planning, and sociology, within the development of an integrated field of leisure and tourism studies.

Also, increasingly, tourism and leisure are regarded as steps in a continuum of human mobility. Inclusion of mobility in the series offers the prospect to examine the relationship between tourism and migration, the sojourner, educational travel, and second home and retirement travel phenomena.

The series comprises two strands:

Contemporary Geographies of Leisure, Tourism and Mobility aims to address the needs of students and academics, and the titles will be published in hardback and paperback. Titles include:

The Moralisation of Tourism
Sun, sand . . . and saving the world?
Jim Butcher

The Ethics of Tourism Development
Mick Smith and Rosaleen Duffy

Tourism in the Caribbean
Trends, development, prospects
Edited by David Timothy Duval

Qualitative Research in Tourism
Ontologies, epistemologies and methodologies
Edited by Jenny Phillimore and Lisa Goodson

The Media and the Tourist Imagination
Converging cultures
Edited by David Crouch, Rhona Jackson and Felix Thompson

Tourism and Global Environmental Change
Ecological, social, economic and political interrelationships
Edited by Stefan Gössling and C. Michael Hall

Routledge Studies in Contemporary Geographies of Leisure, Tourism and Mobility is a forum for innovative new research intended for research students and academics, and the titles will be available in hardback only. Titles include:

Living with Tourism
Negotiating identities in a Turkish village
Hazel Tucker

Tourism, Diaspora and Space
Edited by Tim Coles and Dallen J. Timothy

Tourism and Postcolonialism
Contested discourses, identities and representations
Edited by C. Michael Hall and Hazel Tucker

Tourism, Religion and Spiritual Journeys
Edited by Dallen J. Timothy and Daniel H. Olsen

Tourism, Power and Space
Andrew Church and Tim Coles

China's Outbound Tourism
Wolfgang Georg Arlt

China's Outbound Tourism

Wolfgang Georg Arlt

Routledge
Taylor & Francis Group

LONDON AND NEW YORK

First published 2006
by Routledge
2 Park Square, Milton Park, Abingdon, Oxon OX14 4RN

Simultaneously published in the USA and Canada
by Routledge
270 Madison Ave, New York, NY 10016

Routledge is an imprint of the Taylor & Francis Group, an informa business

© 2006 Wolfgang Georg Arlt

Typeset in Galliard by
Florence Production Ltd, Stoodleigh, Devon
Printed and bound in Great Britain by
Antony Rowe, Chippenham, Wiltshire

British Library Cataloguing in Publication Data
A catalogue record for this book is available from the British Library

Library of Congress Cataloging in Publication Data
Arlt, Wolfgang Georg, 1957–
 China's outbound tourism / Wolfgang Georg Arlt.
  p. cm. – (Routledge studies in contemporary geographies of leisure,
  tourism, and mobility)
  Includes bibliographical references and index.
 1. Tourism – China. 2. Chinese – Travel. I. Title. II. Series.
 G155.C6A75 2006
 338.4′7910951 – dc22   2005034256

ISBN10: 0–415–36536–8 (hbk)
ISBN10: 0–203–96816–6 (ebk)

ISBN13: 978–0–415–36536–9 (hbk)
ISBN13: 978–0–203–96816–1 (ebk)

Contents

Illustrations

Tables

Figures

Preface

China has re-entered the world stage. When the author started his studies in sinology in 1975, Mao Zedong was ruling a closed and very poor country. When he visited the People's Republic of China for the first time in 1978, the policies of 'Reform and Opening' were just beginning to reshape the society and economy in unforeseeable thoroughness. Three decades on, China is the factory of the world, envied for its economic progress and feared for its unpredictability resulting from a halted political development.

In temporal parallel to China's progress, tourism, especially international mass tourism, has reshaped the world, pushing globalization and being pushed by it at the same time, moving towards the touristification of the world.

In the twentieth century the one-way traffic of international tourism mirrored more or less the former colonial divisions of the world in the roles of visitors and visited. The new century ushers in a two-way traffic with the non-Japanese Asian half of mankind entering the global tourism stage as tourists themselves. China and soon-to-follow India are, by sheer size of population, the two potential main players.

'China' and 'Tourism' have been the two foci around which the author's professional and academic life has been organized; fortunately they have now joined into a single focus: China's outbound tourism.

Acknowledgements

Chinese friends, business partners, customers, academic colleagues and unknown persons observed in Chinese cities and villages, they all have over the years taught me many things about China, its people and its changes as a necessary precondition for analysing the integration of outbound travel into the fabric of the Chinese modern culture.

More specifically Chinese tourism experts, among them Zhang Guangrui, Wang Xinjun, Bao Jigang, Wang Ning, Xu Jing and others helped me a lot in my quest to understand the background and policies of China's outbound tourism. Colleagues at the Association for Tourism and Leisure Education (ATLAS), ELRA and at the University of Brighton, namely Peter Burns, gave inspiration and support.

My students and former students at universities in Europe and China engaged in the China Outbound Tourism Research Project and the experts, organizations and companies participating in the ECTW Award also supported my work with ideas as well as with practical help. Kevin Hannam and Alan Crosier kindly polished my non-native English and provided many useful hints. Feng Gequn and Tong Xin provided their assistance with some Chinese language materials, as did Marcel Kelemen, Elias Indrich and Alexander Schuler with some graphics.

The staff members at Routledge publishers, especially Zoe Kruze, and Bethany Dymond and Sarah-Jane Fry of Florence Production, were always accessible and helpful, as was Michael Hall, the editor of this series.

The acknowledgements certainly have to include the unfaltering support of my parents and my wife Judith, whom I first met in China. This book is dedicated to her.

Abbreviations

ADS	Approved Destination Status
APEC	Asia Pacific Economic Council
APB	Asia Pacific Bulletin
ASEAN	Association of South East Asian Nations
ATC	Australian Tourism Commission
ATEC	Australian Tourism Export Council
ATLAS	Association for Tourism and Leisure Education
BAT	British American Tobacco
BEM	Big Emerging Market
BHG	Bayrischer Hotel- und Gaststättenverband (Bavarian Hotel and Restaurant Association)
BISU	Beijing International Studies University
BITE	Beijing International Tourism Expo
BITTM	Beijing International Travel and Tourism Market
BTI	Business Travel International
CAAC	Civil Aviation Administration of China
CAS	Chinese Academy of Sciences
CASS	Chinese Academy of Social Sciences
CATAC	China Association of Tourist Automobiles and Cruises
CATS	China Association of Travel Services
CBITM	China Business and Incentive Travel Mart
CBS	Central Bureau of Statistics (Israel)
CCP	Chinese Communist Party
CCPH	China Cartographic Publishing House
CCT	China Comfort Travel
CCTV	China Central Television
CEN	China Economic Net
CEPA	Closer Economic Partnership Arrangement
CITM	China International Travel Mart
CITS	China International Travel Service
CMT	China Merchants Travel
CNMI	Commonwealth of Northern Marianas
CNNIC	China Internet Network Information Center

CNTA	China National Tourism Administration
CRI	China Radio International
CTA	China Tourism Association
CTC	Canadian Tourism Commission
CTHA	China Tourist Hotels Association
CTS	China Travel Service
CWTS	China Women Travel Service
CYTS	China Youth Travel Service
DMO	Destination Marketing Organization
DPS	Destination Promotion System
DSEC	Direccao dos Servicos de Estatistica e Censos Macao (Macao Statistical and Census Office)
DTA	Domestic Travel Agency
DWIF	Deutsches Wirtschaftswissenschaftliches Institut für Fremdenverkehr
DZT	Deutsche Zentrale für Tourismus (German National Tourist Board)
ECRS	Education, Cultural and Recreational Services
ECTW	European Chinese Tourists Welcoming (Award)
EIU	Economist Intelligence Unit
ELRA	European Leisure and Recreation Association
ETC	European Travel Commission
ETOA	European Travel Organizations Association
EU	European Union
EUCCC	European Union Chamber of Commerce in China
FEC	Foreign Exchange Certificate
FDI	Foreign Direct Investment
FIT	Foreign Individual Traveller
GATS	General Agreement on Trade in Services
GATT	General Agreement on Trade and Tariffs
GDP	Gross Domestic Product
GDR	German Democratic Republic
GITF	Guangzhou International Tourism Fair
GTA	Gullivers Travel Associates
HKTB	Hong Kong Tourism Board
HKTDC	Hong Kong Trade Development Council
HTA	Hawaiian Tourism Authority
ICT	Information and Communication Technology
ITA	International Travel Agency
ITE	International Travel Expo (Hong Kong)
IVS	Individual Visitors Scheme
KNTO	Korea National Tourism Organization
MEK	Matkailun Edistämiskeskus (Finnish Tourist Board)
MFA	Ministry of Foreign Affairs
MICE	Meetings, Incentives, Conferences and Exhibitions

MOFCOM	Ministry of Commerce
MOU	Memorandum of Understanding
MPS	Ministry of Public Security
NBSC	National Bureau of Statistics of China
NGO	Non-Governmental Organization
NTO	National Tourism Organizations
OL	Office Lady
OTTI	Office of Travel and Tourism Industries (US Department of Commerce)
PATA	Pacific Asia Travel Association
PDA	Personal Digital Assistant
PPP	Purchasing Power Parity
ReBAM	Recreational Belt Around Metropolis
RMB	Renminbi
ROCTB	Republic of China Tourism Bureau
SAR	Special Administrative Region
SARS	Severe Acute Respiratory Syndrome
SPTO	South Pacific Tourism Organization
SAT	South Africa Tourism
STA	Swedish Tourism Authority
STB	Scandinavian Tourist Board
STV	Schweizer Tourismusverband (Swiss Tourism Association)
TCL	Travel Career Ladder
TCP	Travel Career Pattern
TBP	Travel Business Partnership
TFC	Tourism Forecast Council
TRCNZ	Tourism Research Centre New Zealand
UNDP	United Nations Development Programme
UNESCO	United Nations Educational, Scientific and Cultural Organization
UNWTO	United Nations World Tourism Organization
VFR	visiting friends and relatives
WFOE	wholly foreign-owned enterprise
WTF	World Travel Fair (Shanghai)
WTO	World Tourism Organization
WTO	World Trade Organization
WTTC	World Travel and Tourism Council

Introduction

The People's Republic of China, long regarded as the last frontier for the tourism industry, has been a latecomer in leisure travel and tourism. Such activities were in fact for decades ridiculed as a wasteful and potentially dangerous bourgeois activity, only to be employed as a tool of foreign politics. Since 1978 however, as a part of the biggest economic miracle heretofore witnessed by mankind, leisure travel, both domestic and outbound, has been embraced by the Chinese people with much greater enthusiasm than the Chinese government anticipated or wished for.

Within the framework of the policy of 'Reform and Opening', tourism as a business was regarded as inbound tourism from hard-currency countries only, with the goal to find an easy and fast way to earn such foreign currency needed to start the modernization process.

Domestic tourism, after some official hesitation, proved to be a useful way to ward off deflation by enticing consumers to spend: 'Farmers who used to save every coin earned, no longer hesitate to spend a few hundred yuan on a tour to "open their eyes"' (Zhang 1989: 55). In 2005, the number of leisure trips for the first time almost equalled the number of inhabitants, turning, at least statistically, every Chinese citizen into a tourist.

Private outbound travel, in the form of visits to Chinese friends and relatives in Hong Kong and Macao and some Southeast Asian countries, was approved as it could be employed by the Chinese government to strengthen the ties with ethnic Chinese living outside Mainland China.

Outbound tourism, leisure trips across the border, paid privately and not connected with visits to friends and relatives, however, was an activity officially ignored until the middle of the 1990s. Demand by the urban elites in China pushed the government step by step to open the door to the world beyond the realm of *Chineseness*.

It has to be kept in mind, however, that only from the beginning of the twenty-first century did outbound tourism involve more than 1 per cent of the citizens of the People's Republic of China per year. By 2006, this figure, which includes travels that already terminate in Hong Kong or Macao, is preparing to clear the 3 per cent hurdle.[1] With only limited transport capacities as a foreseeable hindrance to further growth, major global catastrophes aside, the mantra-like repeated 100 million Chinese outbound travellers prediction by the World Tourism Organization (WTO) for 2020 (WTO 2000) will probably turn out to have been too conservative.

As part of the development and as an agent for the change of the Chinese society, the importance of outbound tourism is bound to increase. Deng Xiaoping's quip against the dangers of contamination at the beginning of the 'Reform and Opening' policy, that 'When you open the window, fresh air comes in and also flies come in' is still popular today (Smith 1997). Whether the prophecy of Li Yiping, '[w]ith its policy of opening the country to the outside world, the regime itself may have unwittingly sown the seeds of its own destruction' (2004: 190), holds any truth or if, in reverse, the yielding to the consumeristic demands of the *nouveaux riches* for outward mobility helps to prolong the lifespan of the last major one-party government, remains to be seen.

If China is a latecomer to tourism, mass tourism itself is a rather recent development, intimately intertwined with the advancement of modernity and globalization. Travel for leisure purposes existed in all developed societies, from Mesopotamia to Imperial China, for a selected few within the respective borders and for an even smaller group of people in the form of border-crossing tourism. Consequently it is often repeated that 'travel for leisure dates back thousands of years' (Wen, Tisdell 2001: 16) and that 'tourist activities in China can be traced to ancient times' (Zhang, Pine, Lam 2005: 11).

This *elite* tourism, however, created almost no impacts on the economy, society, culture or environment of the regions travelled in or to. On the contrary, modern, or rather late modern (Wang 2000), forms of mobility are in the process of achieving no less than a *touristification* (Picard 1996) of our world.

From the industrial production of the division between order and chaos, nature and civilization, work and leisure, the *rationalization* in Max Weber's terms, a new quality of tourism (Urry 2000; Urry 2002) arises. Mirroring and accelerating the development of the world economy in the second half of the twentieth century, mass tourism and especially international mass tourism started in the countries of the former colonial powers in Europe, in North America and in Japan.

As commodification, customization, standardization and mobility, hallmarks of both tourism and globalization, permeate ever deeper in each and every corner of the world in the twenty-first century, so tourism now, in parallel with the flow of goods and information, starts to be a 'two-way-traffic' phenomenon. With a quarter of all international travellers already originating from cultures not based on occidental values, the simple picture of 'white' guests and 'brown' or indeed 'yellow' hosts is becoming blurred.

1 Framework of China's outbound tourism

China's outbound tourism development is a multi-faceted phenomenon. Anything but a multidisciplinary approach to this topic will end up like in the story of the group of blind scholars pawing an elephant, with each confidently announcing what the whole animal resembles on the basis of the specific part that each scholar is touching.[1] Only with the utilization of different perspectives will a more realistic picture emerge.

Tourism studies 'does not recognize disciplinary boundaries' (Jafari, Ritchie 1981: 22) and is therefore a field legitimately occupied by sociologists, economists, anthropologists, geographers, semioticists, scholars of cultural and cross-cultural studies and others. However, 'the political aspects of tourism are rarely discussed in the tourism literature' (Hall 1994a: 1), and more specifically '[s]eldom is tourism considered in terms of the political needs of those who wield power or of the government as a whole' (Richter 1989: 19).

In China, as will be demonstrated in this book, the *political needs of those who wield power* cannot be ignored or hushed up, as they constitute one of two main forces shaping the development of tourism in general and outbound tourism in particular, the other being the want of the Chinese people to travel. These two forces interact in much more complex ways than most papers written on the topic of outbound tourism policies in China recognize or admit.

Rather than the government granting degrees of freedom gradually and organizing 'well-managed, orderly and controlled systems of travel abroad' (Verhelst 2003: 31), the demand of the people to travel and their creative utilization of loopholes and cracks in the system pushed the Chinese government forward in allowing outbound travel in a number of steps.

The Chinese central governments' political approach to use tourism as a tool of China's foreign policy, especially to use it to strengthen relations to overseas Chinese, and the economic and financial approaches to use inbound tourism from western countries as, in the words of Deng Xiaoping's closest ally Chen Yun, 'the export of scenic spots, earning foreign exchange more quickly than the export of goods' (Zhang, Chong, Ap 1999), have been counteracted by the underground growth of domestic travel and outbound leisure travel by the Chinese people. Competing with overseas Chinese and western visitors for space at the attractions and the transport infrastructure inside China and cancelling out the currency import by spending private or even public hard currency outside China, they have successfully elbowed themselves into the guest position in tourism where a host position – or a non-paying guest situation – was prepared for them.

However, the Chinese population is concerned with tourism development in different ways. Only a small minority of the Chinese population is able to partake in outbound tourism activities. Both inbound and outbound tourism are closely related spatially to the coastal areas.[2] Forty-one per cent of the total population live here, but 80 per cent of inbound tourists and 88 per cent of revenue from inbound tourism is received by the coastal areas (Wen, Tisdell 2001: 78). The regional distribution of the place of abode of outbound tourists is almost equally concentrated in the coastal areas, with Beijing, Shanghai and Guangdong alone accounting for two-thirds of the revenue from inbound tourism and for half of all outbound travellers[3] (Wang 2003b).

Different estimates as to the number of Chinese citizens who can potentially afford a trip to a foreign country, i.e. beyond the Special Administrative Regions (SARs) of Hong Kong and Macao, have been put forward, ranging from 50 million (Arlt 2002a) to 65 million (Smith 2003) or even 80 million (Xu 2002). Given the continued speed of economic growth in China, the low prices resulting from price wars in the outbound market and the expected strengthening of the Chinese currency, these figures might even be considered as too low. Nevertheless, they represent well below 10 per cent of the total Chinese population.

Seen in absolute numbers, however, China's outbound tourism is of growing importance for almost anybody earning his or her livelihood directly or indirectly from tourism, customarily described as the world's biggest industry. In 2004, more than 8 million Chinese out of 29 million travelled not only beyond the borders of the People's Republic but also further than Hong Kong and Macao. The figure of these 'real' international travellers is bound to multiply, transporting China into the top ten of global tourism source markets within the next few years and into a position of one of, or even *the* most important source market for quite a few destinations especially in Asia.

In the opposite direction, the figures of international tourism into China are seriously inflated by including same-day visitors from Hong Kong and Macao. Out of the 98 million persons entering China in 2002, only 13 million foreigners and 23 million compatriots[4] actually stayed overnight (CNTA 2004a). In 2004, total arrivals grew to 109 million persons and the overnight visitors to 42 million[5] (CNTO 2005).

Research

In terms of previous research, texts about China and tourism published in the twentieth century often do not mention outbound tourism at all. Gerstlacher, Krieg and Sternfeld (1991) for instance, when presenting 'Historical Aspects' of tourism in China, do not consider travels out of China, but only think about journeys *to* China, not even mentioning the seafarer Zheng He or the monk Xuan Zang. Likewise, otherwise important articles such as those by Gao and Zhang (1983), Choy, Guan and Zhang (1986) and Zhang (1989) remain silent on the topic.

Xu (1999) in his detailed look at the tourism industry in China never even mentions outbound tourism. Wen and Tisdell (2001), in their *Tourism and China's Development*, only offer a one-page 'Note on Outbound Travel', in which they downplay the importance of outbound tourism. Even Lew *et al.* (2003) in the most important text currently available, *Tourism in China*, include only a chapter on outbound tourism to Hong Kong, written from a Hong Kong inbound perspective.

Research focused on China's outbound tourism outside proprietary studies by consulting companies has mainly been initiated by overseas National Tourism Organizations (NTOs) and Destination Marketing Organizations (DMOs) (Roth (1998) for Austria, Strizzi (2001) for Canada, STB (2002) for Scandinavia), or has been conducted with a focus on tourism to Hong Kong by the Hong Kong Polytechnic University (see for instance Zhang, Lam 1999; Zhang, Chow 2004; Zhang, Jenkins, Qu 2003).

In international academic journals, tourism to China and, to a lesser extent, domestic tourism have been but occasionally examined. On China's outbound tourism, mainly to specific countries, just a handful of papers have appeared since the 1990s (Arlt 2002a; Arlt 2004; Cai, Boger, O'Leary 1999; Chen 1998; Dou, Dou 2001; Jang, Yu, Pearson 2003; Junek, Binney, Deery 2004; Li, Bai, McCleary 1996; Laws, Pan, 2002; Ryan, Mo 2001; Yu, Weiler 2001; Zhang, Heung 2001; Zhang, Qu 1996; Zhao 1994), supplemented by region-specific PhD dissertations (Guo 2002; Verhelst 2003).

The joint study of the European Travel Commission (ETC) and the WTO in 2003 finally attempted the first comprehensive global perspective (WTO 2003). The International Forum on Chinese Outbound Tourism and Marketing, organized by the China Tourism Association (CTA) and hosted by China Comfort Travel (CCT) in the same year, provided the first opportunity for intense discussion between academics and the industry from both inside and outside China (Wang 2003a).

As a topic for undergraduate theses,[6] China's outbound tourism has become a 'hot' issue recently.

Articles published on outbound tourism in Chinese academic magazines[7] are seldom recognized in the international discussion and are also limited in number and scope, mirroring the main interest of official and private tourism sectors in China in inbound and domestic rather than outbound tourism. Many articles that do examine outbound tourism are focused more on giving advice for the development of better policies to regulate and control the chaotic market situation than on providing data and analysis. The lack of market research opportunities and published official information is reflected in the limited number of texts that contribute data on a more than local level (Guo 1994; Dai, Zhang 1997; Du *et al.* 2002; Long 2003; Project Research Team 2003; Sun 2003; Sun, Dong 2003; Xiao, Ren 2003; Xu, Chen 2003; Yu, Zhang, Ren 2003; Guo 2004; Du, Dai 2005). Still, the provision of these texts in English would enrich the international discussion and its dearth of detailed information.

The Yearbook of China Tourism Statistics, tellingly, still does not provide any tables on outbound tourism in the main body, just a one-page text in the introductory report[8] (CNTA 2001a–2004a).

The emerging questions of how to deal with the growing flood of outbound tourists and especially the hard-currency drain connected with it might bring more of this discussion to the forefront. The 2005 edition of the quasi-official *Green Book of China's Tourism* (Zhang, Wei, Liu 2005) for the first time included two major articles on outbound tourism (Zhang 2005; Dai 2005) and an invited article by a foreign researcher on this topic (Arlt 2005a). The Beijing International Studies University (BISU) published an *Annual Report of China Outbound Tourism Development 2004* (Du, Dai 2005). A new bilingual magazine *China Tourism Research*, launched by the Hong Kong Polytechnic University, might provide a platform for discussion, even so the Editorial in the first number does not mention outbound tourism at all (Song 2005). The first volume includes accordingly not a single text on outbound tourism.

Overall, studies on China's outbound tourism have been not only limited in number but also in scope. Texts from inside China necessarily discuss developments mainly from a government's point of view, even if reading between the lines yields some insights into the internal discussions going on. Most texts from Hong Kong – before and after 1997 – follow a similar line of what could be regarded as self-censorship. The majority of western authors writing about outbound tourism from China have to rely on these sources and seldom leave a narrow economistic- and marketing-orientated path. This dependence is reflected in an *en masse* adaptation of assertions, such as 'Travel agencies in China started to exist in 1923 only', 'There was no tourism in China in the 1930s and 1940s', 'There was no tourism but inbound tourism before 1978', 'Tourism starts when the average income per year tops 1,000 US$' and so forth, which despite their mantra-like repetition have little relation to reality.

Structure

This text is divided into two main parts: a presentation and analysis of the development of outbound tourism in China and a discussion of the practical experiences of different destinations and the consequences for tourism service providers as well as tourism academics. The closing chapter attempts to catch a glimpse of future developments. Emphasis is placed on the leisure travel activities, but journeys with the motives of visiting friends and relatives (VFR) and business travels are also examined.

The first part of the book starts with a look at the analytical framework under which China's outbound tourism can be studied. It provides some information about the social and economic development of the People's Republic of China with a focus on data relevant for outbound tourism as the background of the coming into existence of the necessary preconditions for international travel: beside sufficient levels of income and leisure time, the permissions to leave the home country and to enter the destination country plus the acquired consumer wish to spend time and money for travelling abroad (Chapter 2).

As tourism in and out of China is still strongly controlled and instrumentalized politically by the government, it is necessary to look closely at China's tourism policies. Given the missing clear separation of inbound, domestic and outbound policies, or rather their expressively stated connection, a complex treatment is necessary. With a government not answerable to electorates, a look at the real travel wishes of the people mirrored in their strategies to fulfil their demand is also necessary if only possible by indirect means. Both views are presented in Chapter 3. Whereas political sciences inform the third chapter, the quantitative aspects of outbound tourism development in China are looked at from an economist's point of view, supplemented by a sketch of the development of outbound tourism in Hong Kong and Taiwan, in Chapter 4. The following fifth chapter scrutinizes the sociological particularities of the tourists themselves, the roots of their behaviour in travel traditions and cultural features, as well as data on their characteristics and demand, especially when compared to 'western' and Japanese travellers.

The second part of this book looks at specific experiences of different continents and countries in the encounter with the guests from China as case studies (Chapter 6). Practical consequences for product adaptation and marketing strategies of destinations wishing to attract more Chinese tourists are presented in Chapter 7. The subsequent consequences for the re-thinking of many theories and beliefs of tourism studies are then discussed in Chapter 8.

Finally, Chapter 9 attempts a glimpse into the likely future of China's outbound tourism.

A note on spelling and place names

To minimize confusion, in this text Chinese words are generally spelt utilizing the official *Hanyu Pinyin* spelling used in the People's Republic of China. For instance, this includes for place names the usage of *Guangzhou* instead of *Canton*, and *Taibei* instead of the locally used *Taipeh*. Hong Kong and Macao, however, are referred to by their English names, not as *Xianggang* and *Aomen*. *China* in modern contexts normally refers to the People's Republic of China within its existing borders. *Mainland China* is also used for the People's Republic to specify the exclusion of the SARs Hong Kong and Macao, which are treated in Chinese tourism statistics as separate entities, and of Taiwan. *Countries* is a term used as shorthand for all political entities that issue passports, including SARs and again Taiwan, regardless of their recognition by the government in Beijing or their membership in the United Nations. *Provinces* in China is meant to include the provinces, autonomous regions and municipalities. In the bibliography authors and editors names are given with first names, to avoid misunderstandings caused by the relative scarcity of Chinese family names.

2 Economic and social development of the People's Republic of China

The People's Republic of China came into existence as the result of the victory of the armies of the Chinese Communist Party (CCP) led by Mao Zedong and Zhu De against the Guomindang forces led by Jiang Jieshi (Tschiang Kai-shek), who with the help of the USA, found refuge on a former Japanese colony, the island of Taiwan. The poor, war-torn country of 400 million people, the majority of them illiterate peasants, started to develop a socialist national economy, isolated by the western countries and with some support from the Soviet Union in the first decade only. Independent attempts to jump-start the progress towards a communist society resulted in the major economic and humanitarian catastrophes of the 'Great Leap Forward' and the 'Great Proletarian Cultural Revolution'. After a period of stagnation in the years before and after the death of Mao Zedong in 1976, the way towards the modern China of today started with the third plenary session of the eleventh congress of the Communist Party in December 1978. Under the slogan of 'Reform and Opening' the modernization of the Chinese economy and society was put on the agenda, ushering in a period of unprecedented economic growth for several decades. Agricultural reforms were followed by the acceptance of foreign investment into industries and services and the integration into the world economy as signalled by gaining membership of the World Trade Organization (WTO) in 2001. As the twenty-first century unfolds, China has developed into the 'factory of the world' – topping the list of manufacturers for such diverse products as cereals, cotton and meat, steel, coal and electricity, textiles, fridges, TV sets and telephones (China Statistical Bureau 2004) – and has become the largest domestic tourism market in the world in terms of travels. A comparison with India shows the different developments paths: in 1982, both countries reported an almost equal level of Gross Domestic Product (GDP), China with 222 billion US$ and India with 195 billion US$. After 20 years, China had reached in 2002 a level of 1,233 billion US$ GDP, compared to 510 billion US$ GDP for India (China Business Weekly 2004). Calculated per head the difference is even more striking, as China's population growth slowed as a result of the one-child policy, whereas India is bound to overtake China as the world's most populated country within a decade.

Life expectancy in China has grown from below 50 years to above 70 years. Whereas in the 1950s less than 80 out of 100 children would reach the age of five, now 96 out of 100 children can celebrate their fifth birthday. Illiteracy has been reduced to below 10 per cent in a country where reading and writing abilities were traditionally restricted to the upper classes. The Human Development Index of the United Nations Development Programme (UNDP), taking into account life expectancy, education and income per person, gives China a middle-ranking eighty-fifth place out of 177 countries with an index of 0.76 of the ideal 1.0, well above the one hundred and twenty-seventh place for India (Economist 2005d).

The rosy picture of modern China, sending *taikonauts* into orbit and hosting the Olympic Games, is however marred by three main problems: a large and growing income gap, mounting ecological problems and the absence of any progress towards a democratic system. Gaps between coastal areas and the western part of China, between cities and rural areas in general, but also within cities and rural areas are bigger than almost anywhere in the world. Especially since the mid-1990s the gaps, now also including the employed and unemployed, are widening. A city dweller in 2002 earned 3.1 times[1] as much as a rural person, up from a ratio of 1.8 in 1985. 'If non-currency factors are taken into consideration, China's urban-rural income gap is the widest in the world' (Li, Yue 2004). The top 1 per cent of the population in 2002 earned 6.1 per cent of the total income, the top 5 percent could claim 20 per cent. Accordingly, the Gini coefficient[2] in 2002 has not only reached 0.454 on the national level, but also within the cities and the countryside, the Gini index readings of 0.32 and 0.37 respectively prove the pronounced differences of wealth distribution in urban as well as in rural areas (Li, Yue 2004; Wu, Perloff 2004). In 2005, the Gini index figure was reported to be above 0.48 and approaching 0.5 (China Daily 2005). India's Gini index is calculated as 0.33, level with most industrialized countries except the USA. With respect to wealth, the *Social Blue Book* of the Chinese Academy of Sciences (CAS) reckons that the top 10 per cent of the Chinese society owns 50 per cent of the total private wealth, whereas the bottom 10 per cent owns just 1 per cent of the total private wealth (Hao 2005).The Qinghua University Center for China Studies divides China into 'four worlds', from the rich 'first world' in coastal areas to areas in West and Northeast China as 'fourth world'. One-quarter of the Chinese live in the first and second world, another quarter in the third and half of the Chinese in the fourth world (Blume 2003). The 'well-to-do-life' (*xiaokang*), as the declared goal of modernization, is provided only to a minority.

Pollution of air, soil and water is the second fact that threatens to stop the forward movement of the juggernaut *Chinese modernization*. A study by the Chinese Academy on Environmental Planning claimed that one-third of all Chinese were exposed to harmful levels of air pollution and blamed air pollution for the premature death of more than 400,000 Chinese in 2003.

Despite efforts to bring more trees and green areas into cities, 16 out of the 20 most air-polluted cities of the world are located in China, with pollution levels still rising due to higher levels of energy consumption and automobile use. Altogether 91 cities did not reach the Chinese official minimum levels of air quality (Li, Li, Huang 2004). Overuse of water supplies lead to regular drying-ups of the Yellow River; 70 per cent of all rivers and lakes are too toxic to be used for drinking water (Watts 2005).

As the early international and iconoclastic socialist orientation of China as the self-proclaimed leader of a Third World and non-aligned movement gave way to the economic orientation of the 'Reform and Opening' policies engineered by Deng Xiaoping, China modernized in what Oakes (1998) called a paradoxical process through which the people produce, confront and negotiate a particular kind of socioeconomic change. The integration into the global market-driven economy was connected to a revaluation of Chinese traditions and heritage; the almost unlimited possibilities of economic activities were accompanied by a freeze in the evolution of political institutions and clear limitations to the development of a civil society. The modernity of the '[s]ocialist economy with Chinese characteristics', which might be called 'post-socialism with Chinese characteristics' to distinguish it from the societies of the former Warsaw Pact (Chen 2002), is a 'false modernity' (Li, Y. 2004) as the constituent element of democratic self-rule is missing and seems to be still far away (Zhao 2000a).

Economic and social development 1978–2004

To provide background to the development of outbound tourism, some data about important factors in the economic and social development of China since the beginning of 'Reform and Opening' are given on the following pages with a focus on the income and consumption of potential outbound tourists. As outbound tourism of any quantitative relevance started not earlier than 1983, no figures for the pre-reform time are included. The qualitative development will however be covered for the whole period in Chapter 3.

A word of caution about statistics is necessary. Chinese statistics are known to have been instrumentalized for many centuries and until this day to serve politics rather than faithfully reporting the reality. Furthermore in a country as vast as China, reliable figures are hard to acquire whatever are the politics involved. Many data are based on surveys with samples that out of necessity are smaller than one per million in most cases. Central statistics are often revised backwards in later editions of statistical yearbooks without explanation. Percentages for growth often do not correspond with the absolute figures given, provincial figures do not always add up to the figures given by the central government. The discussion about how much the Chinese GDP growth has been under- or over-reported in past years fills a library (Sachs, Woo 2003).[3] For example, the National Bureau of Statistics of China (NBSC) announced during a press conference in 2004 that it used for the calculation of the regional GDP per head figures for 2003, for the first time, the real number of inhabitants of a city or area. Before, the Bureau admitted, that only the figure for the officially registered inhabitants was used. Due to the large number of migrant workers, the officially registered inhabitants, however, make up less than half of the total population in some big cities such as Shenzhen. The GDP per head figures for Shenzhen, but also for Beijing or Shanghai, central arguments in many outbound tourism discussions (WTO 2003), turned out to be vastly over-reported for the years up to 2002 (Li, D. 2004).

All figures given here should therefore be treated as indicators of trends rather than hard facts down to the last digit.

The overall development of the Chinese economy since 1978 is given in Table 2.1.

Calculated at Purchasing Power Parity (PPP),[4] the Chinese GDP accounts for 13 per cent of the global GDP, which puts China in the position of the second-biggest national economy after the USA (Economist 2004). A large

Table 2.1 GDP 1978–2003 (100 million Yuan RMB, current prices)[5]

Year	GDP	Primary industry	Secondary industry	Tertiary industry	Per capita GDP (Yuan/person)
1978	3,624	1,018	1,745	861	379
1979	4,038	1,259	1,914	866	417
1980	4,518	1,359	2,192	966	460
1981	4,862	1,546	2,256	1,061	489
1982	5,295	1,762	2,383	1,150	525
1983	5,935	1,961	2,646	1,328	580
1984	7,171	2,296	3,106	1,770	692
1985	8,964	2,542	3,867	2,556	853
1986	10,202	2,764	4,493	2,946	956
1987	11,963	3,204	5,252	3,507	1,104
1988	14,928	3,831	6,587	4,510	1,355
1989	16,909	4,228	7,278	5,403	1,512
1990	18,548	5,017	7,717	5,814	1,634
1991	21,618	5,289	9,102	7,227	1,879
1992	26,638	5,800	11,700	9,139	2,287
1993	34,634	6,882	16,429	11,324	2,939
1994	46,759	9,457	22,372	14,930	3,923
1995	58,478	11,993	28,538	17,947	4,854
1996	67,885	13,844	33,613	20,428	5,576
1997	74,463	14,211	37,223	23,029	6,054
1998	78,345	14,552	38,619	25,174	6,038
1999	82,068	14,472	40,558	27,038	6,551
2000	89,468	14,628	44,935	29,905	7,086
2001	97,315	15,412	48,750	33,153	7,651
2002	104,791	16,117	53,541	35,133	8,184
2003	117,252	17,092	61,274	38,886	9,101

Table 2.2 FDI 1995–2004 (in billion US$)[6]

	1995	*1997*	*2000*	*2001*	*2002*	*2003*	*2004*
FDI	38	45	41	47	53	55	61

part of the increase in Chinese output has been realized with the help of Foreign Direct Investment (FDI). After 1995, each year more than 40 billion US$ of FDI flowed into China, making China the biggest recipient of FDI in most years and responsible for the absorption of about one-third of all FDI outside the industrialized countries.

More than 30,000 companies in China funded by companies from Hong Kong, Macao and Taiwan and from foreign countries in 2002 were responsible for almost 30 per cent of the gross industrial output value of the Chinese industry.

China's foreign currency reserves, mainly in American treasury bonds, increased more than fourfold between 2000 and 2005. From 166 billion US$ in 2000 they moved to 285 billion US$ in 2002 (NBSC 2003) and to 711 billion US$ in mid-2005 (Economist 2005f). China is, with the exception of Japan, which is still ahead, by far the largest holder of foreign currency reserves in the world.[7] These reserves represent also a value higher than the total export figure of 2004, which stood at 593 billion US$, with imports at 561 billion US$ creating yet another trade surplus of 32 billion US$ (EIU 2005).

The majority of China's population is concentrated in the eastern half of the country. In 1949, 89 per cent of all Chinese were dwelling in rural areas (Zhang, K.H. 2004). A major trend connected to the economic development

Table 2.3 Industrial enterprises and their gross output value in 2002[8]

	Number of enterprises	*Gross industrial output value (billion Yuan RMB)*
National total	181,557	11,078
Domestic funded enterprises	147,091	7,832
State-owned industry	29,449	1,727
Collective-owned industry	27,477	962
Cooperative enterprises	10,193	320
Joint ownership enterprises	1,964	94
Limited liability corporations	22,486	2,007
Share holding enterprises	5,998	1,412
Private enterprises	49,176	1,295
Other enterprises	348	14
Enterprises with funds from Hong Kong, Macao or Taiwan	19,546	1,367
Foreign funded enterprises	14,920	1,879

of China is however the increase in the level of urbanization. The proportion of city dwellers moved to 18 per cent of the 960 million people living in China in 1978, 26 per cent in 1990 and 30 per cent in 1996. In 2002, 500 million Chinese lived in urban surroundings, 39 per cent of all the 1.28 billion people living in China in that year according to official statistics. Of these cities, 450 had more than 500,000 inhabitants, 138 of these over 1 million, 23 over 2 million and 10 had even more than 4 million inhabitants. For 2020 an urbanisation rate of 50 per cent is set as a goal by the Chinese government (China Daily 2005).

The income of all households increased considerably since 1978. The Engel coefficient, an indicator showing the percentage of the total expenditure needed for food, decreased to 46 per cent for rural and 38 per cent for urban dwellers.

A close look at the different levels of income is, however, necessary to identify the part of the Chinese society able to engage in long-distance

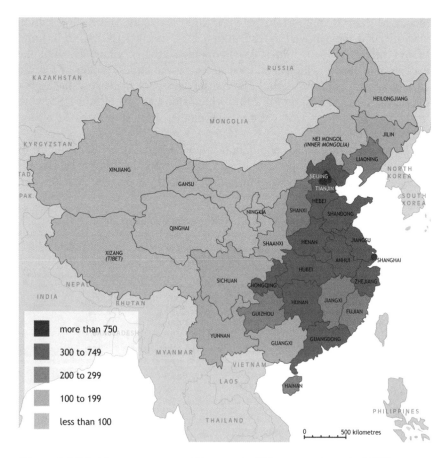

Figure 2.1 Inhabitants per square kilometre in Chinese provinces in 2002[9]

Table 2.4 Development of total population and urbanization 1978–2002[10]

Year	Total population	Urban population	Proportion of total (%)	Rural population	Proportion of total (%)
1978	96,259	17,245	17.92	79,014	82.08
1980	98,705	19,140	19.39	79,565	80.61
1985	105,851	25,094	23.71	80,757	76.29
1989	112,704	29,540	26.21	83,164	73.79
1990	114,333	30,195	26.41	84,138	73.59
1991	115,823	31,203	26.94	84,620	73.06
1992	117,171	32,175	27.46	84,996	72.54
1993	118,517	33,173	27.99	85,344	72.01
1994	119,850	34,169	28.51	85,681	71.49
1995	121,121	35,174	29.04	85,947	70.96
1996	122,389	37,304	30.48	85,085	69.52
1997	123,626	39,449	31.91	84,177	68.09
1998	124,761	41,608	33.35	83,153	66.65
1999	125,786	43,748	34.78	82,038	65.22
2000	126,743	45,906	36.22	80,837	63.78
2001	127,627	48,064	37.66	79,563	62.34
2002	128,453	50,212	39.09	78,241	60.91

Table 2.5 Development of rural and urban households 1978–2002 in Yuan RMB (current prices) and index values (comparative prices) and per cent[11]

Year	Rural annual net income	1978 = 100	Urban annual disposable income	1978 = 100	Engel coefficient rural	Engel coefficient urban
1978	134	100	343	100	67.7	57.5
1980	191	139	478	127	61.8	56.9
1985	398	268.9	739	160.4	57.8	53.3
1989	602	305.7	1,374	182.5	54.8	54.5
1990	686	311.2	1,510	198.1	58.8	54.2
1991	709	317.4	1,701	212.4	57.6	53.8
1992	784	336.2	2,027	232.9	57.6	53
1993	922	346.9	2,577	255.1	58.1	50.3
1994	1,221	364.4	3,496	276.8	58.9	50
1995	1,578	383.7	4,283	290.3	58.6	50.1
1996	1,926	418.2	4,839	301.6	56.3	48.8
1997	2,090	437.4	5,160	311.9	55.1	46.6
1998	2,162	456.2	5,425	329.9	53.4	44.7
1999	2,210	473.5	5,854	360.6	52.6	42.1
2000	2,253	483.5	6,280	383.7	49.1	39.4
2001	2,366	503.8	6,860	416.3	47.7	38.2
2002	2,476	528	7,703	472.1	46.2	37.7

Table 2.6 Participation on total consumption by income quintiles in 2001 (in per cent)[12]

Lowest quintile	Lower middle quintile	Middle quintile	Higher middle quintile	Highest quintile
5	9	14	22	50

Table 2.7 Number of durable consumer goods owned per 100 urban households by level of income in 2002[13]

	Average	Lowest decile	Second lowest decile	Second quintile	Middle quintile	Fourth quintile	Ninth decile	Highest decile
Motorcycle	22	12	17	20	23	24	26	32
Bicycle	143	133	145	149	150	145	142	139
Automobile	1	0	0	0	1	1	1	4
Washing machine	93	81	87	91	94	97	100	102
Refrigerator	87	61	74	84	90	95	98	104
Colour TV	126	102	112	118	126	134	144	161
Video disc player	53	31	41	49	57	59	62	69
Computer	21	3	6	11	18	28	37	54
Camera	44	14	23	32	42	53	66	80
Air-conditioner	51	8	18	29	46	63	85	128
Telephone	94	74	86	92	96	99	101	104
Mobile telephone	63	15	30	46	65	83	99	128

Table 2.8 Urban income and expenditure for ECRS according to household income level in Yuan RMB and per cent in 2002[14]

	Average	Lowest decile	Second lowest decile	Second quintile	Middle quintile	Fourth quintile	Ninth decile	Highest decile
Per capita annual income	8,177	2,528	3,833	5,209	7,061	9,438	12,555	20,208
Per capita disposable income	7,703	2,409	3,649	4,932	6,657	8,870	11,773	18,996
Per capita annual living expenditure	6,030	2,388	3,260	4,206	5,453	6,940	8,920	13,041
ECRS Yuan RMB	902	318	425	577	798	1,046	1,374	2,149
ECRS %	15.0	13.3	13.1	13.7	14.6	15.1	15.4	16.5

Table 2.9 Rural income and expenditure for ECRS according to household income
level in Yuan RMB in 2002[15]

	Lowest 20%	Lower 20%	Middle 20%	Higher 20%	Highest 20%
Income	1,551.79	2,288.34	3,025.17	4,075.60	7,567.22
Expenditure	1,725.44	2,093.55	2,581.46	3,259.33	5,533.63
ECRS and goods	97.36	145.50	193.28	246.81	416.88

leisure travel. The 20 per cent of the total population earning the highest
income is in fact responsible for 50 per cent of the total consumption in
2001.

This consumption can be illustrated by the different levels of durable
household goods possessed by the different income groups. For the highest
income decile of urban households still only 4 cars exist per 100 households, but with all other goods the households are well equipped.

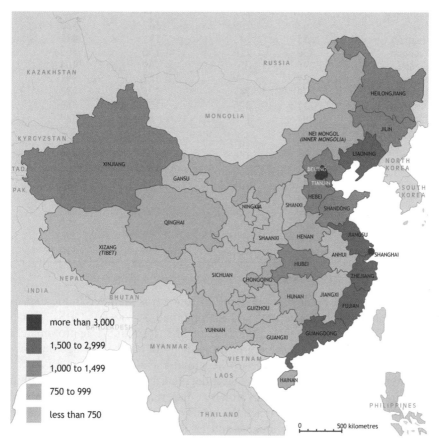

Figure 2.2 GDP per head in US$ in Chinese provinces in 2002[16]

Table 2.10 Urban expenditure for ECRS according to residence in Yuan RMB in 2002[17]

	ECRS total	Cultural and recreational goods	Cultural and recreational services	Education
National average	902	245	162	495
Beijing	1,809	525	404	880
Shanghai	1,668	541	306	821
Zhejiang	1,407	408	264	735
Guangdong	1,386	334	361	691
Tianjin	1,151	349	141	661
Chongqing	1,065	237	195	633
Shandong	929	293	121	515
Hunan	884	231	182	471
Shaanxi	882	220	119	543
Xinjiang	881	260	142	479
Fujian	875	295	156	424
Guangxi	869	267	155	447
Jiangsu	867	262	134	472
Hubei	863	199	142	522
Yunnan	855	233	191	431
Sichuan	830	213	150	468
Gansu	795	251	139	404
Shanxi	782	179	112	491
Hainan	744	147	125	472
Ningxia	711	206	120	386
Liaoning	697	182	81	434
Qinghai	683	183	164	337
Inner Mongolia	679	193	83	402
Guizhou	675	160	144	371
Jiangxi	661	159	103	399
Jilin	656	166	75	415
Hebei	604	182	85	337
Heilongjiang	590	120	88	382
Henan	586	172	135	279
Anhui	480	121	85	274
Xizang (Tibet)	445	104	30	310

The amount of money that is spent for tourism is included in the expenditure for Education, Cultural and Recreational Services (ECRS). Urban households spend more then 2,000 Yuan RMB (Renminbi) (approximately 200 Euros in 2005) on these items, more than double the average but only somewhat higher as percentage of their total spending.

The data for rural dwellers are less detailed, but the differences within the rural income groups are nevertheless discernable. The highest quintile of rural residents does only reach the levels of the middle and lower middle urban quintile in terms of income and ECRS spending. Another figure of the statistical yearbook gives a percentage of rural income households for

2002 above 5,000 Yuan RMB of only 10.1 per cent, which would indicate a strong differentiation of incomes even within the highest quintile or rural households.

Beside the rural/urban divide the differences between the eastern seaboard provinces and the rest of the country are also very pronounced with coastal provinces reaching levels of more than 3,000 US$ GDP per head in 2002, compared to less than 750 US$ in the bigger part of the hinterland. This perspective on the spatial differentiation is confirmed by the differences among urban residents in their ECRS spending. Half of it is spent on the education part of the total spending, leaving on average only 245 Yuan RMB (approximately 25 Euros in 2005) for cultural and recreational goods and 162 Yuan RMB (approximately 16 Euros in 2005) for cultural and recreational services. The urban areas of those provinces that are also the main sources areas for outbound tourists show up on top of the list with spending levels above the national average. However, the fact that the difference from the average in both directions is not much higher than double and half respectively indicates the strong differentiation also within the urban areas of all provinces.

Economic and social relevance of outbound tourism in China in 2005

From a quantitative point of view, outbound tourism is a minor phenomenon in China. Only about 50 million or 4 per cent of the current Chinese population has ever travelled outside China, if the quasi-domestic tourism to Hong Kong and Macao is not counted. Half of all these travels happened in the years 2003 to 2005. The percentage of Chinese travelling beyond Hong Kong and Macao per year has yet to reach the 1 per cent mark of the total population.

If all border-crossings including those to Hong Kong and Macao are counted, less than 200 million or 15 per cent of all Chinese can claim to have touched foreign soil.[18] The foreign currency outflow initiated is also very minor compared to the total foreign currency reserves of China. The international tourism expenditures were calculated as half a billion US$ in 1990, 14 billion US$ in 2001 and 15.4 billion US$ in 2002.[19] Compared to 286 billion US$ reserves, 53 billion US$ FDI and 30 billion US$ trade surplus for the same and even better figures for the following years, the monetary aspect (Lin, Schramm 2003; Zhang, L. 2004) of outbound tourism is not very important. Furthermore, within the tourism industry there are revenues from inbound tourism into China to be taken into consideration, even if the margin between inbound revenues and outbound expenditures is getting smaller over the years[20] and might have disappeared altogether by 2003.[21]

From a qualitative point of view, however, outbound tourism is important for China. The 'Opening' of the 'Reform and Opening' paradigm of 1978 was meant in the beginning as an opening to let foreign investment, new technologies and inbound tourists in, rather than as an opening to let Chinese tourists out. With the fall of the Berlin Wall and the end of the Soviet Union and the bipolar world system of the Cold War the *normality* of forbidding the citizens of a country to leave the country temporarily or permanently disappeared.[22] In China, the government managed to survive the *anni horribilis* for all Communist parties of 1989/1990 and reacted to the strong demand for outbound leisure travel in China with permission to travel, first inside the *Chinese world* in Southeast Asia and after 1997 to other countries also. The response of the top quintile and certainly of the top decile of Chinese population to the post-1989 situation was the embracement of consumerism, the orientation towards personal advancement and high earnings to consume prestigious brands, with many international destinations as brands among them.

Outbound tourism to Hong Kong and Macao, which became quasi-domestic destinations with the return to China in 1997 and 1999 respectively and the step-by-step easing of travel restrictions, is affordable for Cantonese even with only average income as the cost involved is not very high. For border traders the expenses arising from crossing the border are part of their cost calculation. But self-paid outbound tourism to other destinations, with costs starting at 4,000 Yuan RMB for a five-day trip to Thailand, is still out of reach for the vast majority. According to the figures presented above, 4,000 Yuan RMB (approximately 400 Euros in 2005) represents half of the average total disposable annual per capita income of the urban population and more than a year's disposable income for the rural population. Mobility other than inland migration is certainly not within reach for the bottom quintile, which is too poor to afford even basic health services, let alone outbound tourism. For the top income decile however – almost 100 million persons living in 36 million households – 4,000 Yuan is *only* two months disposable annual income per capita.

Outbound tourism is nevertheless setting the tone for the whole society. The acceptance of leisure travel, once officially scalded as a wasteful activity, reflects in the increase in domestic tourism to the level of one trip per person per year, changing China with more then a billion trips into the largest domestic tourism market in numbers of travellers. 'Outbound' tourism for many Chinese tourists is substituted by a trip to one of the many mini-world theme parks, which more or less faithfully (Hendry 2000) reproduce famous buildings from all over the world. Outbound travel is influencing the Chinese society by providing another *distinction* (Bourdieu 1979) tool, a way to show either belonging to the group of people who are well connected enough to travel with public or company money, or to the group of people rich enough to travel with private money. The impressions collected during visits of foreign destinations are also acting as a tool to benchmark China's situation by comparing the economic and social situation perceived in those other countries and by experiencing the different ways the Chinese are treated there. The influence on local Chinese customs and behaviours is harder to judge, being inseparable from other modernization and media influences. Very rich people such as the 265,000 US$ millionaires in China (Merrill Lynch 2005) certainly find new ideas on how to spend their money abroad. Outbound tourism influences the Chinese society in its form as diaspora as well. Business ties are strengthened with the provision of business opportunities to cater for groups by acting as incoming agency or providing restaurant services for them.

According to official figures, China's population number crossed the 1.3 billion mark in January 2005. Demographs foresee stabilization on this level with no further major increases until 2019, followed by a decrease after that date.

The Chinese society is rapidly ageing. In 1994, 56 per cent of the 1.2 billion people were below the age of 30 and only 9 per cent above 60 years of age, with 35 per cent in middle age. In 2004, the ratio for the now 1.3 billion inhabitants of China had changed to 44 per cent below 30 years, 45 per cent between 30 and 60 years and 11 per cent above 60 years. For 2014, the expected ratio is 35:48:17, with a total figure still of 1.3 billion inhabitants. Within two decades the share of under-30s will have fallen from more than half to just more than one-third of the total population (EIU 2005).

Accordingly, the number of persons under 40 years of age peaked already at 800 million and is expected to decline by 250 million persons before the year 2024. From 38 per cent in 2004, the proportion of persons over 40 years is likely to increase to 58 per cent in 2024 (Economist 2005c). Careers as outbound tourists will have shaped the experiences and behaviour of the upper stratum of this new developing generation of older persons in stark contrast to the current generation.

3 Government policies and the development of demand in Chinese tourism

The Chinese character for 'law' *fa* consists of the parts 'water' and 'movement'. Control of the movement of water, keeping within their banks the big rivers and organizing the irrigation systems used for rice production, has been for many centuries the *raison d'etre* of a centralized government in China. Control of the movement of people has been another goal of all Chinese governments, indigenous or foreign. Over most of the more than 2,000 years of history of Imperial China, foreigners were welcomed in China to present wares, tribute, ideas, religions and technologies for possible adaptation in a sinocized form.

Mobility inside China was as cherished by scholars, monks and pilgrims as it was by traders and administrators and was helped by major infrastructure works such as the Grand Canal. 'Chinese people are well known for enjoying travel, particularly travel as a way to broaden one's mind. As the old saying goes "travelling for one thousand Li equals reading ten thousand volumes of books"' (Jang, Yu, Pearson 2003: 89).

The movement of Chinese out of the country, however, was restricted or forbidden for most of the time. Chinese travellers of equivalent worldwide fame as Marco Polo or Ibn Battuta, the fourteenth-century globetrotters from Venice and Tangiers respectively, do not exist in global memory (Arlt 2005a).[1]

However, the desire to move out of the government's sphere of influence if not for pleasure then for economical or political reasons can be found in the movement from northern and central China to the south before the twelveth century and onwards to other parts of Asia and the world afterwards – the creation of a Chinese diaspora of considerable influence.

The control of the mobility of the Chinese people has been a major concern of the Chinese government since 1949 as well (WTO 2003). Until recently, all long-distance movements even within the country needed reasons and permissions. It was only in the 1980s that the government stopped viewing leisure with suspicion. The right to travel outside Mainland China without connection to official business, studies or contacts to Chinese living there was granted by the government even later, with the majority of destinations becoming accessible for leisure travel only since the beginning of the twenty-first century.

In comparison, in the Soviet Union and Eastern Europe, travelling domestically or to the 'brother states' of the Warsaw Pact was meted out as a favour to the common people as a reward for obedience. The inadequacy of these provisions was demonstrated impressively in 1989 with the falling of the Berlin Wall under the chants for *Reisefreiheit* – freedom of travel – a clear warning not to underestimate this demand for the last remaining sizeable communist party in power.

> Outbound tourism has been actively discouraged both for political reasons and because it causes outflows of capital and foreign currency. Nevertheless, the authorities have pursued a policy of progressive opening-up since the nineties, albeit within a very strictly limited framework, as a result of a growing interest in foreign travel coupled with the ensuing pressure exerted by the more affluent (and politically more influential) segments of the population, which can afford such leisure trips.
>
> (EUCCC 2001)

Chinese tourism policies and demand before 1949

Considering the vast territory that makes up China today, travelling from one place to another has, of course, since time immemorial been an important part of life for military, commercial, religious, representative, educational and administrative reasons for the rulers, or out of necessity to flee from war, famine and disasters or to act as soldier or brigand for the ruled.

Between the first unification of China under Qin Shi Huangdi (221 BC) and the end of the Qing Dynasty in 1911, non-Chinese, inducing mobility between the homelands of the conquerors and the conquered territories at least for the rulers, have more often than not ruled all or part of China. China was indeed 'an empire that incorporated different regions and different peoples as it was taking shape and that remained open to outside influence throughout its long history – not a central kingdom closed to foreign influences' (Hansen 2000: 4).[2]

Outbound travel for leisure to areas beyond the realm of the Han culture did not exist, but otherwise a growing number of Chinese ventured abroad. The Buddhist monk Xuan Zang in the seventh century journeyed to the Indian subcontinent to acquire knowledge and collect scriptures to bring back to China without the consent of the ruling Tang emperors (Shen 1996). The Song Dynasty and especially the Mongolian rulers of the Yuan Dynasty supported long-distance contacts, making it possible for Marco Polo and his contemporaries to travel safely overland between the Mediterranean and the Yellow Sea. Admiral Zheng He was sent out by the court at the beginning of the Ming Dynasty to re-establish tributary relations between the Chinese Empire and its Southeast Asian neighbours (Levathes 1994; Arlt, Domnick 1994).

Chinese armies on land ventured into the Korean Peninsula or today's Mongolia or Vietnam to assert the status of overlord, to support factions in power struggles or to ward off invasions, not to conquer.[3] In the north, the Great Wall and the Willow Palisades (Arlt 2001) were built to control the influx as well as the outflow of foreigners and Chinese. In the south, Ming and especially Qing courts actively discouraged maritime contacts with other countries (Arlt 1984; Nyiri 2005b), not always with success:

> In the 1630s, for example, Chinese barbers, having travelled on return-ing Manila galleons,[4] aroused local ire in Mexico City by undercutting prices, and to this day a style of embroidered women's blouse found in Central Mexico bears the name *china poblana*.
>
> (Waley-Cohen 2000: 122)

When Lord Macartney presented the demand of George III for free trade to the Qing Emperor Qianlong at his summer retreat in Chengde in 1793, he was given the answer that China had everything and therefore not the slightest need for products from the outer world.[5] The idea that dwellers of the Middle Kingdom, the centre, could themselves travel abroad, to the periphery, albeit just out of curiosity, never crossed the emperor's mind (Hevia 1995).

In fact, even before 1793, some of his subjects in spite of being forbid-den to do so ventured outside Asia, as sailors or as translators in the Vatican and elsewhere.[6] In the 1830s emigration from China in the so-called 'coolie trade' started the first mass movement out of China, which brought millions of Chinese as workers to Southeast Asia and the USA. Subsequently, as a result of the Second Opium War, the Qing government was forced in 1860 to establish a 'Foreign Ministry' and to sanction the mass emigration of Chinese.

With the earlier establishment of Hong Kong, outbound travel also became easier by bringing a foreign nation-state bridgehead onto Chinese soil. Meanwhile, visitors such as the followers of Hong Xiuquan, self-styled 'younger brother of Jesus Christ' and leader of the Taiping Tianguo, the biggest peasant revolt in the history of mankind (Spence 1997), or Sun Yat-sen, father of the Republic of China, got hold of outlandish ideas that changed the direction of the development of China.

Nevertheless, however widely defined, none of these movements could constitute an 'outbound tourism policy' in Imperial China beyond the cyclical changes between encouraging and discouraging international contacts. However, an underlying desire to venture abroad, if not for leisure reasons, can be detected from early times. As Zhang emphatically puts it:

> The Chinese nation has an age-old tradition of tourism, which is an inseparable part of Chinese traditional culture. The Chinese tourism

tradition emphasizes enlarging one's knowledge, raising one's under-standing of the world and society, enhancing self-cultivation, making friends and conducting cultural exchanges. An old Chinese saying is, 'He who travels far knows much'.

(Zhang 1997: 569)

An especially under-researched period, which is seldom treated in the literature on Chinese outbound tourism, is the time between the fall of Imperial China 1911 and the end of the civil war in China, which led to the establishment of the People's Republic of China in 1949. Before 1927 China was in turmoil and after 1938, war and civil war were waged. Still, as western lifestyle took hold in Chinese cities, students, diplomats and traders went abroad to Japan, the USA or Europe. Some early nouveau riche Chinese copied the leisure travel behaviour they could learn from the growing number of international tourists visiting China, from Bernhard Shaw to Charlie Chaplin, facilitated by new shipping lines and the Trans-Siberian Railway. The statement that 'the prolonged and ruthless wars from the late 1930s through to the 1940s,[7] ... put an end to all pleasure travel in the country' (Zhang, Pine, Zhang 2000: 282) remains to be substanti-ated.[8] For instance, the first periodical on tourism, *Lüxing Zazhi* (*Travel Journal*) was started in 1927 as a quarterly, but later changed to a monthly magazine and continued to be published until 1954.

With the British forces winning the First Opium War, a British entre-preneur opened the first travel agency named *Tongjilong* in 1841, to be followed by an American-owned competitor called *Yuntong* (Shu 1995). In the 1920s Thomas Cook & Sons, as well as other international travel companies, opened offices in Shanghai and Beijing, offering international travel arrangements to foreign and Chinese potential travellers.

At the same time, several small Chinese-owned travel agencies were established, such as *Yusheng* and *Luzhou*, among others. In 1923 these were incorporated into the 'Travel Department', four years later renamed 'China Travel Service', within the Shanghai Commercial & Savings Bank in Shanghai. This travel department is in most Chinese sources presented as the first Chinese travel agency (cf. Qian 2003: 145; Zhang, H.Q. 2004: 369). As a forerunner of today's structures, it combined the functions of retailer and tour organizer, inbound, domestic and outbound tourism and even owned a hotel.

The first outbound tour was organized by the Bank in 1925 as a 'Cherry Blossom Appreciation Tour'. Twenty Chinese tourists visited Nagasaki, Kyoto, Tokyo and Osaka during a three-week journey (Fu 2004: 41). The CTS, however, never succeeded in attracting more than a few hundred outbound customers per year.

The analysis of tourism policies in the republican period remains un-satisfying as it stands. The grandiose schemes of Sun Yat-sen (Sun 1922) to develop railroads and harbours to facilitate national and especially international mobility could not be realized in his time, but with the growing number of Chinese living temporarily or permanently overseas, what could be called VFR outbound tourism thrived.

Chinese tourism policies and demand 1949–1997

The People's Republic of China during the rule of Mao Zedong did not only consider outbound tourism as dangerous but also as an unnecessary expense of scarce hard currency. Maoism, adopting a pre-industrial society point of view, regarded even travelling for leisure within China as wasteful behaviour, as a sign of bourgeois lifestyle 'which one should always guard against' (Zhang 1989) and as a potential source of unrest. Domestic travel was therefore confined to the privileged few or organized for political reasons especially during the Cultural Revolution, when 'the Red Guards . . . combined sightseeing with the arduous work of building a socialist China' (Graburn 2001a: 80).

In post-Maoist China, inbound tourism was favoured as a supposedly easy way to earn money from American, European and Japanese tourists, as before, the few visitors passing the 'bamboo curtain' were used to earn international political support. China International Travel Service (CITS) was at the core of this process. However, from the very beginning of the existence of the People's Republic, the second part of inbound tourism was the responsibility of the Overseas Chinese Travel Service, later called China Travel Service (CTS), and was used to strengthen the bonds with ethnic Chinese living in Hong Kong, Macao, Taiwan and elsewhere overseas, the 'four kinds of persons' (Gao, Zhang 1983: 76).

This specific idea of bonding with ethnic Chinese living outside the current reach of the Beijing government instigated the first moves to allow Chinese citizens, from November 1983 onward, to cross the border to Hong Kong and Macao, and later to some Southeast Asian countries with large overseas Chinese communities for the sole purpose to 'visit friends and relatives'. As the costs were to be borne by the hosts, no financial losses were deemed to have occurred.

The restrictions on domestic and outbound travel were not simply endured by the Chinese people however. 'In Chinese culture, travelling is a way of learning' (Chen 1998). In a situation of very limited access to information, this could prove to be especially interesting, not withstanding the recreational rewards. Like in other socialist countries, domestic leisure tourism was organized between large companies and institutions before the Cultural Revolution during which revolutionary meetings also included a sightseeing component part. In the 1980s, domestic tourism 'often took the form of organized package tours, such as round trips for employees and "travel meetings". . . . Hence it came to be viewed by some as a legitimate excuse for squandering public funds' (Wen, Tisdell 2001: 131). Local Chinese travelled with their rich overseas relatives and 'visited sites they'd never seen before, and ate food they'd never imagined existed' (Richter 1989: 36). 'With more disposable income and available leisure time, and the impact of the growth of international tourism, domestic tourism had become a widespread billion-yuan business with millions of participants even before any actual government involvement in 1984' (Zhang 1989: 58). Put in a nutshell: 'The policy for domestic tourism has been one of "no encouragement" . . . Whether or not domestic tourism is encouraged, it continues to increase' (Zhang 1985: 141). Tourism in China thus developed 'in a seemingly out-of-control gallop. . . . There were considerable gaps between political decision-making, research, planning and implementation' (Gerstlacher, Krieg, Sternfeld 1991: 86).[9]

By the mid-1980s, the growth of the domestic travel demand could no longer be ignored. The size and pace of the emerging tourism industry were unknown before, the participants out-ranged in size and social spread all traditional dimensions. Local authorities in some areas supported the mobilization of resources to cater for domestic tourism for the obvious positive effects on income and regional development. The 'First National Conference on Domestic Tourism', held in Tianjin in 1987, acknowledged the growing importance for the first time and emphasized its possible positive role as a regional development factor. 'In summary, it can be argued that, like policy-making in many other fields, domestic tourism policy was discussed and formulated at a time when much development had already taken place in many parts of the country' (Xu 1999: 75).

After 1989, the attempts of the government to restrict the growing demand for domestic recreational tourism weakened furter. Domestic tourism started to be actively supported with hasty, but hefty, investment at local and provincial levels. The introduction of a five-day working week in 1995 and three collective week-long holidays in 2000 by the central government also underpinned this development. 'The Chinese government . . . has switched its policy from yesterday's "not to encourage domestic tourism for the time being" to today's [1992] "vigorously guide and steadily develop tourism in accordance with local conditions"' (Du 2004b: 29).

In outbound tourism, the efforts to keep developments in line with policies continued much longer. However after the suppression of the Tiananmen movement, VFR travels were permitted to Thailand, Singapore and Malaysia, with Philippines added in 1992 (Ryan, Mo 2001). Again the idea of using outbound tourism for closer bonds with overseas Chinese persisted.

With the start of the Approved Destination Status (ADS) system in 1995 and the 1997 'Provisional Regulation on the Management of Outbound Travel by Chinese Citizens at Their Own Expense' the government began to give way. Across the board bureaucratic procedures with regard to getting passports, exchanging hard currency, getting visas, etc., were simplified and liberalized. After 2003, the speed with which countries were opened to leisure group travel increased.

Deng Xiaoping saw tourism first and foremost as a means for accumulating foreign exchange: 'Developing tourism should first develop those businesses, which could earn more money' (Zhang, Chong, Ap 1999: 473), as an export sector (Xu 1999: 20).

In 2000, five remarks he made on tourism in 1978 and 1979 were published with great fanfare as a book by the Central Committee of the Communist Party and China National Tourism Administration (CNTA) as 'Deng Xiaoping on Tourism' (Literature Research Center 2000). Giving ideological support to the industry, it reiterated Deng's point of running the tourist industry as an economic sector, that there was a lot to do to develop tourism and that tourism must be run in a more prominent way.

A commentary in the party newspaper stressed the fact that the goal set by Deng for 2000, to earn 10 billion US$ in foreign exchange from the tourism industry, had been achieved in 1996, well ahead of schedule (People's Daily 2000). In 2005, however, Chinese tourists spent probably more money outside Mainland China than China earned in foreign currencies from international visitors. 'The development of China's outbound tourism market follows the classic "ripple effect". Just as a pebble dropped into a pond generates ever-widening ripples, so the growth of outbound travel over time becomes more geographically distant' (Zhang, Heung 2001: 8).[10] Within this chapter, this idyllic picture will have to be shattered, as the development of China's outbound tourism needs to be described in more complex patterns.

For the analysis of the development of policies and institutions concerning tourism in the People's Republic of China, most authors follow Zhang's (2003a) temporal structuring into three phases:[11]

1   'politics only' 1949–1978;
2   'politics plus economics' 1978–1985;
3   'economics over politics' since 1986.

This reflects adequately the main government policies towards inbound tourism from western countries – the part of tourism that received the biggest part of attention from the Chinese government. Indeed, when in 1978 it was decided to allow and facilitate 'tourism', when the National People's Congress called for a 'major effort to develop tourism' (Gao, Zhang 1983: 77), the activity in question was always attracting and managing western – including Japanese – inbound tourists only.[12]

A more complete discussion of tourism has to differentiate two aspects twice. First, the policies that were meant to support the political goals of the government on the one hand and the reaction of the governed to realize their demand on the other hand need to be distinguished, as they more often than not will rather contradict than compliment each other.

Second, the tourism policies targeted towards the 'Chinese world', i.e. Hong Kong, Macao, Taiwan and the overseas Chinese living in other countries in Southeast Asia and elsewhere, and the policies targeted towards the 'non-Chinese world' show clear differences, which have to be taken into account.

This differentiation becomes visible with a first look at Chinese tourism statistics and practices. All inbound tourism tables take care to distinguish between *waiguoren*, foreigners, 'outside-country-people', *tongbao*, compatriots,[13] and *huaqiao*, overseas Chinese. The original division of responsibilities between the two established travels services CTS and CITS falls along the ethnic distinction between Chinese and non-Chinese, regardless of the countries of residence or the passport held. This peculiar form of organization of information and services is reported but generally not questioned in texts dealing with tourism in China.

For the Chinese government, tourism before 1978 was considered as 'unproductive' (Wen, Tisdell 2001: 5), the existence of such 'bourgeois indulgence' (Li, Y. 2004: 190) only 'begrudgingly accepted' (Wen, Tisdell 2001: 314). This led western observers to the conclusion that:

> Chinese tourism policy from the 1949 establishment of the People's Republic of China until 1977 can be summarized as cautious at best and characteristically negative in nature – the fewer outsiders the better. This attitude was not unreasonable. For most of the first two decades of the PRC's [People's Republic of China] existence, major tourist-generating countries were unrelievedly hostile to the communist regime. Travel to the PRC was forbidden by the United States and many other western governments.
>
> (Richter 1989: 24)

This statement ignores three important groups of visitors:

- First, overseas Chinese and compatriots who wanted to look for their relatives in their home counties had to be taken care of and – if possible – sent back with a positive impression of the 'New China'.
- Second, a relatively large number of foreign technical personnel, mainly from the Soviet Union, invited by the Chinese government to assist in the implementation of the first Five-Year Plan, which started in 1953, needed to be received (Uysal, Lu, Reid 1986: 113).
- Third, 'foreign friends', invited as part of the 'people-to-people' diplomacy, had to be taken good care of.

After 1978, the need to generate investment sources for the 'four modernizations', namely agriculture, industry, national defence and science and technology,[14] shifted the focus onto more commercial gains but without giving up the differentiation in treatment of foreign and of overseas Chinese visitors.

Overseas and foreign visitors stayed in different hotels and paid different prices[15] even if they received the same services. In the situation of scarce resources, this however had an effect opposite to the one planned:

> Consequently, prices and services varied according to the race of the person rather than the willingness to pay. Very often, overseas Chinese or compatriots were refused by hotels or transport ticket offices, or given less than proper services, because they paid less than foreign visitors.
>
> (Zhang 2003a: 25)

Domestic tourists, who started to grow in numbers with the reform process, had to make do with what was left over by foreigners and overseas Chinese. Domestic tourism had to wait until the late 1980s to be recognized as an important economic force.

This reflected the problem of the undecided goal of tourism policies at the time. In July 1981, a national tourism conference in Beijing discussed the major aims of the development of the tourism industry. The discussions resulted in a State Council document published three months later. It was argued that tourism had to highlight the superiority of the socialist system, as well as to offer the traditional hospitality and historical attractions on the one hand and contribute to the capitalization of modernization on the other. Earning money was not the sole aim of a socialist tourism industry, the State Council emphasized (Gao, Zhang 1983: 78).

A sign of the entrance of China into the tourism sphere was the participation of a Chinese delegation in 1982 at the Second World Conference on Tourism in Acapulco, organized by the WTO, followed by China becoming a member of the WTO in 1983 (Chow 1988).

Also in 1982, the National Heritage Conservation Act changed the iconoclastic attitude of Maoism, which had been responsible for the destruction of a vast number of cultural relics and buildings during the Cultural Revolution. Based on the Act, all heritage objects could be sorted into different categories of national, regional and local importance. At all levels heritage conservation management organizations were set up, many sites were re-opened to the public and developed into tourism products.

Between 1978 and 1985, the number of foreigners entering China multiplied by six from 230,000 to 1,370,000. However, there was a tenfold growth of the number of overseas Chinese and compatriots entering China from 1,580,000 to 16,480,000.

From 1986, tourism was for the first time incorporated into the seventh Five-Year Plan as a 'comprehensive economic activity with the direct purpose of earning foreign exchange for China's modernization' (Zhang, Chong, Ap 1999: 478).

With regard to foreign visitors, the pretence of combining propaganda and business was dropped in favour of earning cash from sightseeing tourists.[16] For Chinese visitors other aspects remained much more important. Overseas Chinese and compatriots became a major source of investment and with the economic growth of China and the new business opportunities awakened, sympathies started to move from Taibei to Beijing. This was also helped by the rescinding of travel restrictions to China by

Southeast Asian countries after 1987 (Lew 1995). Thus, tourism was therefore, even after 1986, not based on direct economic gains from leisure activities alone. Outbound VFR tourism to Hong Kong and Macao and after 1990 to Thailand, Singapore, Malaysia and the Philippines followed the same lines of recognition.

The advent of the ADS system after 1995, meanwhile, signifies the last fight to keep outbound tourism within the boundaries of China's foreign policy interests and of the idea that tourism should bring in and not leak hard currency.

The 1997 'Provisional Regulation on the Management of Outbound Travel by Chinese Citizens at Their Own Expense' acknowledged for the first time that private, non-sponsored, outbound travel existed. With the increased integration of China into the global economy and the loss of the information monopoly with the flow of all kind of news and images into the country via satellite TV and the Internet, the government had to give way. Across the board bureaucratic procedures with regard to getting passports, changing hard currency, getting a visa, etc., were simplified and liberalized.

Australia and New Zealand started to receive ADS groups in 1999. The gates finally opened with ADS agreements coming into force in countries without large overseas Chinese communities, namely Malta, Egypt and Turkey in 2002, to be followed by Germany, South Africa and many more. The flood of outbound leisure tourists seems unstoppable now. In a marked change from previous policies (He 1992), He Guangwei, director[17] of the CNTA, spoke at the National Tourism Conference in 2004 of outbound tourism being the 'desire of the people to see the world' (News Guangdong 2004; People's Daily 2004) and insisted that 'the increase of outbound tourists not only promotes bilateral tourism exchanges, but also enhances mutual friendship and understanding' (People's Daily 2004).

Mutual friendship and understanding is hopefully enhanced, even so watching countries rolling by through the window of a bus as a main form of travelling and the existing language barriers do not necessarily support such an outcome.

Bilateral exchanges might have been originally envisaged by the ADS system with fixed contingents of tourists each year agreed within the rules of a planned economy. With 100 countries open for multinational sightseeing tours and the number of Chinese outbound tourists equalling and soon exceeding the number of tourists coming into China,[18] market realities and the wanderlust of the Chinese people prevail.

Chinese tourism policies towards the 'Chinese world'

'It is impossible to understand the dynamics of tourism development in China without including considerations of the other China' (Lew 1995: 155). What Lew[19] calls *the other China* and what has been called *the*

Chinese Commonwealth by Kao (1993), includes not only the Chinese living in Hong Kong, Macao and Taiwan but also all persons of Chinese descent living in the Chinese diaspora all over the world. These persons have not only been by far the largest group of inbound visitors to the People's Republic and the main investors into the tourism infrastructure there, but also the first 'destination' for outbound tourism from China. In this sense, transcending national and spatial realities, this author prefers to use the term *Chinese world*.

The importance attached to the inhabitants of this imagined community in terms of tourism is illustrated by the fact that within weeks after the official declaration of the creation of the People's Republic of China on 1 October 1949, the first Overseas Travel Services were opened at the border-crossing between Hong Kong and China in Shenzhen (Qian 2003) and in Xiamen (Zhang, Pine, Lam 2005). In 1953, the Beijing Overseas Chinese Travel Service was set up 'to receive the overseas Chinese who came back to the mainland to meet their relatives or friends, or tour the country' (Gao, Zhang 1983: 75), a year earlier than the CITS for non-Chinese visitors. For non-tourism questions, provincial and county level Overseas Chinese Affairs Offices were established in relevant areas after 1949 (Huang 2000).

In 1963, the 'Beijing' in Beijing Overseas Chinese Travel Service was replaced by 'General', using the change of name to show its increased national importance, again a year before the establishment of the China Bureau of Travel and Tourism responsible for non-Chinese under the leadership of the Foreign Ministry. In 1974, the 'overseas Chinese' also made way for 'China' to the now-familiar CTS, four years earlier than the paradigm change on inbound tourism.

To understand the distinctions made and the importance attached to them, a brief look into the concept of overseas Chinese is necessary.

Even within the framework of the western idea of a nation-state, Chinese governments still insist on the special and close relations between the so-called *overseas Chinese*[20] and China, regardless of the citizen status of such persons. Kotkin (1993) identified the Chinese and the Jews as the most important of what he calls *global tribes*, geographically dispersed culture groups that have a strong sense of common origin and maintain relationships to their ancestral homelands through a network of business and cultural ties. Urry (2003) points out that the overseas Chinese are 'societies formed through the global fluid of travelling peoples' (Urry 2003: 62) and calls them 'the most striking non-nation-state society' (Urry 2003: 107).

Communities of Chinese, starting in the twelveth and thirteenth centuries, were increasingly established in the seventeenth century in Southeast Asia but also in Japan, mainly by traders from Fujian and Guangdong. The mass emigration of so-called coolies resulted in the start of most Chinatowns in the Americas, Oceania and Europe that still exist today.[21] Most of these *sojourners* (*qiao*[22]) returned home after working hard and saving money for many years. Some stayed on though in their 'host' countries especially as after the end of Imperial rule in China it became less and less attractive to return to China. Estimates put the number of overseas Chinese at 56 million, of which about 50 per cent live in Hong Kong, Macao and Taiwan, the other half in other places, mostly in Southeast Asia. In the Americas, 3 million can be found and half a million reside in Europe (Lew 2000).

The People's Republic before 1978 did not really welcome returnees, suspecting them of spying or troublemaking. The Cultural Revolution stopped even the possibility for family members of overseas Chinese living in China to leave the People's Republic as they could before. After 1966, ties to relatives living overseas were a reason for persecution, property of the overseas Chinese was often confiscated. Under such circumstances, most overseas Chinese communities became affiliated with the government on Taiwan and experienced a reduction in their ties with Mainland China.

Three conferences in Beijing in 1977 and 1978 signalled a turn-around in overseas Chinese policies, with 'a United Front of revolutionary struggle giving place to the United Front of modernisation construction' (Barabantseva 2005: 9). A large array of institutions was started and overseas Chinese relatives and returnees were given privileges such as easier university access. From 1985 onwards, no visas were necessary anymore for overseas Chinese to enter China:

> Since 1978, and especially since the mid-eighties, the Chinese govern-ment has shifted its discourse of the nation from a territorially orien-tated 'peoples of China' (*Zhongguo ge minzu*) to a culturally or racially focused 'Chinese people' (*Zhongguo minzu*) and to traditional culture and 'values'.
>
> (Duara 2005: 52)

The appeal towards Chinese living outside Mainland China to use the new investment opportunities was very successful. The overseas Chinese have played an important role in the modernization of China, not the least by providing approximately two-thirds of all FDI entering China between 1979 and 2000, a sum of over 220 billion US$. The importance of Chinese FDI was even more pronounced in the earlier part of the integration of China into the world economy, with 76 per cent of all FDI between 1979 and 1993 originating in Hong Kong and Macao and 9 per cent in Taiwan (Fan 1997: 148).

The Chinese government strongly influenced and supported the strength-ening of the overall sense of Chineseness among the Chinese commun-ities with special government institutions, media and conferences and the sponsorship of associations, schools and events abroad. The wave of 'new migrants'[23] after 1978 resulted in the fact that today in western countries and in Japan the majority of Chinese communities are made up by persons who 'grew up under the red flag' (Barabantseva 2005: 19).

In Europe, Japan and Australia respectively more than 30 newspapers, radio and TV channels are provided for and by overseas Chinese. Chinese satellite TV stations and Internet websites are received worldwide. Sun (2002: 9) argues that in the global age the emergence of a single 'media-tized' Chinese community can be witnessed.

Since the early 1990s, these policies of strongly connecting Chinese living outside China with the homeland (Lew, Wong 2003) were extended from the *huaqiao*, Chinese citizens, to the *huaren*, ethnic Chinese with other nationalities. A regulation of the State Council about 'strengthening the work towards Overseas Chinese and Foreigners of Chinese Descent' stipulated that ethnic Chinese should not be treated like *common foreigners*. Whereas students studying abroad had been strongly encouraged to return to China in the 1980s, this not very successful policy was changed towards the establishment of strong ties with those who did not return.

In 1996 the State Council stated that 'people who left Mainland China to reside abroad ... will become a backbone force friendly to us in the United States and some other developed Western countries' (Barabantseva 2005: 14).

Premature exposure to Western culture by sending abroad primary and secondary students is, however, not approved of, in opposition to the support of the going abroad of mature and nationally aware 'ripe' citizens.

To sum up: by claiming the Chinese identity of all ethnic Chinese and their 'natural' affiliation towards the People's Republic, the government in Beijing has quite successfully used a transnationalistic approach (Nyiri, Breidenbach 2005) to utilize the Chinese living in Hong Kong, Macao, other countries and even Taiwan to support the modernization of China economically and the status of China politically. This happened within the framework of the long-terms goals of regaining control over Hong Kong, Macao and Taiwan, winning the support of overseas communities in com-petition to the Taibei government and turning the brain drain of Chinese students not returning, but staying outside of China, into an advantage.

Tourism has been a major tool in achieving these goals. The task of CTS before 1978, to show overseas Chinese visitors the splendour of socialist development was not an easy one with a limited number of guests arriving and little to show to them, hampered by inadequate tourism infra-structure on top. Citizens of Hong Kong and Macao who did not even need a visa but just a travel identity card, would visit relatives living not too far away from the border more frequently before the Cultural Revolution. Some also crossed just into Shenzhen for political as well as recreational reasons:

> In the 1950s and 1960s, when China lay secluded behind its bamboo curtain, ... troupes of artists, dancers and theatrical performers would come here to perform and people from Hong Kong would be invited to attend those performances as special guests. ... For those who had

time to stay longer, the major tourist attraction centred around the Shenzhen Reservoir, a scenic spot of green hills and water and with an art exhibition hall.

(Wong 1982: 68)

For the years 1956 to 1965 approximately 200,000 overseas Chinese from 80 different countries and 1 million Hong Kong and Macao Chinese are thought to have been handled by CTS (Fu 2004: 43).

Since 1978 the number of border-crossings of Chinese without a Chinese passport has increased almost year by year, with 1989 (Tiananmen upheaval) and 2003 (Severe Acute Respiratory Syndrome (SARS))[24] being the only exceptions. For 2002, almost 85 million border-crossings were recorded for compatriots from Hong Kong, Macao and Taiwan. This figure is heavily inflated by day trippers, Hong Kong citizens having their second or even first home across the border, etc. The WTO figure of 23 million international travellers, which is not disputed by CNTA, is still a sign of the successful opening of China to this group.

Furthermore, a good part of the 13 million foreign-passport holders entering China in that year are in fact ethnic Chinese with Singaporean, Malaysian, American or other nationalities. Lew (1995) gives an estimate of 15 per cent for 1991. This percentage may have risen with the larger number of naturalized Chinese living abroad and closer business ties but a precise figure remains unknown.

Eric Cohen (1979) coined the term 'existential tourism' to describe tourists who travelled in search of a personal or spiritual 'centre' that is located beyond their immediate place of residence, distinguishing them from the experiential and the experimental tourists who only observe or take part without commitment in what is considered as authentic life in the place to which they travelled.[25]

With travel into China becoming an option after 1978 after decades, many overseas Chinese choose to engage in existential tourism back to their *existential home*. Whereas foreign visitors flocked to Beijing, Shanghai and Xian, overseas Chinese would rather travel, laden with gifts, to the rural areas of Guangdong and Fujian, home of the majority of overseas Chinese (Lew 1995).

Overseas Chinese and compatriots also faced no restrictions in their travels or contacts within China, whereas for foreigners cities and provinces were 'opened' only gradually; spontaneous contact of local inhabitants with foreigners was still discouraged in the 1980s.

Outbound policies towards the Chinese world did not differ from those to the rest of the world before 1983. International travel was prohibited and only a few official delegations had the chance to get a glimpse of the outside world. The only large group of Chinese citizens leaving China temporarily before that date was probably the survivors among the 260,000 so-called Chinese People's volunteer troops fighting in the Korean War in 1950–1953 (Chen 1994) – not exactly a leisure activity.

In 1983, it had become clear that the British colony of Hong Kong would have to be given back to China after the lapse of the lease to the New Territories in mid-1997 and that Portuguese Macao would follow soon. The treaty putting paid to British rule was signed in December 1984, the Portuguese government agreed in 1987 to hand over Macao in December 1999. In this situation, the demand for outbound travel and the policies of strengthening the ties with the inhabitants of the soon-to-be SARs coincided and prompted the organization of 'visiting relatives' tours in November 1983 to Hong Kong and from 1984 also to Macao. These tours took the form of bus tours for a few days only, with all costs paid by the relatives who had to officially invite the participants. A quota on the number of visitors was imposed by the Hong Kong government, from just a few persons per day to 1,500 in 1998, 2,000 in the year 2000 and finally the abolishment of all restrictions in 2002 (Zhang, Jenkins, Qu 2003).

A growing number of Chinese, adhering in their own way to the policy of 'Some getting rich first' of Deng Xiaoping, became more affluent and the number of official sightseeing trips paid with public funding increased during the 1980s. Since 1986, the law on the administration of entry and exit of Chinese citizens officially granted every citizen the right to obtain a private passport if an invitation letter and sponsorship from abroad could be provided (Pieke 2002).

These developments increased the pressure to open further travel opportunities outside official delegations. Only 350,000 people, much less than one per million of the population, had been able to join a 'meet relatives' tour between 1979 and 1986 (Verhelst 2003).

The 1990 'Provisional Regulations on Management of Organizing Chinese Citizens to Travel to Three Countries in South-East Asia' opened Thailand, Singapore and Malaysia for VFR group travels. The Philippines was added as the fourth country in 1992. The process of acquiring the necessary permission for a passport was cumbersome and invitation letters and sponsorship were needed from the country chosen as the destination. These travels still served the purpose of using outbound tourism for closer bonds with overseas Chinese and operated under the precondition that no losses for the Chinese economy occurred. Over time, the obligation for the hosts to pay for the trips was less and less strictly enforced, and the character of many trips was at least partly inspired more by sightseeing and shopping than by a yearning to meet relatives became apparent. Beside the channel of official, public-paid business, study and investigation tours to other countries, a new possibility to see at least some corners of the world was added.

Until 2001 the system of ADS, installed in 1995, still concentrated on Asian and Oceanic countries, most of them with sizeable Chinese communities. However, with the official acknowledgement of the existence of privately paid holidays outside China in 1997, the VFR aspect could no longer solely justify recreational journeys.

This does not mean however, that tourism ceased to be a political tool within the Chinese world. Visits to summer camps in China for young people are important parts of strengthening the links between Mainland China and the next generation of the diaspora, especially by parading the economic success of China vis-à-vis Taiwan. Delegations from China are still sent out to visit overseas Chinese and influential individuals:

> These are not always government delegations but can take the form of informal groups from mass organizations, business delegations, or even tour groups . . . In Europe in particular, delegations almost invariably engage in sightseeing trips, typically accompanied by representatives of local overseas Chinese associations.
>
> (Nyiri 2005b: 151)

In 1998, 270 official delegations were received by the Federation of Chinese Associations in Rome alone. In the other direction, the overseas Chinese bureaux in Fujian Province in 1995 received more than 500 overseas Chinese delegations comprising 70,000 people (Nyiri 2005b). Chinese outbound tourism is also serving the strengthening of ties. A large, if not the largest, part of official and unofficial inbound tourism from China is organized by overseas Chinese. The business acquired from the growing outbound service links them in a stronger way to the current development of China and makes affiliation with Beijing, instead of Taibei, an economically sensible decision.

Chinese tourism policies towards the non-Chinese world 1949–1997

'Tourism' as a concept did not apply to China before 1978 and was treated predominantly as international tourists' sightseeing visits to China for many years afterwards. Given the economic situation and the isolation of the People's Republic from the western world, outbound tourism to such countries was not an option. However, with the Maoist view of tourism as an anti-socialist form of behaviour, even possible visits to 'brotherly countries' before 1960 are not recorded.

The former premier and foreign minister of China, Zhou Enlai, allegedly encouraged the idea of establishing an official organization to deal with foreign visitors (Gerstlacher, Krieg, Sternfeld 1991). These were, beside technical personnel from the Soviet Union, selected 'friends' of China, in the 1950s mainly also from the Soviet Union and Eastern Europe as well as from North Korea and Vietnam.

In 1954 the CITS was set up, modelled after 'Intourist' style monopolistic organizations in the Soviet Union and elsewhere with the notable difference that the majority of visitors, the overseas Chinese and compatriots, actually were handled by another organization, the CTS:

The limited foreign visitation, which existed, was sanctioned on the grounds that the successes of communism could be paraded before a selected international audience. Tours focused on the material achievements of communism such as factories, communes and revolutionary peasant, and worker communities. Heritage was not promoted. Contact between tourists and locals were strictly regulated. Segregation in hotels into the categories of foreign tourists, overseas Chinese, Hong Kong and Macau Chinese, and locals, was rigidly enforced.

(Sofield, Li 1998a: 369)

Even though international understanding and friendship were the topics of many a toast at banquet tables, Chinese and tourists were kept apart under a system of what Richter (1989) called 'portable apartheid'.[26]

The reasons were spelt out in a paper on 'Socio-cultural Impacts of Tourism' as late as 1984:

Tourists from capitalist countries bring capitalist concepts like 'democracy and freedom' to developing countries, which may lead to social problems and unrest in peaceful, stable societies. The tourists bring western capitalist ethics like 'love money more than everything' and egoism to the host countries. This may lead to frauds and rivalry in the modest host countries. Moreover, western life-style introduced by tourism, brings the social evils of the western societies such as gambling, prostitution, drug addiction and alcoholism. The younger generation may very easily be influenced by the capitalist way of life from abroad. This is a great danger for the traditional life-style. Tourism may promote the unreflected worship of everything foreign. The tourists' wealth and comfort may corrupt those employed in tourism and the local people and create desires, with the effect that they may lose their national pride and get an inferiority complex. This is a very serious form of pollution caused by tourism.

(Xu, Shao 1984: 10, in Gerstlacher, Krieg, Sternfeld 1991: 70)

With Mao Zedong temporarily losing much of his influence after the famines of the early 1960s, the Central Committee of the Communist Party approved a 'report on tourism development', which also included gaining some foreign currency for the state as one of the goals of tourism. To increase the number of foreign visitors, the China Bureau of Travel and Tourism was established under the direct leadership of the State Council in 1964. In the following year an increase in the number of foreign visitors to 13,000 could be reported, 11,000 of them coming from western countries (Fu 2004: 43).

Within a year, however, the Cultural Revolution made tourism a 'non-issue' (Zhang, Chong, Ap 1999: 473), destroying the established structure and almost completely closing off the country.

Following the visit of US President Nixon to China in 1972, a limited growth of inbound tourists from the USA, Europe and also from Japan set in, reaching 56,000 visitors in 1977. In the same period Vice-Premier Li Xiannian nevertheless still insisted that China was not planning to expand tourism in the foreseeable future (Richter 1989: 25).

The 'First All China Tourism Conference' convened in Beijing in January 1978, two months before the third session of the eleventh Central Committee officially declared the start of 'Reform and Opening'. It called for the improvement of tourism infrastructures and the reduction of travel restrictions for foreigners and overseas Chinese. A second conference in September 1979 in Beidaihe set an optimistic goal to reach a number of 3 to 5 million international tourists by 1985.[27] The need to separate the local population from unhealthy and uncivilized influences was emphasized again: 'Some weak-willed Chinese, particularly youngsters, could not withstand such influences and blindly pursue the way of life of the foreigners' (Gao, Zhang 1983: 78). Development was needed, not only to earn money from western tourists, but also to facilitate business visits of potential investors.

During the 1980s the rapid growth of foreign visitors meeting an under-developed infrastructure, and an organizational structure overwhelmed by the influx but still trying to closely control the movement of all visitors, led to all kinds of service problems and complaints. The inexperience of the personnel working in hotels and as guides, etc. was aggravated by the lingering Maoist attitude that it is demeaning or inferior to serve others (Huyton, Sutton 1996; Tsang, Qu 2000). After 1980 foreign tourists were forced to use the newly introduced 'Foreign Exchange Certificates' (FECs) for all payments in China instead of the RMB, emphasizing the idea of foreign tourists moving in a 'tourist bubble' separete from the life of normal Chinese citizens. FECs acquired a status of quasi-hard currency with a black market exchange rate, giving access to products and services otherwise limited to foreigners only up to their final abandonment in 1994 (Pan 2002).

With the inclusion of tourism into the seventh Five-Year Plan, the CNTA, which replaced the State Administration for Travel and Tourism, prepared the first 'National Tourism Plan' again setting a goal of 5 million international visitors, this time for 1990.

The right to deal directly with foreign tour operators was decentral-ized and recentralized several times. Beside CNTA, an inter-ministerial Tourism Coordination Group from 1986 to 1988 and the National Tourism Commission after 1988 were set up to rectify the chaotic situation by, for instance, for the first time, establishing a horizontal channel of com-munication to forward CNTA's arrival forecasts to the Civil Aviation Administration of China (CAAC).

Also in the 1980s, under the supervision of CNTA, the CTA and the China Tourist Hotels Association (CTHA) were set up to promote the growth of the tourism industry. The China Association of Tourist Automobiles and Cruises (CATAC) started in 1988, but became officially approved by CNTA only in 1991. Star ratings for hotels were also introduced in 1988. The fact that the chairmen of the tourism industry associations are normally the retired leaders of tourism administration institutions, illustrates their limited power and main function as mere advisory bodies to the local and central tourism administrations (Zhang, Pine, Lam 2005).

The growth of international arrivals slowed down in 1987 and 1988, to experience a sharp reduction in the wake of the suppression of the Tiananmen movement. This gave some breathing space for the infrastructure to catch up with demand and to try to establish more rational organizational structures.

In 1991, the recovery of the international inbound tourism started the tenfold increase of arrivals from 1.7 million in 1990 to 17 million in 2004. In 1992 tourism was declared to be a key industry among tertiary industries and the ninth Five-Year Plan of 1995 marked tourism as the first industry to be vigorously developed among the tertiary industries.

High-level recognition of tourism was underlined by the reception of the delegates of the 2001 National Conference on Tourism Development by Premier Zhu Rongji. In the same year, he presented the outline of the tenth Five-Year Plan: 'Develop consumer-orientated industries such as real estate, community service, tourism, catering, recreation and fitness services, and increase service contents' (Du 2004c: 345).

Outbound travel to western countries remained the field of official delegations and students pursuing their education abroad in increasing numbers after 1985. The growing gap between the socialist ideology and the market economy reality was answered especially after 1989 with an increased nationalism[28] from the government side and with consumerism especially from the urban elites.

The 'normality' of travel restrictions in a Cold War world disappeared together with the Soviet Union. The export successes and the massive amounts of FDI flowing into China made it even harder to argue against outbound leisure travel outside the idea of invited VFR trips in the 1990s.

With hard currency easier to obtain by government institutions and big companies, more or less open leisure trips to western countries paid by public funds increased. In the second half of 1993, an austerity campaign tried to stop this phenomenon but the outbound figures only decreased for some months, so that the 1994 number of border-crossings out of China reached the same level as the figure for 1993.

Private outbound tourism as official tourism policy 1997–2005

The general policy of the Chinese government for many years has been that 'inbound tourism should be developed rapidly, domestic tourism should be developed actively, and outbound tourism should be developed moderately' (Wei, Wei 2005).[29]

In practical terms, this meant the discouragement of outbound tourism and, if this was not possible, its channelling into VFR tours within the 'Chinese world'.

This did not reflect the existing demand:

> Domestic tourist demand is enormous, and growing at a rapid pace. The rise of mass domestic tourism has documented many ongoing economic, social and political changes in contemporary China. Improving standard of living, certain degree of individual freedom, urban growth and changing rural-urban relations, booming business and social activities as well as changing lifestyle have stimulated increasing mobility and rising demand for leisure and tourism of a considerable part of the Chinese population. Travel begins to take a higher place in consumers' scale of preferences.
>
> (Xu 1999: 211)

An example of the ingenuity of the would-be travellers to non-Chinese destinations is the 'barter tourism' that developed after 1988 from the border-town Heihe in Heilongjiang across the Amur River to the city of Blagoveshchensk on the opposite side:

> In September 1988, the Heihe government originated the idea of tourism trade with its counterpart. This idea received the full support by the Chinese people because of their urge to see the outside world. It was a good compromise to see Russia if it was not possible to visit the United States or other western countries.
>
> (Zhao 1994: 402)

Because hard currency was missing on both sides, it was agreed that groups of 40 tourists would be exchanged for one-day excursions with each side charging its own tourists the cost of meals, transportation and guiding services plus profits. Half of the day was reserved for sightseeing, the other half for shopping, the Russians buying Chinese textiles and the Chinese buying Russian fur coats and hats. After the first 520 tourists were exchanged in 1988, the number swelled to 49,000 in 1992.

Very swiftly, the established contacts were used to extend offers. From 1989, package tours of four or seven days exchanged visitors to Khabarovsk with tourists going to Harbin or Jiamusi. In 1990, round-trips were started to Vladivostok and Qiqihar respectively. Even trips to Moscow were successfully negotiated with one Chinese tourist going to Moscow 'worth' four Russians going to Harbin.

After the end of the Soviet Union, the Russian side replaced person–person barter with payment in trucks, boats and other goods of unspecified former ownership and later with payment in US dollars. Besides providing outbound tourism opportunities to Chinese, Heihe City even experienced an economic boom with a railway line and several first-class hotels constructed and Harbin Airport getting daily international flights to serve the 67,000 Russians arriving in 1992 alone.

It was not that 'the government's relaxation of the rules for travelling overseas encouraged people to travel abroad' as Zhang, Chong and Ap (1999: 481), quoting Zhang and Qu (1996), assume, but the urgent wish of the people to travel abroad brought about the relaxation of the rules, granted by a government wise enough not to show the inflexibility of their erstwhile Eastern European comrades.

Activities like the Amur River border tourism signified the urgency of the establishment of organized outbound tourism being brought under the control of the central government. The ADS system was developed to achieve this goal.

In the decade following its introduction in 1995, the ADS system was a major instrument of the Chinese government for the attempted control of the development of outbound tourism. At the end of the decade, almost all important destinations were covered by this agreement between the Chinese government and the destination countries; its demise has been predicted for the near future (Arlt 2004; TBP 2004), even if CNTA publicly insists[30] that 'the ADS-system is only young and should therefore not be repealed any time soon' (WTO 2003: 20).

The ADS system is based on bilateral[31] tourism agreements whereby a government allows self-paying Chinese tourists to travel for pleasure to its territory within guided package groups and with a special visa. Only ADS countries can openly be promoted as a tourism destination in Chinese media.

The main interest of the Chinese government in introducing the ADS system was initially a means of restricting the number of Chinese travellers spending hard currency abroad. For the receiving countries, the focus was on an increase in inbound tourists as well as finding ways to reduce the risk of travellers not returning to China but attempting legal and especially illegal immigration in the host country.[32]

ADS agreements involve mainly three government departments in China: the Ministries of Foreign Affairs (MFA) and of Public Security (MPS) as well as the CNTA. In the receiving country, similar ministries and state-run NTOs are involved, Ministries of Immigration, Foreign Affairs, the Economy or others responsible for international relations and tourism.

According to Chinese usage, ADS is put into operation in several stages: in the first round, after preliminary discussions, a country officially hands in an application, which is discussed by CNTA, MFA and MPS, resulting in a report sent to the State Council. The State Council approves the application and announces the conferment of ADS status. The announcement triggers a second round of discussions of the details of the agreement.

Finally, a Memorandum of Understanding (MOU) is agreed and signed by both parties. The agreement is considered operational. This is then followed by further discussions about the setting up of organized tours. As a result, with the commencement of the first organized Chinese tour, the agreement is finally consummated. At each of these steps, delays and interruptions are possible. In some instances, most prominently Australia and Japan, the validity of the agreement has been limited at the beginning to travellers from selected provinces only, i.e. Beijing, Shanghai and Guangdong.

The practical consequences of the ADS status for the potential Chinese travellers are mainly that:

- ADS visas can be obtained by designated travel agencies directly, the travellers do not have to visit the consulates and to do all the paper-work themselves;
- for ADS countries private passports can be used and hard currency exchanged;
- the time for the procurement of visas can be reduced;
- there is no need anymore to pretend to visit business partners or fairs, etc.

For the receiving countries positive effects are:

- the opportunity to receive a larger number of Chinese tourists especially if the destination is given the status earlier than competing destinations;
- the reduction in workload for the consulates in China, and release from the pretence that 'business' visa applications were clearly from travellers going on pleasure trips;
- the possibility of actively communicating their country as a destination for Chinese tourists.[33]

The biggest problem for both sides remains the question of how to minimize the risk of unwanted illegal immigration, given the fact that the consulates are not in direct contact with the travellers and the travel agencies can be more interested in a larger number of customers than in strictly controlling the background of the participants in ADS groups.

As an important part of each MOU, an annex with a list of authorized companies is included, which are in theory the only companies able to handle ADS tourism groups, both in China as travel agencies that can send leisure tour groups as well as the inbound agencies in the receiving destinations.

ADS covers only Chinese travelling as part of a leisure tour group of at least five persons[34] out of China. Business travel, government other official travel, and some other types of travel – such as students going abroad for education, sports travel or incentive trips – are not included.

The restriction regarding the promotion of tourism to non-ADS countries has an effect in practice also on group tourism only. An individual Chinese passport holder, provided that he or she has the financial resources, has no restrictions on travelling abroad, as long as the country of destination issues individual visas for tourists.

With the ADS system and the increase in outbound tourism, the number of passports issued in China has increased while the difficulties in getting a passport became less.

Not counting the special passports for Chinese living in Hong Kong SAR or Macao SAR and for seamen, there are two groups and several different types of passport. A general distinction can be made between Public and Private Passports.

Public Passports are issued and renewed by the MFA or the respective Foreign Departments of the municipalities or provinces (*wai-ban*). There are three different kinds of Public Passport: Diplomatic Passport, Service Passport and Ordinary Passports for Public Affairs. Diplomatic Passports are issued to senior officials of the Communist Party, the government and the People's Liberation Army, principal leaders of the National People's Congress, the Chinese People's Political Consultative Conference, and all democratic parties, diplomats, consular officers and their accompanying spouses, underage children, diplomatic couriers, etc.

Service Passports are issued to officials of governmental offices ranking division chief or higher, staff members of diplomatic and consular missions in foreign countries, members of the United Nations' organizations and its specialized agencies, and their accompanying spouses and underage children, etc. Diplomatic and Service Passports are valid for five years.

Ordinary Passports for Public Affairs are issued to staff in the governmental offices at all levels and to employees of state-owned or joint-venture enterprises or institutions. They can also be used by attendants of international conferences, visiting scholars, contract workers of state-owned enterprises, representatives of Non-Governmental Organizations (NGOs), journalists, exhibitors, athletes, aircrew members, etc. They are valid for two years (Verhelst 2003).

Official Passports can be used individually or in a group for business and professional trips or for study tours to all countries. Voyages undertaken with them are typically not paid with private money but from government or company budgets. The issue of a Public Passport still involves rather complicated and time-consuming processes. Official Passports cannot be used for openly tourist type travelling, even if paid privately, and therefore not for the application of ADS visas.

Private Passports are issued by the offices of the MPS in the form of Ordinary Passports for Private Affairs and Tourist Passports. Ordinary Passports for Private Affairs are issued to those who go abroad for personal affairs, mainly for VFR or studying. The Ordinary Passport for Private Affairs is valid for five years and can be extended once for another five years.

Before the year 2001, it was necessary for the application for a Private Passport, if it was not issued for studies abroad, to present an invitation letter from the friend or relative together with a copy of the identification card of this person to the Public Security Bureau. Until 2002, Private Passports had an exit-card attached, which indicated the destination of the first trip. Once the passport holder returned from the indicated destination, the exit-card was taken off. This meant that after the first exit and return to China, subsequent trips to other destinations did not need further approval. Application for Private Passports used to be hard work, with approval of the work unit (*danwei*), the police and the municipal authorities necessary as a part of a process of prolonged investigation. In the early 1990s, six months would be required to obtain a passport. In 1995, this was shortened to one month, in 2000 to 15 days (CTS 2001). From 2005, ID and the proof of legal residence (*hukou*) is enough for the citizens of most bigger cities for the application for a Private Passport.

The holder of a Private Passport can travel to any ADS country for leisure, as long as s/he joins an organized group. For other destinations, for leisure trips, and for business trips of owners or employees of private companies who could not obtain a Public Passport, evidence of friends or relatives had to be produced.[35]

Since the advent of the ADS, a new type of Private Passport was created in order to facilitate travel to these ADS destinations: the Tourist Passport. These passports are valid only for one year and for one exit and one entry, stating the destination of the travel they are issued for. They are issued by the MPS after application by the travel agency, so there is no need for the tourists themselves to apply and for tourism trips to ADS countries there is no need to produce invitation letters.

Tourist Passports are valid for travels to ADS countries only. However, in the late 1990s, groups of Chinese tourists already travelled to Europe on Tourist Passports, which were regarded as valid travel documents by some European embassies and consulates. As leisure trips to Europe at that time were not approved by the Chinese government, 'the departure to Europe on tourist passports . . . (was) only possible through, to say the least, "special connections" of the travel agency with Chinese passport control officers' (Roth 1998: 12).

For Hong Kong and Macao, a travel permit can be used instead of a passport.

In the first years after the introduction, ADS agreements were the result of long and protracted negotiations. Complicated rules of reciprocity and other limitations were included, with seven preconditions to be met by the applying country:

a. The country should be a source market of outbound tourists to China.
b. The country should have good political relationships with China.

   c.   Tourism resources should be attractive and facilities suitable for
        Chinese travellers.
   d.   Safety and freedom from discrimination for the Chinese travellers
        should be guaranteed.
   e.   The country should be easily accessible from China.
   f.   The expenditure of tourists from the country visiting China should
        be balanced with the expenditure of Chinese tourists travelling to
        that country.
   g.   The number of tourists from the country and from China should
        only increase reciprocally.

(Guo et al. 2005b)

These kinds of restrictions clearly were impossible to meet for most countries in the world, which could neither guarantee total safety nor force their own tourists to visit China and to spend a specific sum of money there. If China had insisted on these rules, ADS could have probably only have been awarded to North Korea and Cuba.

With the rapid development of the Chinese outbound tourism, however, the speed of successful discussions increased. Many countries that were awarded ADS do not meet several of the seven limitations, being either no source market for tourism to China, or having no direct air link, or being short of Chinese restaurants and Chinese-speaking guides, or all of these. The increase of Chinese visitors to many countries and the amount of money spent there is in reality not connected to the numbers of visitors or the amount of money spent in China by visitors from those countries.

In some cases, ADS has been used by the Foreign Ministry as a kind of 'gift' to countries that in terms of tourism business are insignificant.[36] It might also, as Guo remarks, 'be used in future *against* some governments for political reasons' (Guo 2002: 8, emphasis added).

The idea of reciprocity of numbers of tourists from both parties of an ADS agreement has been given up, as has the rather naive idea that in market economies the governments of the receiving countries could restrict the incoming business activities of private companies through an MOU with China.

The crucial part of the open acceptance of private outbound leisure tourism can be seen in the 'Provisional Regulations on the Management of Outbound Travel by Chinese Citizens at Their Own Expense', which came into force in mid-1997, in parallel with the return of Hong Kong and – unexpectedly no doubt – with the start of the Asian financial crisis of 1997/1998, bringing a welcomed drop in the exchange rates of many Asian currencies vis-à-vis the Yuan, which stayed pegged to the US$. With the 'Provisional Regulations', the main goal of tourism development, repeated again and again since 1978, to earn hard currency from international tourism exchanges, had been abandoned.

Wen and Tisdell, using 1998 figures, sound like they are whistling in the dark when stating:

> It seems possible that there is some suppressed demand for outbound travel from China due to the regulations on such travel and administrative obstacles to outbound travellers. Greater liberalisation and continuing rises in income could result in a substantial rise in the number of China's outbound travellers. Nevertheless, China's international tourism account is likely to remain in positive balance for some time to come. Consequently, tourism and China's tourism industry will continue to make a net positive contribution to China's foreign exchange earnings.
>
> (Wen, Tisdell 2001: 40)

In 2003, Zhang Jianzhong, director of the Policy and Regulation Department of CNTA, stated during the First International Forum on Chinese Outbound Tourism and Marketing in Beijing that tourism consumption is a mark and symbol of the well-off society China is striving to become. He argued that outbound tourism development should, however, continue to be developed 'moderately' and only in a way that guarantees the balance between revenue and expenditure of the international service trade of China. 'The outbound tourism of Chinese citizens is provided with a long term potential, which shall form an integrated system together with the inbound tourism and internal travels' (Zhang 2003b: 79).

The emphasis on earning foreign currencies is also still visible in the *Yearbook of China Tourism Statistics*: outbound tourism is only mentioned in the text preceding the main table part; for inbound tourism a whole chapter provides detailed tables on 'International tourism receipts', whereas for domestic tourism only the 'expenditures' are mentioned (CNTA 2004a).

The currency control policy relaxed in line with the general development. Before 1994, just 60 US$ could be exchanged for overseas travel per year, when the amount was raised to 1,000 US$. From 1997 until 2001, a total of 2,000 US$ could be exchanged for overseas travel per year for places other than Hong Kong or Macao, after April 2001 this sum could be exchanged per trip. Since October 2003, the amount increased to 3,000 US$ per trip and in July 2005 it increased further to 5,000 US$ for trips shorter than six months and 8,000 US$ for trips longer than six months.[37] In Hong Kong, Macao and in some Southeast Asian countries RMB are also accepted using an acceptable exchange rate.

The ADS system was introduced by China in 1995, when agreements were made with Hong Kong and Macao, co-existing for a while with the 1984 rules of VFR trips, which supposedly had to be paid by the relatives there. The 'Three (plus one) Southeast Asian' countries of Thailand, Singapore, Malaysia and the Philippines were similarly upgraded.

Table 3.1 Nations with approved ADS as of January 2006[38]

Country and date of ADS MOU signed	Date[a]	Regional openings
Hong Kong[b]	1983	1983 Guangdong; 1984 Beijing, Shanghai,
Macao[b]	1983	Fujian, Zhejiang, Jiangsu; 1996 26 provinces; 1998 all provinces except Xizang (Tibet); 2003 International Visitors Schemes for 8 cities in Guangdong, Beijing, Shanghai; 2004 all Guangdong, 11 more big cities
Thailand[b]	1988	
Singapore[b]	1990	
Malaysia[b]	1990	
Philippines[b]	1992	
South Korea (1998)	1998	1998 Beijing, Shanghai, Tianjin, Chongqing, Guangdong, Shandong, Anhui, Shanxi, Jiangsu; 2000 all provinces
Australia (1999)	1999	1999 Beijing, Shanghai, Guangdong;
New Zealand (1999)	1999	2004 Tianjin, Hebei, Shandong, Chongqing, Jiangsu, Zhejiang
Japan (2000)	2000	2000 Beijing, Shanghai, Guangdong; 2004 Liaoning, Tianjin, Shandong, Jiangsu, Zhejiang
Vietnam (2000)	2001	
Cambodia (2000)	2001	
Myanmar (2000)	2001	
Brunei (2000)	2001	
Nepal (2001)	2002	
Indonesia (2001)	2002	
Malta (2001)	2002	
Turkey (2001)	2002	
Egypt (2002)	2002	
Germany (2002)	2003	
India (2002)	2003	
Maldives (2002)	2003	
Sri Lanka (2002)	2003	
South Africa (2002)	2003	
Russia (2002)	Agreement outside ADS in 2005	
Croatia (2003)	2003	
Hungary (2003)	2003	
Pakistan (2003)	2003	
Cuba (2003)	2003	
Jordan (2003)	2004	
Mongolia (2003)	2004	
Greece (2004)	2004	
France (2004)	2004	

a Date of agreement coming into force.
b Until July 1997 officially only for VFR tourism.

Country and date of ADS MOU signed	Date[a]	Regional openings
Netherlands (2004)	2004	
Belgium (2004)	2004	
Luxembourg (2004)	2004	
Portugal (2004)	2004	
Spain (2004)	2004	
Italy (2004)	2004	
Austria (2004)	2004	
Finland (2004)	2004	
Sweden (2004)	2004	
Czech Republic (2004)	2004	
Estonia (2004)	2004	
Latvia (2004)	2004	
Lithuania (2004)	2004	
Poland (2004)	2004	
Slovenia (2004)	2004	
Slovakia (2004)	2004	
Cyprus (2004)	2004	
Denmark (2004)	2004	
Iceland (2004)	2004	
Ireland (2004)	2004	
Norway (2004)	2004	
Romania (2004)	2004	
Switzerland (2004)	2004	
Liechtenstein (2004)	2004	
Ethopia (2004)	2004	
Zimbabwe (2004)	2004	
Tanzania (2004)	2004	
Mauritius (2004)	2004	
Tunisia (2004)	2004	
Seychelles (2004)	2004	
Kenya (2004)	2004	
Zambia (2004)	2004	
Laos (2004)		
Fiji (2004)		
Commonwealth of Northern Marianas (CNMI) (2004)		
Tonga (2004)		
Cook Islands (2004)		
Vanuatu (2004)		
Argentine (2004)		
Brazil (2004)		
Chile (2004)		
Peru (2004)		
Mexico (2004)		
Barbados (2004)		
Antigua and Barbuda (2004)		
United Kingdom (2005)	2005	
Israel (2005)		

The first real ADS agreements with new destinations, having Chinese communities but with 'western' cultural backgrounds, were signed in 1997 with Australia and in 1998 with New Zealand, and came into force in 1999. After a number of Asian countries were added, the year 2002 saw the first European (Malta) and the first African (Egypt) countries added. From here, the pace increased, with the en bloc agreement with most countries of the European Union (EU) member countries in April 2004 forming a remarkable milestone.

In many countries the first ADS tourist groups were greeted as the 'first tourists from China' (Finck 2004; Williamson 2003). In fact Chinese citizens have been the customers of incoming agencies all over the world for 20 years already, albeit in the disguise of business or VFR travellers.[39] ADS has brought these phenomenon into the open by establishing the opportunity to participate in group leisure travel with private money and Private Passports. This helped to substitute political influence and *guanxi* as the major tools to catch an opportunity to travel abroad with the common market economy instrument of exchange: money.

The temporary distortion of the flow volumes from a supply and demand pattern, which was caused by the ADS system for the decade starting in 1995, is coming to a close. By the end of 2005, more than 100 countries have completed at least the first round of the ADS agreement procedure, with even the USA entering the process.[40]

Domestic travel services' involvement in Chinese outbound tourism

As seen above, before 1978 tourism was officially only organized with CITS and CTS as the main government organizations handling the few foreign visitors and the overseas Chinese and compatriots respectively.

The concentration of hard-currency earning was reflected after 1978 in the development of inbound tourism before domestic – or even outbound – tourism, a very unusual policy, an 'extraordinary strategy' (Du 2004c). The duopoly of the centrally organized travel service, to which China Youth Travel Service (CYTS) was added in 1980, was unable to handle the increasing numbers of foreign visitors and faced a growing number of other organizations run by regional government offices, state- and collectively-owned companies and even private initiatives, which were involved in the unregulated domestic tourism sector (Choy, Gee 1983).

As result of several steps of decentralization, recentralization and anewed decentralization and in acknowledgement of the evolved complex situation, the 'Provisional Regulation on the Administration of Travel Agencies' was issued in 1985, establishing a three-tier system of travel agencies:

1   Category I agencies were permitted to receive all types of tourists and conduct direct sales and marketing overseas. Their number increased from 17 in 1987 to 136 in 1992 and 352 in 1996.

2 Category II agencies could handle domestic as well as inbound Chinese tourists, i.e. overseas Chinese and compatriots, and organize ground services for international inbound tourists. The number of this kind of agencies remained more or less stable with 677 in 1987, 701 in 1992 and 607 in 1996.[41]

3 Category III was open to private and collective ownership. These agencies were only allowed to handle domestic tourists. Their number grew in accordance with the development of domestic tourism from 551 in 1987 to 1,755 in 1992 and 3,995 in 1996.

In this system, outbound tourism still was absent (Choy, Guan, Zhang 1986). Outside the three-tier system, CNTA licensed originally the three headquarters of CITS, CTS and CYTS plus the Guangdong and Fujian branches of CITS and CTS to handle overseas tours for Chinese nationals (Zhang, Pine, Lam 2005), with three more agencies added before 1992 (TBP 2004).

Within the three-tier system, it developed into a common practice for inbound business-starved second category agencies to operate illegally in the inbound business by luring foreign tour operators with low prices and other privileges. On the other hand, to participate in the more profitable and easier to undertake domestic travel business, both first and second category agencies established their own domestic tourism departments to the disadvantage of third category agencies, making 'an already competitive domestic tourism market even more competitive' (Du 2004d: 67).

In 1996, with the 'Regulations on the Administration of Travel Agencies', the three-tier system was replaced by a simpler distinction between 'International Travel Agencies' (ITAs) and 'Domestic Travel Agencies' (DTAs), in effect merging the first two categories into one. Within the ITAs, however, still only the minority was licensed to handle domestic, inbound and for the first time also outbound tourism, whereas the majority was only allowed to handle inbound and domestic tourism. DTAs were confined to servicing domestic tourists. The China Association of Travel Services (CATS) was founded the following year as a trade association.

To be awarded a licence, deposit payments to a newly established Tourism Quality Control Institute of the CNTA were now required.[42] For a full ITA licence, beside certain minimum requirements for turnover and number of inbound customers, this deposit reached the sum of 1 million Yuan RMB (approximately 100,000 Euros in 2005). According to the volume of inbound business, ITAs were given a passenger quota of how many outbound travellers they were allowed to handle each year.

The number of fully licensed international ITAs was restricted to an elite group of 67 agencies until 2002. These were mainly the headquarters and the provincial branch offices of the big companies CITS, CTS, CYTS plus China Women Travel Service (CWTS) as well as CCT and China Merchants Travel (CMT). Thirty-seven, i.e. more than half of all fully licensed ITAs, were concentrated in Beijing, Shanghai and the coastal provinces of Liaoning, Jiangsu, Zhejiang, Fujian and Guangdong, with only one or two such ITAs in all other provinces.[43]

The location of the ITAs was not merely important for practical reasons as provincial and local government tried to protect local travel agencies by building barriers for the establishment of out-of-area companies on their local turf (Zhang, Pine, Lam 2005: 132). The scarcity of fully licensed ITAs against an increasing demand let provincial governments often to turn a blind eye towards practices such as the unlicensed organization of outbound travels, the use of licenses of ITAs for their business purposes upon payment of a so-called 'management-fee', or the reselling of parts of the outbound quota of ITAs to smaller agencies on a commission basis (Roth 1998: 8), the so-called *chengbao* system (Qian 2003: 147).

With the 'Management Regulation on Chinese Outbound Tourism', issued in July 2002, the quota system was abandoned and the number of fully licensed ITAs was raised to 528 (Wang 2003b) and then 529[44] (CNTA 2004c). In 2005, 5 ITAs' licences were revoked, while 143 others were newly licensed, bringing the total to 672 (DPS 2005a). These agencies are the official partners for group travel to ADS countries; even so for each destination a distinct list exists, grouping the ITAs together that are empowered to organize groups to the specific country.[45]

In 2001, the outbound market share of CITS, CTS, CYTS and CCT was reported to have reached 56 per cent (CTC 2001). With the licensing of a greater number of companies, shares of the market are becoming more widely spread. As the provincial organizations of the big companies are operating more or less independently, calculation of aggregated market shares are of limited use only.

Still it is quite common that agencies that do not have the full ITA status will nevertheless sell outbound tours to their local customers and then pass

Table 3.2 Licensed travel agencies 1997–2005[46]

Year	(1) ITAs fully licensed	(2) ITAs inbound and domestic only	(3) DTAs	Total (2) + (3)
1997	67	991	3,995	4,986
1998	67	1,312	4,910	6,222
1999	67	1,256	6,070	7,326
2000	67	1,268	7,725	8,993
2001	67	1,310	9,222	10,532
2002	528	1,349	10,203	11,552
2003	528	1,364	11,997	13,361
2004	529	1,319	n.d.	n.d.
2005	672	1,472	13,867	15,339

Note: n.d. = no data available.

their customers on to licensed agencies. ITAs are also known to exchange customers among themselves; even so such activities are outside the legal requirements. This might even be done by individual staff that work within a travel agency, but on their own initiative (Hu, Graff 2005).

The blurred responsibilities are connected to a characteristic of the Chinese tourism industry that is common to many Asian countries: the absence of an independent retail channel. As quality control and customer complaints are the responsibility of the companies that sell travel products, no company is willing to sell products over which they cannot exert quality control. In the same way, Chinese customers tend to think that being transferred to other travel operators is tantamount to being sold around for a profit by 'vicious' travel dealers (Guo 2002: 120). The main reason, however, is the chaotic competition based solely on price cuts, which result in a small margins for retailing thus preventing a retailing function from developing independently.

According to their historical roots, the companies responsible for organizing outbound tourism can be divided into two distinct groups: on the one hand, there are large travel companies created by the central government that were started to facilitate inbound tourism. They later expanded into

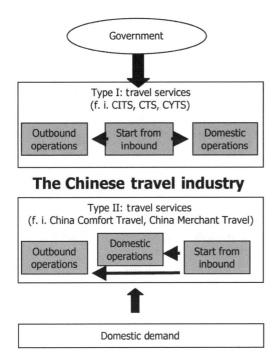

Figure 3.1 Two ways to the development of outbound departments in Chinese travel services[47]

domestic and outbound operations and owe their influence to their long history and their strong connections within the Chinese power structure. On the other hand, companies exist that developed with the growth of the Chinese domestic tourist market. Some of them are today among the largest travel companies in China. Most of them expanded into inbound and outbound operations as opportunities presented themselves with the changes in tourism policies.

These policies, which until now emphasize a 'balance' between inbound and outbound tourism and the specific history of tourism companies in the People's Republic of China, are responsible for the fact that there are no solely outbound operators in China.

The ITA structure is, however, mainly responsible for leisure travel paid by the travellers themselves – with or without subsidies by their *danwei*. For official and business outbound tours, a very different picture emerges. These tours include investigation tours, business discussions, study and training tours, visits to exhibitions and conferences and other such activities as well as trading missions.[48] All these travels are done with Official Passports and are paid by the responsible government institution or state-owned company. They are not restricted to ADS countries, but can be organized to all countries.

The main destinations are – in line with the business contacts of China as well as the perceived attractiveness as a destination – Japan, Australia, the USA and Europe. The percentage of time and enthusiasm that is dedicated to sightseeing, shopping and other leisure activities rather than the official duties, on which the public sponsorship of such tours is based, varies. For seasoned experts visiting a business or academic partner frequently, it may be close to zero, for cadres having their first-ever chance to go abroad it can reach 99 per cent.

For official and business groups, six different organizing bodies can be identified:

1. State-run Service Centres for foreign contacts on different national, provincial and local levels
2. State-run exhibition companies
3. Commercial consulting companies for international contacts
4. Private service companies for outbound travel
5. Travel agencies with or without ITA status
6. Representative offices of foreign travel agencies

(Ling, Wang 2003)

The role of the travel agencies is rather minor in this business segment. Among the reasons for this are the fact that ITAs are not necessarily knowledgeable about non-ADS countries and may fear being punished for organizing tours to such countries. In addition, the controlling department of the organization, which pays for the tour, often does not accept invoices from travel agencies as they are seen as organizers of leisure tours only. Another reason is the fact that such groups tend to be small (six to seven participants) for business trips or very big (up to several hundred participants) in the case of fairs and conferences. They cannot be combined with members of other travel groups to achieve economy of scale or the minimum required size of a group – a common practice.

Further, ITAs are not able to organize the professional contacts needed to foreign companies, institutions, training providers, etc., and are totally dependent on the receiving inbound organization.

The 'foreign offices' of the cities and provinces and the organizations responsible for organizing visits of international fairs, etc. have been in business much longer than many ITAs. Their strong position and good connections to facilitate the necessary approvals and permissions in many cases result in a virtual monopoly for the organization of outbound tours paid with public money in their specific region or branch of government.

To complicate matters further, some representative offices of foreign travel agencies in China are also known to organize Chinese official and business outbound tour groups by using their Hong Kong branches or other ways to circumvent the restrictions that officially disallow such activities. All this results in a situation described as simply 'chaotic' (Ling, Wang 2003: 144).

Ling and Wang (2003) give the figures for Shanghai for the relative importance of these six providers of outbound non-leisure trips:

1. State-run Service Centres 35%
2. State-run exhibition companies 25%
3. Private service companies 15%
4. Travel agencies with or without ITA status 10%
5. Representative offices of foreign travel agencies 10%
6. Commercial consulting companies 5%

> The Chinese tourism authorities try to control the operations of Chinese and foreign travel agencies with great efforts, but often they are not able to keep pace with the rapid developments of the travel trade. Outbound travel is particularly prone to illicit practices and corruption. As long as the growing demand for outbound travel from Chinese citizens is not met with relevant regulations, a sometimes rather 'chaotic' market will try to satisfy this demand.
>
> (Roth 1998: 5)

In 1998, He Guangwei, chairman of CNTA, described China's travel agencies as small, fragmented, weak and inefficient. Recent newspaper reports as well as academic studies illustrate the neck-breaking, price-cutting competition that leads to deteriorating service and decreasing profits especially in outbound operations. 'The situation faced by China's travel agencies appears bleak' (Qian 2003: 145).

The list provided by Wei and Wei is an insight into the strategies used by travel services to survive in the market as of 2005:

> The travel agencies are not allowed to do any business beyond their business scope, nor organizing Chinese citizens to travel to destination countries without permission from the State, nor selling at a price lower then the cost, nor accepting the quoted price of a reception agency abroad, which is lower than the cost, nor coercing and inducing travellers to do shopping or to take part in items at their own expense and nor taking tourists to porno centres.
>
> (Wei, Wei 2005: 709f.)

International tour operators' involvement in Chinese outbound tourism

The modernization process under the 'Reform and Opening' programme eased the restrictions on travel within China, with the possibilities for involvement by foreign companies in the travel business following suit, especially after China's accession to the World Trade Organization. Outbound tourism however remains officially off-limits for foreign companies.

With the start of inbound tourism in 1978, the inadequacy of just 203 suitable hotels in the whole country became apparent. Foreign capital mainly from overseas Chinese and hotel chains were invited to invest in joint-venture hotels, with tax breaks thrown in as an incentive. By 1985, 45 foreign-invested hotels were operating (Zhang, Chong, Ap 1999: 476), representing 85 per cent of the total investment in the tourism industry until this year (Pine, Zhang, Qi 2000). The opening in 1982 of the Jianguo Hotel with management by Hong Kong's Peninsula Hotel Group and the China Hotel in Guangzhou introduced international tourism practices to China for the first time (WTTC 2003). Amidst hotel construction that led to an over-supply of hotel capacity in the 1990s, the number of foreign-invested hotels rose to 464 hotels in 1997, to recede to 267 in 2003. Hotels with investment from Hong Kong, Macao and Taiwan on the other hand moved almost exactly in the opposite direction, with 270 hotels in 1997 growing in numbers to 411 by 2003 (CNTA 1998a–2004a).

Beside the hospitality sector, from 1992 onwards aviation operations were opened to foreigners resulting in foreign-invested aircraft maintenance and ground services cooperations and investment in Chinese airlines (Zhang, Chong, Ap 1999: 481).

Travel agencies were the final sector of the tourism industry to open up to foreign participation.[49] After decades of denial, the first projects of agencies with minority foreign funding were approved in 1998 with a total number of 15 until the end of 2002 (WTO 2003).

The main impulse originated in the prolonged negotiations about the accession of the People's Republic of China to the WTO (WTO[50]), which led to the signing of the agreement granting China entry into the WTO in November 2001. As part of the membership, China agreed on the General Agreement on Trade in Services (GATS). This opened the way for wholly foreign-owned enterprises (WFOEs) to be established in the hotel sector in China after 2003[51] and for the establishment of joint ventures for the running of attractions and the production of souvenirs. China also allowed 'qualified' foreign service providers to establish joint-venture travel agencies in Beijing, Shanghai, Guangzhou, Xian, Shenzhen and some tourism resorts, having to choose one only of these locations. As Chinese partners only fully licensed ITAs were qualified to apply for partnership in a joint-venture travel agency.

From 2003 such companies were to be organized with foreign partners as majority shareholders and no later than the end of 2005 travel agencies were to be permitted to be run as WFOEs and to have branch offices. Outbound tourism was, however, not included in the permitted business scope and no date was given for an end to this restriction.

The first such foreign joint venture under WTO rules was Beijing Morning Star (CTC 2001). American Express agreed to establish the travel services joint venture CITS & Amex Co. Ltd with CITS in January 2002. Business Travel Centres were opened in Beijing, Shanghai and Guangzhou within a year, specializing on travel management services in China to multinational corporations, regional companies and domestic companies, as well as selling air tickets (WTTC 2003). Another joint-venture partnership between CCT and Rosenbluth fell apart when Rosenbluth and American Express themselves merged in 2003 (TBP 2004).

In June 2003, two years earlier than required by GATS rules, CNTA and the Ministry of Commerce (MOFCOM) published 'Interim Experimental Provisions on Setting Up Wholly Foreign Owned Enterprise or Joint-Venture Travel Services With More Than 50 % Shares', to be implemented as of 15 July 2003.

The provisions erect some high hurdles for market entry by stating that:

- the registered capital of the Joint Venture should not be below RMB 4 million (approximately 400,000 Euros in 2005)
- the company must have the form of a limited company
- the foreign tour operator should have annual gross sales exceeding US$ 40 million to set up a Joint Venture with foreign majority or exceeding US$ 500 million to register a WFOE

- the Joint Venture cannot start subsidiary companies and
- the business scope of such a Joint Venture or WFOE is strictly limited to inbound and domestic travel only, with outbound travel, i.e. Chinese nationals travelling abroad (including Hong Kong, Macao and Taiwan) still being completely off limits.

(DZT 2005)

Within a week of the 'Provisions' coming into force, the Japan Airlines-owned Jalpak International China Ltd became the first WFOE travel agency, beating TUI AG of Germany, which on 1 December 2003 started a joint venture with CTS[52] in 2003, the signing of the official agreement for the TUI China Travel Ltd in Beijing witnessed by the German Chancellor Schröder. A cooperation agreement had been signed in November 2002 and the contract signed in September 2003. The Hong Kong branch of CTS meanwhile took Synergi as a partner to form CTI Sunshine Business Travel Management Co.

While the EU protested that the very high requirements for annual sales volume, which were not mentioned in the protocol of the first round of WTO negotiations, shut out the majority of European tour operators (EUCCC 2003), many international big players followed Japan Airlines and TUIs lead in 2004.

Joint ventures with a foreign majority were started by UK-based Business Travel International (BTI) in cooperation with Shanghai Jinjiang International (Group) Ltd and by Shenzhen Skal Travel Service Ltd. Kuoni Travel acquired full control of the Hong Kong-based tour operator P&O Travel Hong Kong, which already operates in Mainland China. Flight Centre Ltd, the largest travel agency group in Australia, established the joint venture 'Flight Centre Comfort Business Travel Service' with CCT in August 2004. Cendant Travel Distribution Services entered a joint venture with CYTS Tours Holding. Other joint-venture companies include Diethelm Travel, Beijing BTG-Accor Travel and Wing On International Travel (MOFCOM 2005).

Besides Japan Airlines (Jalpak), All Nippon Airlines also started a WFOE company called ANA Tours China Ltd. Singapore-based Star Cruises with Star Cruises Travel Agency (Shanghai) did likewise. Gullivers Travel Associates (GTA), a major European incoming agency, launched two China subsidiaries with representative offices in Beijing, Shanghai and Chengdu for operations in inbound and domestic travel services (TBP 2004).

With the outbound market beckoning, the temptation to bend the rules by 'close cooperation' with their joint-venture partner, other Chinese ITAs or the usage of own companies in Hong Kong or elsewhere outside China is very high (often seemingly too high) for the foreign tour operators engaged in China. This is certainly true for incoming agencies from the main destinations, which can only set up representative offices in China. Their business scope is officially restricted to liaison services to travellers for business, educational purposes or training. As a representative office they are also forbidden to hold direct negotiations, to sign contracts, to collect payments or even to hire staff directly.

In a situation where the market is characterized by no-name products and price wars among the Chinese tour operators, all but one quadrant of the price-quality portfolio are vacant. Under current circumstances, for most international tour operators the premium segment of high price and high quality would be of the biggest interest.

Unlike other global brands in different consumer product branches such as Nike, L'Oreal or Nokia, the names of international tour operators however mean nothing to Chinese consumers. The smiling face icon of 'Tu Yi', the Chinese name of TUI China, is not familiar, and with no independent retail system existing will not be as easily distributed on catalogue titles as in other countries. The global selling power of the big companies therefore will not necessarily lead to the domination of the premium segment by international tour operators in China after an official widening of their business scope.

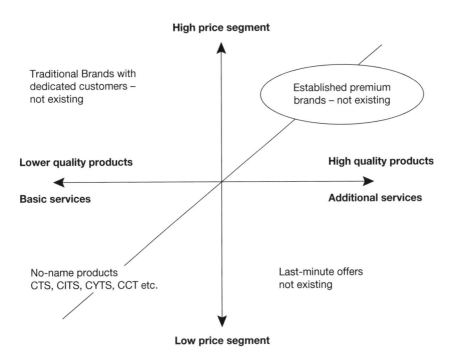

Figure 3.2 Position of outbound products in a price-quality portfolio[53]

Guo and Turner argue for foreign travel companies to hold back from their engagement in China until the 'inevitable shake out into a divided wholesaling and retailing structure' (Guo, Turner 2001: 63) of the Chinese tourism industry. They envisage lower costs for entry, better economies of scale and a more innovative environment for new entrants in a more flexible system with reduced entry barriers, even if this is counterbalanced by missing the opportunity of acquiring in-depth knowledge of the market-place early on.

Outbound tourism research and education policies and institutions

Compared to the volume of business and the influence on the daily life of the majority of the world's population, tourism worldwide is a relatively under-researched subject and an industry with a comparatively small percentage of university-degree holders working in it. For China, this is also true, with outbound tourism especially almost completely neglected as a topic for research or higher education.

Before 1978, tourism research or education did not exist in China. As foreign visitors were seen in the context of foreign policy and domestic and outbound tourism discouraged, it seemed unnecessary to study tourism as such (Zhang 2003b).[54] The personnel working for CITS including the guides were educated in the Foreign Language University under the guidance of the Foreign Ministry from a diplomatic rather than an economic viewpoint.

With the acceptance of inbound tourism as a way to earn foreign currency, tourism research started from scratch, mainly with the translation of some western standard monographs, stressing the positive economic effects of tourism development. In the 1980s and early 1990s, strategies for a 'tourism industry' and planning of numerous new attractions reflected the 'boosterism' and 'industry approach' stages of tourism (Hall 2000; Hall 2005)[55] China was going through at that time, catering first of all for the needs of governmental investors on all levels. 'Chinese officials . . . were essentially only exposed to an industry perspective. Area specialists, anthropologists, sociologists, and political scientists (were) seldom advisors' (Richter 1989: 32).

With the growth of international and especially domestic tourism, the impacts of many unsustainable developments on the economy, culture and not least the environment became obvious and a topic for research. 'As a consequence, some tourism research is now becoming more closely linked with the contemporary issues and concerns that are studied in the West' (Zhang 2003b: 74).

With the exception of the BISU in Beijing (formerly Second Foreign Languages Institute), the few existing tourism research institutes within

Chinese universities concentrate on project-orientated tourism planning projects rather than basic research. Most such research is conducted by the Chinese Academy of Social Sciences (CASS) and CAS. CASS established its dedicated Tourism Research Centre in 2000. CNTA as well as other government agencies have only very limited dedicated research facilities. Private and consulting companies conduct limited research but most findings are not made public.

The weakness of tourism research in quality and quantity research is uniformly lamented in all texts dealing with this subject. Blame is put onto the lack of funding of scientific research and the absence of awards in this field (Du 2004e), the mix up of responsibilities between many government branches (Lam, Xiao 2000) and the low pay of academics in universities that encourage them to engage in paid consulting and planning work rather than in basic research (Zhang 2003b).

Giving these unfavourable conditions and the government's longstanding reluctance to acknowledge or even support outbound tourism, it comes as no surprise that only very recently research into the outbound tourism phenomena has started in China, and non-Chinese organizations and researchers have organized the bulk of studies.[56]

The establishment of the Nanjing Tourism School in 1978 (Lam, Hong 2000), the Yixing Tourism School in 1979 (Gerstlacher, Krieg, Sternfeld 1991) and the Shanghai Institute of Tourism in 1979 (Wang, Mai 1983; Du 2004d) are variously seen as the beginning of higher tourism education in China. From 1980, the universities in Hangzhou, Dalian and Xian also opened departments for tourism; in 1983, 200 students were enrolled (Wang, Mai 1983). In 1986, there were 27 universities and colleges offering hospitality and tourism programmes with 4,800 students. With the maxim of producing experts who are 'red and professional'[57] still prevailing and with inadequate teaching materials and lecturers, the establishment of an acceptable tourism education system started only after that date (Zhang, Chong, Ap 1999: 477).

Until 1994, progress was slow with 109 tourism higher education institutes counted in 1994 (Xiao 2000), but from then the development acquired a faster pace. In 1998, 192 higher educational institutes of tourism and colleges with travel and tourism departments with 30,000 students existed (Lam, Hong 2000); in 2003, these figures had gone up to 494 and 200,000 respectively, with even higher numbers for institutions and students of professional courses (CNTA 2004a). Nevertheless, lack of expertise and lack of service quality are still seen as major bottlenecks of tourism development in China, especially on the level of higher and middle management (Huyton, Ingold 1997; Zhang, Lam, Bauer 2001; Zhang, Lam 2004; Zhang, Wu 2004).

At the end of the 1980s, regional tourism education departments and tourism programmes in several universities were set up by CNTA directly, which still plays a bigger role than the Ministry of Education, leading to a 'pattern of crude expansion' and a 'blurriness' (Du 2004e) of educational objectives.

Among the many problems besetting the fast-growing but substandard quality tourism education sector in China are a lack of funding, a lack of qualified educators, and old-fashioned or questionable curricula.[58] As a reflex to former government policies, inbound tourism is still the major topic, with less regard to domestic tourism or questions of leisure products. Outbound tourism seems not to be a topic in Chinese higher education tourism programmes at all.

4 Quantitative development of China's outbound tourism

The limits to the existence, publication and reliability of statistical information in China have been discussed in Chapter 2 already. For China's tourism statistics, even stronger reservations have to be made. Calculations about the revenue from international visitors to China were hindered before 1994 by the double standard created by the FECs in use (Mok 1985). Besides tinkering with results for political reasons, the legacy of the pre-reform era also resulted in structural problems to obtain reliable results in the first place. 'By no stretch of imagination can PR [People's Republic of] China's statistical system be regarded as sophisticated, ruined as it was between 1958 and 1978. PR China's statistical concepts and methodology may also differ from standard international practices' (Mok 1985: 279).

A major reform of tourism statistics took place in 1994. For domestic tourism, the first national statistic was published only in this year. Xu (1999) reports that the database for the statistics was very shaky before that time. Estimates by the central administration had to be based on unsystematically collected and rather anecdotal information at the local level such as the number of tickets sold for major tourist attractions (Xu 1999: 76). The calculation of tourism receipts also started only in 1994 to follow 'international standards' (CNTA 2004a). For outbound tourism the statistical range was changed in 1997 and projected backwards to 1995, resulting in an old figure of 4.5 million outbound travellers and a revised figure of 7.1 million outbound travellers.[1] 'CNTA changed its methodology for calculating outbound trip volumes in the mid-1990s, which complicates analysis of trends. It is further complicated by the fact that some annual counts have been restated more than once' (TBP 2004: 10).

Number of participants and expenditure

Domestic and outbound tourism activities existed in China before 1978 only in a number of guises, making it rather inaccessible for quantitative representation; the number of inbound visitors was very small. Therefore no attempt is made in this chapter to cover that period of time. After the beginning of the 'Reform and Opening' process however, because of the close political and economical interrelation between inbound, domestic and outbound tourism, all three forms have to be presented and analysed quantitatively together. This is done here as a general overview, with a special focus on outbound tourism in the new millennium. Specific characteristics and more detailed data for single continents and countries are given in Chapter 6.

Using its satellite accounting system, the World Travel and Tourism Council (WTTC) arrives at impressive figures for the overall performance of travel and tourism in China. The directly involved travel and tourism industry was, according to these calculations, directly responsible in 2002 for providing 14.1 million jobs and 253 billion Yuan RMB (approximately 25.3 billion Euros in 2005) revenue. However, if the indirect effects of travel and tourism are included, the travel and tourist *economy*, as it is called by WTTC, was responsible for 53 million jobs and 1,061 billion Yuan RMB (approximately 106 billion Euros in 2005) or more then ten per cent of the national GDP[2] (WTTC 2003). Chinese official statistics arrive for 2002 at a total revenue from inbound and domestic tourism activities of 557 billion Yuan RMB (approximately 55.7 billion Euros in 2005) or 5.4 per cent of the national GDP. After a decline in 2003 to 448 billion Yuan RMB (approximately 44.8 billion Euros in 2005), the result for 2004 climbs in the Chinese statistics to 684 billion Yuan RMB (68.4 billion Euros in 2005).

Looking at the development of tourism in China, the coveted international tourists came to China in ever growing numbers following the opening of the country in 1978, with decreases in the number of overnight visitors only in 1983, 1989, 1995 and 2003. The share of 'foreign' visitors, persons who did not reside in Hong Kong, Macao or Taiwan, hovered around 10 per cent only for most years, growing to more than 15 per cent in 2004 for the first time. Less than half of the persons crossing the border into Mainland China stay overnight, the rest being border tourists and day-trip visitors from Hong Kong and Macao. In 1992 China became one of the top ten international tourism destinations. In 2004 it pushed Italy from the fourth position, being behind only France, the USA and Spain. Without the visiting compatriots, the size of the inbound market and the international ranking would, however, shrink considerably.

For the development of the domestic tourism in China, no reliable figures are available for the 1980s and even the figures for 1990 through 1993 are more guestimates than data. From 1994 to 2004 the number of trips of more than six hours and 10 kilometres doubled to more than a billion trips, while the total expenditure more than quadrupled in the same time. Since 2003 every second rural and each urban resident undertakes more than one trip per year on average. Even though the number of trips undertaken by countryside dwellers is still higher than the number taken by urban residents, the total expenditure of urban tourists is more than double that of the rural inhabitants. Some 80 per cent of all rural domestic tourism trips

Table 4.1 Inbound visitors to China 1978–2004 (in million persons and per cent)[3]

Year	Total visitors	Compatriots including overseas Chinese	Foreigners	Share foreigners (in %)	Overnight visitors (compatriots and foreigners)	Share overnight visitors of total (in %)	Growth overnight visitors (in %)
1978	1,809	1,579	0.230	12.7	716	39.6	–
1979	4,204	3,842	0.362	8.6	1,529	36.4	113.5
1980	5,703	5,174	0.529	9.3	3,500	61.4	128.9
1981	7,767	7,092	0.675	8.7	3,767	48.5	7.6
1982	7,924	7,160	764	9.6	3,924	49.5	4.2
1983	9,477	8,604	873	9.2	3,791	40.0	–3.4
1984	12,852	11,718	1,134	8.8	5,141	40.0	35.6
1985	17,833	16,463	1,370	7.7	7,133	40.0	38.7
1986	22,819	21,337	1,482	6.5	9,001	39.4	26.2
1987	26,902	25,174	1,728	6.4	10,760	40.0	19.5
1988	31,694	29,852	1,842	5.8	12,361	39.0	14.9
1989	24,501	23,040	1,461	6.0	9,361	38.2	–24.3
1990	27,462	25,242	2,220	8.1	10,484	38.2	12.0
1991	33,349	30,639	2,710	8.1	12,464	37.4	18.9
1992	38,115	34,110	4,005	10.5	16,512	43.3	32.5
1993	41,527	36,871	4,656	11.2	18,892	45.5	14.4
1994	43,685	38,503	5,182	11.9	21,070	48.2	11.5
1995	46,387	40,500	5,887	12.7	20,034	43.2	–4.9
1996	51,128	44,384	6,744	13.2	22,765	44.5	13.6
1997	57,588	50,160	7,428	12.9	23,770	41.3	4.4
1998	63,478	56,370	7,108	11.2	25,079	39.5	5.5
1999	72,796	64,364	8,432	11.6	27,047	37.2	7.8
2000	83,444	73,284	10,160	12.2	31,229	37.4	15.5
2001	89,013	77,787	11,226	12.6	33,167	37.3	6.2
2002	97,908	84,469	13,439	13.7	36,803	37.6	11.0
2003	91,662	80,259	11,403	12.4	32,971	36.0	–10.4
2004	109,038	92,105	16,933	15.5	41,761	38.3	26.7

Table 4.2 Domestic tourists in China 1990–2004 (in million persons and Yuan RMB)[4]

	Total persons (in millions)	Urban dwellers (in millions)	Rural dwellers (in millions)	Total expenditure (in million Yuan RMB)	Urban dwellers (in millions)	Rural dwellers (in millions)	Expenditure per trip (in Yuan RMB)	Urban dwellers (in millions)	Rural dwellers (in millions)
1990	280								
1991	300								
1992	330								
1993	410								
1994	524	205	319	102,351	84,821	17,530	195.3	414.7	54.9
1995	629	246	383	137,570	114,010	23,560	218.7	464.0	61.5
1996	640	256	383	163,838	136,836	27,002	256.2	534.1	70.5
1997	644	259	385	211,270	155,183	56,087	328.1	599.8	145.7
1998	695	250	445	239,118	155,113	87,605	345.0	607.0	197.0
1999	719	284	435	283,192	174,823	108,369	394.0	614.8	249.5
2000	744	329	415	317,554	223,526	94,028	426.6	678.6	226.6
2001	784	375	409	352,237	265,168	87,069	449.5	708.3	212.7
2002	878	385	493	387,836	284,809	103,027	441.8	739.7	209.1
2003	870	351	519	344,227	240,408	103,819	395.7	684.9	200.0
2004	1,102	459	643	471,071	335,904	135,167	427.5	731.8	210.2

in 2003 were same-day visits, not including an overnight stay away from home. The total revenue from domestic tourism in 2004 was already twice as high as the total revenue from inbound tourism.

For outbound tourism again there are no reliable figures existing for the 1980s, when the only form of official non-business trip was the visit to friends and relatives in Hong Kong and some border tourism into the Soviet Union. With the reform of tourism statistics in 1994, the way of calculating was changed to include service personnel, sailors and other persons not included before in the outbound statistics. A meaningful comparison of outbound trips can therefore only start from 1995. The growth rates accelerated after the introduction of ADS travel in 1999 and with the easier entry into Hong Kong and Macao. For 2005 a slower growth of about 10 per cent only was recorded.

Tourism acted the role as the earner of hard currency it had been assigned to with the start of commercial tourism in Chinas in 1978. Whether or not this function is still fulfilled, is hotly contested. Up until 2002, each year saw a surplus of inbound revenue over outbound expenditures, even if the significance of these amounts became less and less important in the light of the growing success in international exports of other goods and growing currency reserves in China. For 2003 and 2004 however, two different estimates can be presented. Travel Business Partnership (TBP) estimated in a study of the Chinese outbound market expenditures of 18 and 25 billion US$ for 2003 and 2004 respectively. Compared with the published inbound revenue, such expenditure levels would cancel out the revenue completely, with even a small loss in 2003. The Economist Intelligence Unit (EIU) offers for 2003 an estimate of 16.8 billion US$ outbound expenditure, which would leave a positive balance of just 0.8 billion US$ (EIU 2005). Internal

Table 4.3 Chinese outbound tourists 1991–2004 (in million persons)[5]

1990	Old statistical series	New statistical series	Growth (in %) (new series)
1991	2,134.2	–	
1992	2,929.7	–	
1993	3,740.0	–	
1994	3,733.6	–	
1995	4,520.5	7,139.0	
1996	5,060.7	7,588.2	6.3
1997	5,323.9	8,175.4	7.7
1998	–	8,425.6	3.1
1999	–	9,232.4	9.6
2000	–	10,472.6	13.4
2001	–	12,133.1	15.9
2002	–	16,602.3	36.8
2003	–	20,221.9	21.8
2004	–	28,852.9	42.7

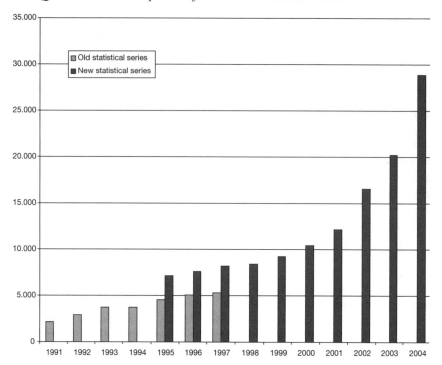

Figure 4.1 Chinese outbound tourists 1991–2004 (in million persons)[6]

data from the Data Centre of China Economic Information and from the People's Bank of China, quoted in Guo *et al.* (2005a) put the outbound expenditure levels much lower for 2003 and 2004 at 15.2 and 16 billion US$ respectively. For 2004, this would add up to a positive result of 9.7 billion US$, the biggest ever surplus in international tourism.

A look at the expenditures per outbound traveller can shed some light on this dispute. After 1996 the expenditure per trip and person jumped from less than 600 US$ to more than 1,200 US$, as more long-distance travels became possible. With the growing importance of short-distance trips to Hong Kong and Macao and fierce price wars in the market, from 2001 the expenditures fell again to below 1,000 US$ per trip. The total figures cited by Guo *et al.* (2005a) however, would bring the expenditures down to 550 US$ per trip, which is unlikely. Even with the TBP and the EIU figures, a decrease to below 900 US$ is envisaged. Total expenditures at the same level for 2002 and 2004 against a growth in travellers from 17 million to 29 million for the same years also seems unrealistic.

Comparing the total domestic and outbound expenditures of the last decade, a proportionate growth can be seen with an average ratio between outbound and domestic spending of 1:2. As the total domestic expenditure

Table 4.4 Inbound revenue and outbound expenditures of China's tourism sector 1978–2002 (in billion US$ and per cent)[7]

Year	1978	1990	1995	1996	1997	1998
Inbound revenue in bn US$	0.3	2.2	8.7	10.2	12.1	12.6
Outbound expenditure in bn US$	0	0.5	3.7	4.5	10.2	9.2
Balance in bn US$	0.3	1.7	5.0	5.7	1.9	3.4
Difference in %	100	77	58	56	16	27

Year	1999	2000	2001	2002	2003[a]	2004[a]
Inbound revenue in bn US$	14.1	16.2	17.8	20.4	17.4	25.7
Outbound expenditure in bn US$	10.9	13.1	13.9	15.4	18.0/ 15.2/ 16.8	25.0/ 16.0
Balance in bn US$	3.2	3.1	3.9	5.0	−0.6/ 2.8/ 0.8	0.7/ 9.7
Difference in %	23	19	22	24	−6/ 14/4	3/61

a   See note 7 for details on these figures.

Table 4.5 Chinese outbound tourism expenditures in absolute values and per trip 1995–2004 (in million persons and billion US$)[8]

	(1) Total outbound travellers (in million persons)	(2) Total outbound expenditure (in bn US$)	Expenditure per trip (2) ÷ (1) (in US$)	(3) Alternative estimates of total outbound expenditure (in bn US$)	Expenditure per trip (3) ÷ (1) (in US$)
1995	7,139	3,688	516.60		
1996	7,588	4,474	589.60		
1997	8,175	10,167	1,243.61		
1998	8,426	9,205	1,092.50		
1999	9,324	10,864	1,176.73		
2000	10,463	13,114	1,253.42		
2001	12,134	13,909	1,146.25		
2002	16,602	15,398	927.46		
2003	20,222	15,180	751.02	18.0/16.8	890.12/830.78
2004	28,852	16,000	554.55	25.0	866.49

Table 4.6 Comparison of domestic and outbound expenditures 1995–2004 (in billion US$)[9]

	(1) Total outbound tourism expenditure	(2) Total domestic tourism expenditure	Ratio (1):(2)
1995	3.7	16.6	0.22
1996	4.5	19.8	0.23
1997	10.2	25.5	0.40
1998	9.2	28.9	0.32
1999	10.9	34.2	0.32
2000	13.1	38.4	0.34
2001	13.9	42.5	0.33
2002	15.4	46.8	0.33
2003a (TBP)	18	41.6	0.43
2003b (Guo *et al.*)	15.2	41.6	0.37
2003c (EIU)	16.8	41.6	0.40
2004a (TBP)	25	56.9	0.44
2004b (Guo *et al.*)	16	56.9	0.28

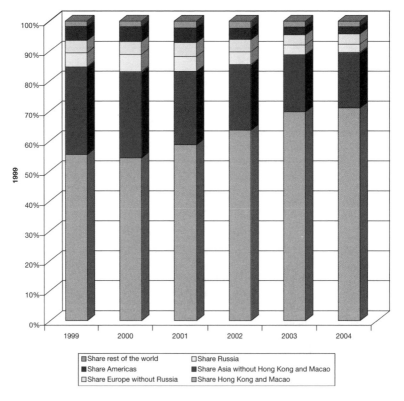

Figure 4.2 Share of continents and the Hong Kong and Macao SARs of the total development of China's outbound tourism in comparison (in per cent)[10]

Table 4.7 Chinese outbound destinations 2000–2004[11]

	2000	2001	2002	2003	2004	% share in 2003	% share in 2004
Asia							
Hong Kong	4,412,191	5,320,446	7,771,047	9,310,103	13,001,635	46.0	45.1
Macao	1,644,421	1,800,432	2,783,082	4,790,589	7,490,491	23.7	26.0
Japan	595,660	608,548	760,073	804,687	1,021,325	4.0	3.5
Vietnam	n.d.	n.d.	267,900	606,600	785,682	3.4	2.7
South Korea	400,958	459,374	551,431	559,120	697,023	2.8	2.4
Thailand	707,456	652,375	688,748	527,835	682,475	2.6	2.4
Singapore	262,776	281,361	289,237	262,138	429,258	1.3	1.5
Malaysia	86,696	123,971	231,035	244,087	337,173	1.2	1.2
North Korea	194,970	171,974	247,876	175,696	295,738	0.9	1.0
Taiwan	86,154	105,380	112,488	126,544	144,526	0.6	0.5
Myanmar	n.d.	n.d.	40,800	74,500	n.d.	0.4	–
Philippines	33,647	51,327	68,180	72,925	102,197	0.4	0.4
Mongolia	63,044	59,640	91,839	64,828	122,499	0.3	0.4
Kazakhstan	44,226	38,333	49,398	57,362	91,122	0.3	0.3
Indonesia	19,963	24,738	45,316	56,867	84,745	0.3	0.3
Cambodia	n.d.	n.d.	25,100	26,500	n.d.	0.1	–
Nepal	n.d.	n.d.	15,500	24,900	n.d.	0.1	–
Pakistan	n.d.	n.d.	n.d.	15,700	n.d.	0.1	–
India	n.d.	n.d.	n.d.	13,400	n.d.	0.1	–
Other Asia	583,709	391,824	156,602	125,652	534,752	4.4	1.9
Total Asia	**8,845,908**	**10,089,723**	**14,195,652**	**17,940,033**	**25,820,641**	**88.7**	**89.5**
Americas							
USA	395,107	418,128	418,585	345,566	443,873	1.7	1.5
Canada	100,178	134,995	147,871	150,244	183,607	0.7	0.6
Cuba	n.d.	n.d.	n.d.	1,500	n.d.	0.0	–

Table 4.7 (cont.)

	2000	2001	2002	2003	2004	% share in 2003	% share in 2004
Other Americas	27,796	31,955	51,745	33,406	52,690	0.2	0.2
Total Americas	**523,081**	**585,078**	**618,201**	**530,716**	**680,170**	**2.6**	**2.4**
Europe							
Russia	606,102	606,981	691,128	661,231	809,606	3.3	2.8
Germany	112,824	138,371	165,687	165,168	222,878	0.8	0.8
France	96,485	114,435	136,692	135,407	201,533	0.7	0.7
UK	61,129	88,440	128,000	134,088	177,601	0.7	0.6
Other Europe	202,549	228,883	276,154	255,215	395,757	1.3	1.4
Total Europe	**1,079,089**	**1,177,110**	**1,397,661**	**1,351,109**	**1,807,375**	**6.7**	**6.3**
Oceania							
Australia	126,852	166,321	199,356	197,621	270,473	1.0	0.9
New Zealand	18,288	38,583	58,741	59,379	61,805	0.3	0.2
Other Oceania	5,091	4,829	7,230	8,155	28,542	0.0	0.1
Total Oceania	**150,231**	**209,733**	**265,327**	**265,155**	**360,820**	**1.3**	**1.3**
Africa							
South Africa	n.d.	20,500	25,800	25,400	n.d.	0.1	—
Egypt	n.d.	n.d.	n.d.	6,300	n.d.	0.0	—
Other Africa	n.d.	38,147	48,562	57,712	n.d.	0.3	—
Total Africa	**47,521**	**58,647**	**74,362**	**89,412**	**115,381**	**0.4**	**0.4**
Unidentified	3,625	12,806	51,144	45,514	68,461	0.2	0.2
Total	**10,649,455**	**12,133,097**	**16,602,347**	**20,221,939**	**28,852,850**	**100.0**	**100.0**

Note: n.d. = no data available.

Table 4.8 Development of China's outbound tourism to continents and the Hong Kong and Macao SARs in comparison (in '000 visitors and per cent)[12]

	1999	2000	2001	2002	2003	2004	Growth 1999–2004 (in %)	Growth 1999–2001 (in %)	Growth 2002–2004 (in %)
Total	9,232	10,649	12,133	16,602	20,222	28,853	313	131	174
Hong Kong	3,571	4,142	5,320	7,771	9,310	13,002	364	149	167
Macao	1,551	1,644	1,800	2,783	4,791	7,490	483	116	269
Hong Kong and Macao	5,122	5,786	7,120	10,554	14,101	20,492	400	139	194
Total without Hong Kong and Macao	4,110	4,863	5,013	6,048	6,121	8,361	203	122	138
Total Asia	7,813	8,846	10,090	14,196	17,940	25,821	330	129	182
Asia without Hong Kong and Macao	2,691	3,060	2,970	3,642	3,839	5,329	198	110	146
Total without Asia	1,419	1,803	2,043	2,406	2,282	3,032	214	144	126
Europe with Russia	824	1,079	1,177	1,398	1,351	1,807	219	143	129
Russia	437	606	607	691	661	810	185	139	117
Europe without Russia	386	473	570	707	690	998	259	148	141
Americas	429	523	585	618	531	680	159	136	110
Rest of the world	166	201	281	390	400	545	328	169	140

Table 4.9 Share of continents and the Hong Kong and Macao SARs of the total
development of China's outbound tourism in comparison (in per cent)[13]

Share Hong Kong/Macao	55.5	54.3	58.7	63.6	69.7	71.0
Share Asia total	84.6	83.1	83.2	85.5	88.7	89.5
Share Asia without Hong Kong and Macao	29.1	28.7	24.5	21.9	19.0	18.5
Share Russia	4.7	5.7	5.0	4.2	3.3	2.8
Share Europe without Russia	4.2	4.4	4.7	4.3	3.4	3.5
Share Americas	4.6	4.9	4.8	3.7	2.6	2.4
Share rest of the world	1.8	1.9	2.3	2.3	2.0	1.9

grew by only 21 per cent between 2002 and 2004, but the number of outbound travellers grew by 80 per cent, a more than proportionate growth of outbound expenditures is very likely, again putting the low expenditure figure cited by Guo *et al.* (2005a) in doubt.

Shares of visited destinations

The choice of destinations for official leisure travels increased rapidly after 1999. The figures given for the distribution of the Chinese outbound tourists, as reported in the statistics of the WTO, CNTA and the national statistics of the destinations, are however almost as diverse as the destinations themselves. Japan, for example, reported for 2002 the arrival of 452,420 Chinese inbound visitors (Koldowski 2003). The official outbound figure given in the CNTA *Yearbook of China Tourism Statistics*, however, states the departure of 760,073 Chinese citizens to Japan (CNTA 2004a). The same sources give the number of Chinese visitors to Vietnam as 724,385 from the receiver's point of view, but as only 267,900 according to the Chinese authorities exit statistics. It is difficult to reconcile such discrepancies, as the obvious difference between the first port of call recorded in the Chinese statistics only and the possibility to travel to several countries in one trip cannot explain the big differences in many cases.[14]

The figures given in Table 4.7 do use the data of the Chinese outbound statistics available from different sources.

The destination data show the clear predominance of Asia. A look at the relative weights of the continents illustrates the fact that the bigger part of the growth of China's outbound tourism after 1999 has been a growth of the quasi-domestic tourism to the Hong Kong and Macao SARs. Compared with the overall growth figure of 313 per cent between 1999 and 2004, Hong Kong and Macao outperform with 400 per cent growth, whereas the rest of Asia, Europe and the Americas can all achieve only an approximate doubling of the number of Chinese visitors in the same period. Especially between 2002 and 2004, an increase of almost double within two years for the two SARs contrasts with the 25 per cent increase for non-Asian destinations.

China's outbound tourism development clearly outperformed the global outbound tourism development, which managed only about 12 per cent between 1999 and 2004; for non-Asian destinations the growth has however been less pronounced in absolute numbers than many worldwide newspaper reports during that time proclaimed it to be.

Outbound tourism from Hong Kong and Taiwan

In a not-too-distant past, when outbound travel was a rare treat for citizens of the People's Republic of China, many inhabitants of Hong Kong and Taiwan were already sophisticated international travellers. To secure some long-term perspective of the travel patterns and behaviour of Chinese[15] outbound tourists, it seems conducive to examine the outbound travel from the SAR and former British colony Hong Kong and from the island of Taiwan, locally governed if not internationally recognized as the remaining part of the Republic of China. This may also benefit the discussion about the question of whether distinct travel patterns are signs of cultural differences or just temporary markers of an immature market.

Outbound tourism from Hong Kong

Two-thirds of the 7.5 million citizens of the SAR Hong Kong travelled across its borders in 2004 to destinations other than Mainland China or Macao and 20 per cent of these even to destinations outside Asia. Adding the eight trips per person per year to the northern homeland and the half trip to Macao to this figure gives an impressive insight into the high level of mobility of the diligent Hong Kongers, even under the circumstances of the prolonged crisis of the Hong Kong economy that started in parallel with the return of Hong Kong to Chinese rule in 1997. Citizens with a SAR Hong Kong passport do not need a visa for most countries, including the main Asian destinations and the Schengen area.

Asia is by far the main region visited,[16] with Europe, the USA, Canada and Australia and New Zealand following. The proportion of the demand for all regions has been more or less stable over the last several years.

According to research by TourismAustralia, in terms of aspiration and intention for leisure travel, Japan is clearly preferred, followed on both counts by Europe, the USA and Australia (TourismAustralia 2004).

Over 3 million (78 per cent) of the trips to 'foreign' destinations in 2003 were undertaken for holidays, 0.5 million (11 per cent) each for business and VFR travels. The most popular form of holidays for Hong Kongers (55 per cent) was a round trip, followed by city trips (15 per cent) and beach holidays (9 per cent) (DZT 2005). Given the fact that almost two-thirds of the population travelled abroad, it is no surprise that the socio-demographic structure of the travellers more or less mirrors the structure of the adult Hong Kong society.

Table 4.10 Outbound travel from Hong Kong 1999–2004[17]

Year	Number of outbound departures of Hong Kong citizens ('000s)			
	To destinations other than Mainland China and Macao	To Mainland China	To Macao	Total
1990	2,400			
1995	3,400			
1997	3,758			
1998	4,197			
1999	4,175			
2000	4,611			
2001	4,799			
2002	4,709	55,648	4,182	64,540
2003	4,428	52,555	3,952	60,936
2004	5,003	59,675	4,223	68,903

Table 4.11 Outbound departures from Hong Kong to destinations other than Mainland China and Macao 2003 and 2004 ('000s)[18]

Region	Year	
	2003	2004
South and Southeast Asia	1,556	1,681
North Asia	700	806
Taiwan	415	536
Europe	323	373
USA and Canada	279	334
Australia and New Zealand	207	222
Middle East	22	30
All others (including unidentified)	926	1,021
Total	4,428	5,003

Several studies have analysed different aspects of the travel behaviour of Hong Kong Chinese tourists. Prestige and status boosting is found to be an important motive (Ap, Mok 1996), realized by patronizing five-star hotels and places such as Disneyland (Mok, DeFranco 1999). Zhang, Qu and Tang (2004) concluded that Hong Kong people perceive safety as the most important attribute when choosing their travel destinations, confirming an earlier finding of Mok and Armstrong (1995), which saw safety as major concern, followed by scenic beauty, price of the trip, and the service quality of hotels and restaurants. Safety was again identified by Wong and Kwong (2004) as the main concern of package tourists from Hong Kong.

Mok and Armstrong (1995) found that sport facilities were not considered important, whereas Hsieh, O'Leary and Morrison (1992) as well as Wong and Lau (2001) identified sport activities as highly valued by Hong Kong tourists together with novelty and authenticity of attractions, visits to historical places, sampling local food, and experiencing different cultures. These results tally only partially with the statement of Kaynak and Kucukemiroglu (1993) that Hong Kong Chinese tourists often repeat travel to the same vacation destination.

Family and friends' recommendations and other word-of-mouth communication are other major sources for decision making processes, especially for first-time travellers (Yau 1988; Kaynak, Kucukemiroglu 1993; Heung, Chu, 2000).

Looking more specifically to travels to Europe, a higher percentage than average of richer and better educated travellers is evident, reflecting first of all the higher cost of such trips, but also an unbalanced gender ratio with a majority of female travellers (58 per cent). Taking the example of Germany, the main interests of Hong Kong holidaymakers, who tend to be older than average, are centred on local festivals landscape and small cities, with shopping also an important activity. Significantly when compared to Mainland Chinese and Taiwanese travellers, Hong Kong Chinese are more interested in eating non-Chinese food when travelling (DZT 2005).

In Taiwan, a study by Hsieh and Chang (2006) looked at the reasons for 600 Hong Kong visitors to visit night markets, finding 'novelty-seeking', 'experimenting local culture and customs' and 'exercising'[19] as the main motivations. As far as tourists' preferred leisure activities in tourist night markets were concerned, eating out overwhelmingly dominated (88.5 per cent), followed by everyday shopping (56 per cent), novelty-seeking (32 per cent) and entertainment (23 per cent).

Outbound tourism from Taiwan

Outbound travel started with the relaxation on the former ban of overseas travel in 1979 and took off when residents of Taiwan were permitted to visit Mainland China from 1987 (Huang, Yung, Huang 1996), in the first few years officially only for VFR trips (Lew 1995). In 2002, one out of three of the 23 million inhabitants left the island temporarily.

Before 1979 the tense political situation as well as the lack of foreign currency informed the official support of inbound tourism as opposed to the almost total restriction on outbound travel, creating tensions in the growing and increasingly affluent society on Taiwan.

After 1979 overseas travel developed swiftly. Early problems with unreliable travel agencies, unsophisticated behaviour of Taiwanese in foreign countries and bureaucratic problems were solved during the 1980s with official quality control measures and training courses on international

Table 4.12 Outbound travel from Taiwan
1979–2004[20]

Year	Number of outbound departures of Taiwanese ('000s)
1979	312
1982	641
1985	847
1988	1,602
1989	2,108
1990	2,942
1991	3,366
1992	4,215
1993	4,654
1994	4,744
1998	5,912
1999	6,559
2000	7,329
2001	7,189
2002	7,507
2003	5,923

etiquette offered by Tourist Information Service Centres. In 1989, further travel restrictions were withdrawn with the abolishment of the so-called 'Leave Country Permit Document' exit visa, the issuing of a general passport and the end of the ban of men of military age to travel for leisure purposes. This was followed by the abolition of embarkation cards and currency restrictions in 1990 (Prideaux 1996). Consequently in 1989, for the first time, the number of outbound departures was higher than the number of inbound arrivals by foreign visitors to Taiwan (Zhang 1993), tripling within a few years.

With the possibility of visiting relatives in Mainland China from 1987 and the growing investment across the Taiwan Strait, Mainland China soon became the most important destination (ROCTB 1993). The economic boom of the 1980s and 1990s put holidays in the surrounding Asian and Oceanic countries or in the USA well within financial reach of the majority of Taiwanese, a situation not changed even after the Asian financial crisis of 1997/1998. Already in 1992 Taiwanese ranked twelfth in the world in tourism expenditures, with more then 80 per cent of the total expenditure of 7.3 billion US$ spent abroad (Karwacki, Deng, Chapdelaine 1997). In 2002, 7.5 million Taiwanese residents went abroad, a figure only temporarily dented in 2003 by the SARS problem.

Huang, Yung and Huang (1996) report a number of traits of Taiwanese outbound travellers:

- *Gifts and shopping*: Taiwanese travellers are expected to bring back gifts to their friends or relatives in similar fashion to the *omiyage* custom in the former colonial power Japan. As a result shopping, especially for branded goods, is an important part of tourism activities overseas. Shopping activities are considered more important than comfortable accommodation.
- *Food*: Chinese cuisine is favoured also when travelling abroad. Taiwanese travellers do not expect to try 'exotic' cuisine every day.
- *Group travel*: more than three-quarters of the Taiwanese tourists prefer package or semi-package tours, with more young people preferring independent travel.
- *Travel type*: theme tours, island resort trips and city tourism were each identified by approximately 30 per cent of travellers as preferred type of travel. From round-trip package tours the market is diversifying into more specialized products for different target groups according to age, travel experience, income and special interests.
- *Destinations*: Europe was named as the favoured destination, even so only about 1 per cent of all travellers go there. Approximately one-third of all travels go to Hong Kong and then Mainland China. Exact figures vary, as Taiwanese travellers still cannot fly to Mainland China directly and therefore are required to change flights in Hong Kong, Macao or use earthbound transportation from there, ostensibly with the single purpose of VFR or business travel. Japan, the USA and Southeast Asia are the next significant destinations. As Taiwan only manages to maintain diplomatic relations with a very small number of countries, visa problems make travel to many countries cumbersome and expensive.

Facilities and safety are the most important factors in selection of travel destinations, followed by attractions, practical accessibility, quality and symbolic accessibility, according to the research results of Lai and Graefe (2000).

Jang and Wu (2005) analysed the push and pull motivations of Taiwanese seniors. According to their study of 353 still actively travelling elderly Taiwanese, the main push motivations were, in hierarchical order, to see new things, to go to places that the participants always wanted to visit and to see how other people live. Laziness, visits to the old home area and fashionable destinations were the least important push motivations.

Pull motivations were concerned with safety and security, hygiene and cleanliness and the environmental quality of air, water and soil. For the elderly shopping facilities, facilities for physical exercise and special events were not important pull factors. Women were more eager to acquire new knowledge than men, whereas richer participants were more strongly concerned with choosing prestigious destinations to talk about afterwards.

Reflections on Hong Kong and Taiwan Chinese outbound tourism

Looking at the information available for the outbound tourism of Hong Kong- and Taiwan-based Chinese, a number of similarities with the situation of the outbound tourism of Mainland China are already evident. Both societies are populated by keen outbound travellers,[21] willing to spend considerable sums of money for leisure and shopping activities outside their normal place of abode. Even after decades of gathering travelling experiences, group travel plays an important role for safety as well as comfort reasons. Activity, learning and excitement rather than relaxation are sought, with prestige and status playing important roles especially in the pre- and post-trip phases of the journey. Western destinations are the most sought-after, even so time and money and – in the case of Taiwan – visa constraints result in neighbouring Asian countries being the most visited destinations.

The review of available information on the travel motivations and behaviours do however also confirm the four main problems of tourism research identified by Cooper (2003b), namely 'conceptual weakness and fuzziness, a spread of topics and a lack of focus, a predominance of one-off atheoretical case studies, and difficulties with access to quality large-scale data sources' (Pearce 2004: 58). What we encounter in the case of Hong Kong and Taiwan and will encounter again in the following chapter in the case of the People's Republic of China illustrates the dire need for stronger, focused, soundly theoretically based research done over a longer period of time and with a broad database. Law, Cheung and Lo (2004), after criticizing the methodological shortcomings of studies on Hong Kong's outbound tourism, accordingly state that 'there is an absence of published articles in the tourism and hospitality context that scrutinise the importance of travel activities as perceived by Hong Kong travellers' (Law, Cheung, Lo 2004: 356). Nevertheless, for the moment anybody interested in Chinese outbound tourism has to make do with the existing data, but has to bear in mind that they mainly offer positivistic snapshots, where a much deeper analytical approach would be needed.

5 Chinese travellers

After looking at the economic, political and quantitative background of China's outbound travel, the focus has to be shifted to the subjects of the activities: the Chinese travellers themselves. Obviously, for a country which has a fifth of mankind living within its borders, the warning of Pizam and Sussman (1995) is especially valid:

> Nationality is only one variable that should be considered in predicting variation in tourist behaviour and should never be used as a *sole* explanatory variable. Certainly, not all tourists of the same nationality, regardless of demographic category, motivation, and life-style, behave the same.
>
> (Pizam, Sussman 1995: 917)

Historical and cultural backgrounds are, however, to be included in any analysis of outbound tourists' behaviour, especially if tourism is seen as a tool to create national identities in domestic travels (Palmer 1998) and to sharpen the perception of cultural differences in overseas destinations (Jameson 1993; Robinson 1998; Robinson 2001). For Chinese outbound travellers there are only a very limited number of historical precedents or role models to inform modern tourism, but the 'cultural distance' (Williams 1998; Bowden 2003) is clearly felt in most encounters between Chinese and non-Chinese actors in the play of international tourism.

Even though more than a billion people are involved in domestic tourism of some kind in China and 30 million Mainland Chinese travel abroad per year, the knowledge about their characteristics, behaviour and motives is still very sketchy, as research in China is concentrated on inbound tourism (Cai, Hu, Feng 2001). Similarly, the information sources for the tourists themselves remain limited and of uneven quality. Using theoretical tools of cross-cultural and semiotic analysis and comparison with other cultural traditions of tourism behaviour will hopefully help to put into perspective the little data available.

There are several reasons why the Chinese outbound tourists are so under-researched and remain a largely unknown entity. First, research is hampered by the general sorry state of basic research into outbound tourism. The point of view of not only the Chinese tourism industry to a large extent is still production-orientated, looking at destinations and attractions rather than applying a consumption-orientated view to the source markets and the consumers (Meethan 2001). Second, as a group of customers only recently appearing on the market outside of Asia in larger numbers, interest from the receiving destinations is just beginning. Deeper insights into the characteristics of a sociological group as big and diverse as China's outbound travellers and as different to other, especially non-Asian cultures, require more than superfluous observations of how name-cards are to be handed over and the like. Third, in China itself outbound as well as domestic tourism has long been ignored or hushed up, so that all but superficial research is sparse and furthermore handicapped by strong ideological self-images. Foreign studies addressing tourism impacts on local communities in China (Oakes 1995; Oakes 1998; Swain 1995; Wall, Xie 2005; Xie, Wall 2002), focused mainly on the ethnic minorities in remote regions. 'Chinese researchers noted tourism's impacts on the majority population, but under the Chinese Government's censorship they could not sufficiently address issues that were socially and politically sensitive in China' (Li, Y. 2004: 189).

This chapter can therefore only attempt to give an overview into what research has been done until now and what limited results it has yielded. The study of the cultural roots of Chinese outbound tourism and the adaptation to a globalized world as a part of the sociology of the Chinese, or what Tu (1994a) called 'The Changing Meaning of Being Chinese Today', will be an ongoing concern for research.

In looking at the Chinese travellers, it becomes immediately clear however, that *the* domestic or *the* outbound tourist does not exist. Very different activities with very different inputs of time, money and mobility are occurring and not all participants climb the tourism career ladder starting at the same time or with the same speed. Domestic tourism, according to Chinese statistics, includes all activities that involve an absence of more than six hours and that take place at least 10 kilometres away from the usual place of residence (Zhang, Pine, Lam 2005: 53). With this very generous definition, peasants visiting a local fair are mixed together with Cantonese businessmen spending their time in a skiing resort in Northeast China, resulting in an average spending equalling just 50 US$ per trip.[1] Similarly large are the differences between the 70 per cent of outbound tourists who just travel to Hong Kong or Macao, the 20 per cent who travel to other Asian destinations and the top 10 per cent who can afford to travel to other continents. In 2004, a year without SARS and with ADS in force for many countries, this last group amounted to only 3 million people.

Travelling in China

Mobility as a leisure activity, travelling for other than military, administrative, commercial, disaster-induced or religious reasons, is not a necessity.
Unlike eating, housing and clothing, tourism is not necessary for the
physical survival of human beings and indeed the vast majority of people
living in China – as in other countries – lived until very recently a life
without indulging in it, considering it as 'spending much money to receive
nothing tangible in return' (Dou, Dou 2001: 47).

> Tourism, of course, existed long before modernity came into being.
> However, it is only under modernity, particularly late modernity,
> that tourism has become a mass phenomenon. . . . It is modernity that
> 'pulls' people away from home in a quest for pleasure in other
> places, due to advances in technology, living standards, social welfare
> . . . and international relations. However, modernity also 'pushes' people
> away from home to tourist destinations for relaxation, recreation, the
> experience of change, novelty, fantasy, and freedom.
>
> (Wang 2000: 214)

In China, the rush to modernity within the fast economic development
of the 'Reform and Opening' policies had to overcome the notions of
frugality and 'working for the next generation' (Xiao 2003). For the current
generation of pensioners, not only is their average income low (Xu, Chen
2003) but their pre-reform era ideas of thrift are preventing tourism
consumption (Yu, Zhang, Ren 2003). Xiao and Huyton (1996) name five
characteristics of traditional leisure as impediments to modern leisure
behaviour:

1 the lower intensity of work minimizing the need for recreation;
2 the devotion of any discretionary income to building or purchasing
   houses and to the welfare of their child instead of spending it on leisure
   and recreation;
3 the Confucianist virtue of filial piety and the idea that one shows
   respect to one's family, consequently spending time with the family
   instead of engaging in outgoing leisure;
4 the tendency to include business or VFR into one's leisure time, leading
   to a predominance of regular family/business; and
5 the lack of leisure and recreation offers.

To implement the 'travel bug' (Yatsko, Tasker 1998) into the Chinese
society, more was needed than just the increase of time and money,
however necessary these are. Without the move away from agrarian – and
Maoist – frugality and thriftiness and official internationalism towards re-
definition of the collective identity by embracing globalized consumerism

and enhanced nationalism at the same time, the upsurge in tourism activities cannot be understood. Domestic and outbound tourism serve these goals in different ways.

Chinese domestic and outbound travellers before 1949

> Historically speaking, travel for pleasure-seeking purposes belongs to the privilege of a few social groups in China, such as the ruling classes, the wealthy and the educated. With regard to the mass Chinese people, traditional travels appeared primarily in two forms, namely travel to cities and travel to religious sites.
>
> (Xu 1999: 72)

As discussed in Chapter 3, travel was seen in Imperial China first of all as a way to broaden one's mind, an intellectual undertaking, as expressed in the saying 'He who travels far knows much'. 'In ancient Chinese history, travel had an extremely high intellectual value among Chinese scholars' (Guo 2002: 19).[2] Educated persons were even advised 'to seek ultimate truth from the landscape' (Petersen 1995). The classical Confucian writings are based on experiences collected during travels through China, the *Analects* opening paragraphs includes the often-quoted sentence 'Is it not a joy to have friends come from afar?' (Lau 1989).[3] As administrators (*Mandarins*) had to move throughout their career several times to the venue of the imperial examinations for different grades and were normally employed in a different province every few years, a lot of opportunities for long-distance travel arose during a bureaucratic career. 'Both China and Japan ... have long had very extensive circuits of internal travel, pilgrimages often akin to tourism' (Graburn 2001a: 72).

Beside commerce, internal and external wars, disasters and famines also provided other reasons for unwanted long-distance journeys. Some emperors travelled themselves on 'inspection tours' or on hunting trips, on pilgrimages to Taishan Mountain or to summer retreats. However, the main obligation of the emperor was to remain in his Forbidden City and to carry out the prescribed rituals at the right time.[4] Roaming the country was left to poets such as Li Bai or Su Dongpu and scientists such as Xu Xiake, whose works influenced the perception of later generations of travellers (Wang 1998; Guo, Turner, King 2002). Monks wandered throughout the country, visiting the four holy mountains of Buddhism and the five holy mountains of Daoism. Guest houses for travelling officials and for commoners existed throughout China 2,000 years ago (Shu 1995). Religiously motivated outbound travellers were the Buddhist monks Fa Xuan in the fifth century and Xuan Zang in the seventh century, whereas other monks brought Buddhism and with it Tang-Chinese culture to Japan (Shen 1996). Admiral Zheng He at the beginning of the Ming Dynasty famously undertook seven journeys that brought him as far as Mombasa[5] with a fleet far superior to the Portuguese ships arriving there a century later (Levathes 1994).

For the vast majority, the agrarian population, the only forms of willingly undertaken travels consisted of visits to towns and cities in the vicinity to participate in festivals and fairs, often for a mixture of economic and leisure reasons, and of pilgrimages, which tended to be short in duration and spatial reach. This was helped by the proliferation of temples and monasteries in the country. 'The sites multiplied over the centuries and as Buddhism became established, even more sacred sites were added' (Sofield, Li 1998a).

The century between 1850 and 1950 increased the travel opportunities with the introduction of modern transportation infrastructure such as railways and steamships for the affluent part of the society.[6] For the *laobaixing*,[7] the common people, this time was marked by the unprecedented displacement of millions of people. Inside the country this was caused by the uprisings of the Taiping and the Yihetuan ('Boxer'), the opening up of Manchuria, the civil wars before and during the Chinese Republican era and the fight against the Japanese occupation as part of the Second World War. At the same time, a growing number of Chinese left the country to work as 'coolies' in Southeast Asia and the Americas. Their number reached a million per year in 1889 and even 2 million per year after the fall of the Qing Dynasty in 1911 (Zhang 1998).

Domestic travellers in the People's Republic of China

The Maoist position of seeing tourism as a wasteful, bourgeois lifestyle and a potentially dangerous possibility of the formation of networks outside the control of the party perpetuated the demise of domestic tourism in China after 1949. 'In the 1950s and early 1960s even local Chinese were easily suspected of being a spy when they travelled. The desire to travel abroad was practically committing treason or defecting to the enemy' (Gerstlacher, Krieg, Sternfeld 1991: 54).

The systems of household registration (*hukou*) and ration coupons made it almost impossible to travel outside the normal place of residence for Chinese citizens before the 1980s. All travels had to be approved by the *danwei*, the work unit every Chinese belonged to, otherwise a train or bus ticket could not be obtained legally. The ration coupons necessary to buy food and other items were only valid locally, so travellers had to apply at the Public Security Office for special coupons with nationwide validity. The traditional visits to local fairs and temples were stopped by the abolition of free peasant markets and the attacks against religion and the *four olds* – old ideas, customs, culture and habits – and all the local customs connected to it (Sofield, Li 1998b). Pilgrimages, which still could be practiced more or less openly in the 1950s, ceased during the Cultural Revolution, when countless religious sites were destroyed.[8]

A primary cornerstone of the ideology propounded by the CCP since its inception in 1921 concerned the need to reject the cultural past as a whole and its replacement with a new Chinese socialist culture. Under this policy, massive destruction of China's rich and varied built heritage occurred and there were sustained attacks on its cultural (living) heritage.

(Sofield, Li 1998a: 364)

For members of the administrative staff of the government or the party and of big state-owned enterprises, *gongfei lüyou*, publicly financed tourism[9] offered a chance to visit the capital Beijing or tourist spots such as Kunming or Hangzhou for national meetings (Xu 1999). The national leadership followed the Qing Dynasty tradition of retreating in August from hot Beijing to a summer resort, mostly to the coastal city of Beidaihe. As an incentive for 'model workers' or for recuperating workers, large companies developed tourism facilities attached to their organizational structure. 'Admission to these tourism facilities was based on the connections between the work units' (Guo 2002: 42).

A Chinese tradition that was left untouched was the family reunion during the Chinese New Year, now renamed the spring festival, in January or February. At this time, students and persons sent to work in other provinces overloaded the underdeveloped transportation system.

During the first years of the Cultural Revolution millions of young 'Red Guards' travelled to take part in political campaigns and meetings. Many intellectuals and students were later 'sent down' (*xia fang*) to the countryside of remote areas, being forced to stay there in some cases for more than a decade. During that time leisure tourism was an unthinkable concept:

Because there are no statistics on tourism during this period [before 1978], we can only speculate as to the exact figures. However, if there [was] any [privately organized] pleasure travel in China, it was in very small numbers. The economic and physical conditions necessary for large scale tourism in China did not exist. A low living standard, an inadequate supply of commodities, a shortage of food and accommodation facilities, and a limited transportation system prevented the growth of such an industry.

(Wen 1997: 565)

Guo states that 'the significance of travel as a life style was always regarded by the general public as desirable. The popularity of travel has never abated' (Guo 2002: 43). In any case, after 1978 the number of domestic tourists grew from probably up to 300 million during the 1980s to half a billion in 1994 and more than a billion in 2004. New tourism activities such as visiting theme parks or going on honeymoon trips developed. The central government tried in vain to contain the travel urge and finally turned it into an instrument of regional development and fiscal policy.

However, with the inclusion of all trips of more then six hours duration and 10 kilometre distance, these immense figures must be seen as rather illustrating the development of new leisure patterns than of tourism development specifically. In 2003, 33 per cent of the domestic travellers engaged only in day trips, another 28 per cent stayed only one to two nights. Also, 48 per cent stayed with friends and relatives and 28 per cent in guest houses, with only 18 per cent using hotels (Wang, Qu 2004). The average expenditure per trip in 2004 represented the equivalent of about 43 Euros, with city dwellers spending 74 Euros and rural inhabitants 21 Euros (Liu, D. 2005: 93). The average expenditure for day trips of rural residents in 2003 was as low as 98 Yuan RMB (approximately 10 Euros in 2005) (CNTA 2004a). The upward market movement to more sophisticated inland travels, as described by Wu, Zhu and Xu (2000), or the household expenditure pattern, with personal computers, houses, cars and leisure travel as typical goods purchased in the 1990s (Doorn, Ateljevic, Bai 2003), are relevant only for the top 5 per cent of the population, which earned 20 per cent, or even the top 1 per cent, which earned 6 per cent of the total national income in 2002 (Li, Yue 2004; Wu, Perloff 2004).[10]

The often-quoted (Zhang 2003c; Huang, Hsu 2005) view of an 'automatic' travel wish triggered by the reaching of a certain GDP per capita level cannot explain the increase in travel activities after the beginning of the 'Reform and Opening' period in 1978:

> Generally speaking, when the per capita income reaches US$ 300–400, people will have the desire to travel in their own countries. When it reaches US$ 800–1,000, they will wish to go to the neighbouring countries. When it is more than US$ 3,000, they would have the desire of travelling to faraway places. . . . In 2000, the per capita disposable income was US$ 1,253, which means that Beijing residents are financially capable of travelling to neighbouring countries.
>
> (WTO 2003: 60)

This mechanistic approach cannot convince even if it uses fitting *post-fact* barriers: 'According to international experience, people generate travel motives after GDP per capita reached about Yuan 3,500 (US$ 1 = ¥8.3). In 1985, the GDP per capita of China exceeded that amount' (Wu, Zhu, Xu 2000: 296). Furthermore, even within the logic of such an argument, employing national averages makes little sense in a country with the widest rural–urban income gap in the world and Gini index readings as high as 0.32 and 0.37 respectively within the cities and the countryside.[11]

Rather, five necessary preconditions for the development of different forms of domestic travel had to be met: a sufficient levels of income, a sufficient amount of leisure time, the freedom to travel within the country, the existence of transport infrastructure and last but not least the consumers wish to spend time and money for travelling.

The development of the income of rural and urban Chinese, moving upward, if at different speeds and uneven distribution, has been described in Chapter 2. Connected to the income, the high private saving rates changed the consumer attitude towards tourism (Guo 2002).[12] For the growing part of the population not working in the field of agriculture, the leisure time account grew, especially after 1990. For bigger companies and government institutions a week-long holiday was introduced in 1992. The average working hours per week were reduced from 48 to 44 in 1994 and to 40 hours in 1995, bringing about the five-day working week as a normal regulation. In 1999[13] the introduction of three 'Golden Weeks' in January/February, May and October extended the non-working days to about 120 days[14] (Guo 2002; Schuler 2005). The introduction of more individual paid holidays is under discussion, as is the more fanciful idea of moving to a system of three weeks of six workdays followed by a free week every month. In the countryside, the increases in population combined with a higher level of mechanization led to a 'surplus' rural population of about 200 million people, 'which constituted the tourist flow in domestic market on one hand, and source of labour in the market of domestic tourism supply on the other' (Du 2004b: 31).

The travel restrictions disappeared step by step with the abolishment of food and textile rationing in the early 1980s and the easing of the *hukou* system in the 1990s. With a 'floating' population of more than 100 million migrant workers, little pretence of controlling the movement of people within the country[15] is possible. The construction of highways, railways and airports and the provision of convenient transport services using this infrastructure developed quickly from very limited beginnings during the 1990s. Infrastructure is no longer the 'bottleneck' of domestic tourism as before, except during the 'Golden Weeks', the introduction of which resulted in chaotic scenes especially in the first new festive weeks in October 1999 (Cai, Hu, Feng 2001) and May 2000 (Latelinenews 2001). For the spring festival 2005, the transport system coped with 70 million tourists on top of the passengers travelling towards family reunions (Lin, Li 2005). Beside city visits, outdoor tourism also grew, with a focus on nature parks. 'Relative to Western cultures, Eastern cultures tend to favour human manipulation of nature in order to enhance its appeal, in contrast to preservation of nature in a pristine state' (Lindberg, Tisdell, Xue 2003: 119). The use of nature reserves as a favoured domestic tourism destination results, however, in heavy overuse or even destruction (Deng *et al.* 2003). This is mainly due to the fact that the emphasis in touristic infrastructure development is put on the generation of financial and economic benefits in tourism, including ecotourism. 'Ecotourism in China involves greater levels of visitation and infrastructure development than one typically finds in other countries' (Lindberg, Tisdell, Xue 2003: 123). Whereas discussions about sustainable tourism tend to concentrate on ecological questions of minimizing the footprint of human interference in Europe and on social development issues for the rural host populations in Japan, sustainable development research in China often tries to find the maximum possible level of economic exploitation of an area just short of destruction (Arlt 2006b). Community involvement is equally almost never included in the discussion (Jim, Xu 2002; Stone, Wall 2003). 'Little appears to be happening in China to ensure the direct participation of residents, . . . in the development of these assets as tourism attractions in a way that is beneficial to them and the heritage assets' (Du Cros *et al.* 2005b: 192).

The fast development already created new divisions betweens different kinds of domestic travellers:

> While natural and historical attractions appeal to the majority of domestic tourists, the more experienced, especially those from the coastal urban regions, already show diversified needs and preferences beyond sightseeing in consuming tourism products and selecting destinations . . . However there is an apparent lack of product diversification and innovation. Tourism products and services are often packaged in the 'one-fit-for-all' approach.
>
> (Cai, Hu, Feng 2001: 72)

Accommodation shortages eased during the 1990s, but even so hotels still tend to prefer international guests (Cai, Hu, Feng 2001), regardless of that fact that the service quality is seen more critically by international hotel guests (Tsang, Qu 2000). The majority of hotels in China lose money, according to another study, which besides badly qualified management blames the reluctance to concentrate on the domestic market (Gu 2003). Altogether, mobility developed from a political and physical problem to a financial and motivational question.

'Western-style sightseeing and travel in China came about as a result of influence from the west' says Shu (1995: 157). Zhang Guangrui, as quoted before, named three influences: disposable income, available leisure time, and the impact of the growth of international tourism (Zhang 1989: 58), hinting on the model-setting influence of western tourists. Consumerism, transported not only by inbound tourists but by the whole modernization process of China since 1978, is a major force in today's China, having taken the place of socialist ideologies especially for the urban elites. Consumerism, 'the active ideology that the meaning of life is to be found in buying things and pre-packaged experiences' (Bocock 1993: 50), includes the consumption of places (Urry 2005) in the form of leisure and tourism products. Buckley (1999) showed in a survey conducted in 1994 the emergence of the consumer-orientated 'nouveau riche' in Beijing, with more recent illustrations provided among others by Tuinstra (2003) of what Mao Zedong would have called the impacts of the *sugar-coated bullets of the bourgeoisie*. 'There is nothing natural about modern consumption; it is something which is acquired, learned; something which some people are socialized into desiring' (Bocock 1993: 54).

A special Chinese form, a *consumerism with Chinese characteristics* emerged from

> the hybridity of revolutionary culture and consumer culture, which resulted from the appropriation of revolutionary culture within the context of Chinese postmodernity. . . .
>
> Since the beginning of the 1990s, Chinese revolutionary culture has been largely commercialised – appropriated by consumer culture – in the context of consumerism. By the term 'appropriation', I mean a rhetoric strategy that makes use of revolutionary symbols or mean a rhetoric strategy that makes use of revolutionary symbols or images – such as Red Guards, Five-Star red flag, Red Book, etc.
>
> (Tao 2005: 70)

Tao quotes examples where, for instance, a Red Guard holds a Personal Digital Assistant (PDA) instead of the Little Red Book in his hand in an advertisement. As a reaction to this appropriation, in domestic tourism can be seen the strong development of the so-called 'red tourism', officially sponsored tours that include visits to sites connected to the history of the Communist Party and the People's Republic of China for old cadres and for young people alike (Gao 2005). Red tourism can be interpreted as an attempt to regain the control over the symbols of the revolution and as a way to bring back the pre-modern forms of pilgrimages and the bourgeois behaviour of leisure trips into the folds of Communist ideology by painting tourism 'red'.[16] China Radio International (CRI) provides a well-developed website with detailed descriptions – quite unnecessarily even in English – of dozens of 'red tours' in China.

Zhang Xiqin, Vice-President of CNTA is quoted on the website as saying 'Red Tourism is an economic project, a cultural project and at the same time, a political project' (CRI 2005).

Another important basic force supporting the increase of domestic tourism in China can be found in the production of the *imagined community* China. Following Benedict Anderson's (1991; 1998) classical analysis of nationalism as a form shaped by a collective imagining enabled by modern technology, the People's Republic of China and their citizens had to invent themselves again after 1978 as a nation based on a thousands of years of history.[17] The end of the official Maoist rejection of the cultural past and the attempt to replace it with a new Chinese socialist culture as well as the end of the reduction of nature to a tool for production alone gave rise to the need of the reassessment of the *Chineseness* of China.

In 1982, the National Heritage Conservation Act 'provided the foundation for tourism to embrace heritage in its development' (Sofield, Li 1998a: 371). Not only could foreign tourists visit the highlights of the achievements of Chinese civilization, but also overseas and domestic tourists could see with their own eyes and relate to what they were proud of as part of their *national* – or, in the case of overseas Chinese, *transnational* – identity. 'The conservation and presentation of traditional culture were also approved because of its perceived contribution to enhance national unity and to develop the country's tourism product' (Sofield, Li 1998a: 363). In the same year, the CNTA Director Han Kehua announced that China would employ in future a 'Chinese' architectural style for the construction of hotels and restaurants (Han 1982; Gee 1983), newly embracing heritage also for the tourism infrastructure.

In parallel to the historical heritage the landscapes of China were also transformed into consumable national products. Nature reserves, mostly under the administration of the State Forestry Administration and the State Environmental Protection Administration, increased from only 34 in 1978 to 573 in 1989 and 1,146 in 1999, covering almost 9 per cent of the total area of China (Han, Ren 2001). In what Palmer (1998) calls 'Tourism as Identity', both architectural and natural heritage is strongly connected in China to this identity-building exercise. The Chinese traditional taste of viewing nature and landscapes is closely associated with the nations many sacred mountains and rivers, where century-long religious sites are still popular tourist attractions (Peterson 1995):

> When Western tourists look at the Yangtze, they see a river; the Chinese see a poem replete with philosophical ideals. Part of the 'common knowledge' of Chineseness is to recognize representations of the picturesque hills of Guilin, the sea of clouds of Wu-shan (Mount Wu), the Three Gorges of the Yangtze River, and the Yellow Crane Terrace pagoda. These images bring spiritual unity even if the people have never visited them; but when they do visit the importance of these images is reinforced.
>
> (Sofield, Li 1998a: 367)

The Chinese experience of nature as a row of sights, as 'pictures' rather than a 'film' of a hiking experience, supports Selwyn's analysis 'that metaphors drawn from the landscape constitute part of the moral discourse which is used in the wider distinctions made between *us* and *them*' (Selwyn 1995: 119). For both nature and heritage consumption, in the 'management of the national imagination',

> tourism's use of identity goes far beyond the commercial it goes to the heart of a people because it serves to deepen their cultural identity and to make this visible, both to themselves, and to *others*. Furthermore, cultural identity underpins national identity as it communicates the past and present traditions and mores of a people, thus enabling them to be identified as a distinctive group.
>
> (Palmer 1998: 316)

Beside the obvious attributes of nations, such as the national flag, anthem, border or frontiers, hidden aspects, such as national recreations, the countryside, popular heroes and heroines and fairy tales, all connected to touristical experiences, are shared by the members of a community (Smith 1991). They act as signifiers of the nation as a community with common beliefs, an historic homeland and as a common culture. This common culture was strengthened during the course of the modernization of China by stronger economic relations between the provinces of China, the spreading of Putonghua as standard Chinese language through schooling and television, improvements in transport infrastructure, etc. By travelling throughout China – or at least to the next bigger city – domestic tourists reinforce the imagined community spatially. 'Tourism plays a vital part in both the "imagining" – i.e. bringing into awareness – and the "re-creation" of national cultures in Asia and Oceania' (Graburn 1997: 201).

The slogan of 'learning to love our motherland, to love rivers and mountains of our country and to make acquaintance with our nation', criticized by Guo as 'a common Chinese excuse for travelling at government expense' (Guo 2002: 45), finds it's justification in its acquisition by the Chinese private domestic travellers.

Characteristics and motivations of Chinese outbound travellers

Information about the characteristics of Chinese visitors to different destinations is recorded by the statistical services of those countries; some of these sets of data will be introduced in Chapter 6. Only a few overall studies are available, however, and these are based on rather small survey samples or sometimes cover only one source region. For instance, the WTO study of 2003 can only offer some figures from Southeast Asian destinations for such general items as the gender and age of outbound tourists. Border tourists are not covered except those to Hong Kong and Macao, no differentiation is made between travellers to these SARs and other long-distance travellers or between self-paying travellers and those who travel on public or company expenses. The basic answer to questions about the characteristics and motivations of Chinese outbound travellers can therefore only be: nobody knows. Nevertheless, the available information is presented on the following pages.

Geographical and sociographical characteristics of Chinese outbound travellers

China is the home of 1,300,000,000 people and therefore in size and diversity rather a continent than a country. Accordingly the differences between the inhabitants are as varied as those between Europeans or Africans from different countries. This is true not only for the 9 per cent of the population who belong to one of the recognized minorities, but also for the 91 per cent Han Chinese who live in areas that have variable climates and natural conditions, which have been integrated into the Chinese culture at different times in history and that have very different local customs, languages and histories. These differences are masked for the outside world by the status of China as a single country and by the perceived similarity in appearance. From the inside, the similarities in the geographical as well as in the historical dimension are also much more stressed by the Chinese government and public with the insistence on a common history and heritage.[18] Furthermore the use of Chinese characters, unifying the culture beyond dialects and historical development of the spoken language functions as the glue that holds the Chinese identity together regardless of the big regional differences.

China is a market, and many regional differences can be expected in any Big Emerging Market (BEM),[19] as has been documented by a study of values and lifestyles by Cui and Liu (2000). East and especially South China, the most developed parts of the country, show a clearly more pronounced interest in 'living one's own life', 'seeking opinions' and favouring foreign brands. Satisfaction with life as it is, hard work and honest living is more highly regarded in other parts of the country. Not surprisingly, travelling is an activity liked the most by Eastern and Southern Chinese.

The statistics about the geographical source areas for China's outbound tourism are biased by the fact that a high number of travellers are engaging in border tourism. Accordingly, the figures for the major outbound travel generated province, Guangdong, is inflated by the large number of Hong Kong and Macao visitors, some of them not even staying overnight. The border provinces of Yunnan, Heilongjiang, Guangxi, Inner Mongolia and Liaoning are also reporting high numbers of outbound travellers, putting Yunnan, Guangxi and Heilongjiang on positions two, three and

Table 5.1 Attitudes and values in different Chinese regions[20]

	South China	East China	North China	Central China	South-west China	North-east China	North-west China
Satisfaction with life	66.6	67.5	81.3	80.2	66.8	81.3	67.6
Work hard and get rich	33.3	31.6	30.7	34.6	42.3	44.2	67.1
Live one's own life	41.7	42.1	40.4	38.1	33.6	31.9	20.0
Pure and honest	8.3	10.5	12.7	11.9	9.3	9.3	4.3
Opinion seeking	66.6	56.4	58.8	57.1	46.8	59.7	47.9
Favour foreign brands	36.4	27.5	22.2	24.1	19.6	28.1	35.7
Like travelling	45.5	47.5	42.1	42.0	38.2	37.6	16.9

five respectively as most important provinces for outbound travels in 2002.[21] In 1997, for instance, 24,924 out of the 25,264 travellers (99 per cent) from Inner Mongolia travelled to either Russia or Mongolia; of 54,009 Heilongjiang travellers, 47,425 moved to Russia (88 per cent); of the total number of residents of Yunnan crossing a border of 345,000, 289,000 went to Myanmar (84 per cent), 27,000 to Vietnam (8 per cent) and 20,000 to Thailand (6 per cent) (Dou, Dou 2001). Border tourism and trade is responsible for the increase of Chinese visitors to Myanmar from just 5,600 in 1994 to 472,000 only two years later (Roth 1998).

For outbound leisure tourism beyond Hong Kong and Macao and beyond neighbouring border areas, the most important provinces are Beijing, Shanghai and Guangdong, also where the main international airports of Mainland China are located, Hong Kong being another important airport acting as a hub for medium- and long-term travel. For some ADS destinations only inhabitants of these three provinces are allowed – or were allowed in the first years of ADS in operation – to receive tourist visas under the ADS regime. Other provinces of secondary importance are Zhejiang, neighbouring Shanghai, and Fujian, home of many overseas Chinese, and Tianjin. These provinces and municipalities also top the list of the regions with the highest spending on living expenditures per head.[22] The relative share of the major generating areas differs with different destinations. One source gives the overall structure of outbound regions as 27 per cent for Shanghai, 25 per cent for Beijing and Guangdong 19 per cent, with 29 per cent for the rest of the country (DZT 2005).

Table 5.2 Most important generating self-supported outbound tourism provinces 1995–2002 (in '000s)[23]

Year	1995		1996		1999		2002	
Rank in 2002	Province	No. of tourists	Province	No. of tourists	Province	No. of tourists	Province	No. of tourists
1	Guangdong	677	Guangdong	719	Guangdong	582	Guangdong	728
2	Yunnan	275	Yunnan	524	Yunnan	293	Yunnan	533
3	Guangxi	105	Guangxi	121	Guangxi	189	Guangxi	303
4	Heilongjiang	52	Heilongjiang	46	Liaoning	55	Shanghai	205
5	Fujian	32	Fujian	42	Heilongjiang	54	Heilongjiang	77
6	Inner Mongolia	20	Liaoning	28	Fujian	47	Beijing	76
7	Liaoning	19	Shanghai	28	Shanghai	44	Fujian	71
8	Shanghai	19	Beijing	27	Beijing	38	Zhejiang	63
9	Beijing	16	Inner Mongolia	23	Inner Mongolia	25	Hubei	60
10	Zhejiang	10	Sichuan	23	Zhejiang	22	Jiangsu	60

The differences in the characteristics of outbound travellers connected to the locality are often discussed within the tourism industry and also in studies (WTO 2003; Blok 2002; Lommatzsch 2004, Hu, Graff 2005). The results are based mostly on anecdotal evidence rather than serious market research. It is an open question how much influence the coming from a specific province has in comparison to the sociographical segmentation, taking also into account that many people living in Beijing, Shanghai and the cities of Guangdong actually moved there from other areas.

The inhabitants of Beijing, the capital, and therefore containing a large number of government officials and administrators, are supposed to be more conservative and sensitive to tradition than people in Shanghai or Guangzhou. They are therefore seen as more interested in culture and history and to have a love for socializing, giving high value to relations with their family and friends.

Shanghai is China's most populous urban area with more than 16 million inhabitants and is the largest commercial city in China. It has developed into the industrial and financial centre of China and is the richest area in terms of GDP per capita. Shanghai people are described as economically practical and as not easily becoming enthusiastic, paying attention to rational cost and benefits calculations. The large number of white-collar workers employed by Chinese and foreign company headquarters represent – and perceive themselves – as the most urbanized strand of the Chinese society. In outbound tourism they are said to act their image as being shrewd business people by looking for value for money and shopping opportunities and to be more risk averse than others. Information from the media is supposed to play a more important role than in other source regions.

Guangdong Province is home of the majority of overseas Chinese and has been the area with the closest contact with western culture through the colonies of Hong Kong and Macao. Guangdong is the most urbanized province with 21 cities having more than one million inhabitants (NBSC 2004), connected into what regional planners call the 'Megacity Pearl River Delta' (Ipsen 2004). Shenzhen is the most important gateway to Hong Kong (Chi 2003), as Zhuhai is for Macao. 'Internationalization and globalization is particularly influential in these cities, making the inhabitants more liberal and cosmopolitan than the comparatively conservative Beijing' (Blok 2002: 11). The traditional distance from the capital in the north[24] is supposed to support the taste for new experiences, entertainment, good food and nightlife. Being located in the sub-tropics, warmer climates hold less appeal for Cantonese compared to travellers from northern parts of China. Good service is said to have higher value for travellers from Guangdong than lower prices.

The information given about the gender of Chinese outbound travellers also varies wildly in different sources. The German National Tourist Board (DZT) reports a ratio of 52 male to 48 female travellers (DZT 2005). Guo (2004) finds in a study conducted in Shanghai a ratio of 54 male to 46 female travellers, almost identical with the figures given by TBP (2004) of 55:45. In a nationwide study, the Canadian Tourism Commission (CTC) gives a ratio of male to female of 58 to 42 (CTC 2001); Zheng (2004) reports an even higher number with 69 per cent male vs 31 per cent female travellers. The current study of the BISU (Du, Dai 2005) however, arrives at a slight advantage for female travellers at 47 male to 53 female. A TFWA-ACNielsen study (Appleton, Yu 2005) reverses the figures of Zheng (2004) completely with 31 per cent male and 69 per cent female travellers. When trying to make sense out of these figures, a look at the national figures of different receiving countries confirms that destinations that welcome a large proportion of business and Meetings, Incentives, Conferences and Exhibitions (MICE) travellers show a bias towards male visitors, whereas leisure tourism and VFR destinations show a more even weight of both sexes. For Hong Kong and Macao, a growing number of female compared to male visitors can be expected with the easier accessibility. The example of the gender structure of the Chinese visitors to Singapore, implied to be representative by WTO (2003), shows a trend toward equality from a 80 per cent male to 20 per cent female ratio in 1993 towards a 56 per cent to 44 per cent ratio in 2000 that should support the assumption that the prominence of male travellers decreases with the growth of the proportion of leisure tourism within all arrivals.

Age structures also vary among different sources. The very young and the old are under-represented when compared to the age structure of China, which at the population census of the year 2000 showed 23 per cent under the age of 14 and 7 per cent above the age of 65 (NBSC 2004). National statistics show that the average age of travellers to predominantly leisure destinations is lower than the average age for destinations with a high proportion of business travellers.

Table 5.3 Age structure of Chinese outbound tourists (in per cent with age groupings in parentheses)[25]

Du, Dai 2005	Zheng 2004	TBP 2004	Guo 2004	Appleton, Yu 2005	CTC 2001	DZT 2005
6 (0–18)	5 (0–18)				4 (15–19)	
15 (18–25)	10 (18–25)	30 (0–24)	39 (0–30)	29 (20–29)	7 (20–24)	17 (15–24)
24 (26–35)	35 (26–35)	38 (25–34)	40 (31–39)	16 (30–39)	31 (25–34)	39 (25–34)
28 (36–45)	30 (36–45)	15 (35–44)	18 (40–49)	19 (40–44)	31 (35–44)	19 (35–44)
17 (46–55)	15 (46–55)	18 (45+)	3 (50+)	36 (45+)	27 (45–59)	19 (45–54)
10 (55+)	5 (55+)					6 (55+)

Table 5.4 Occupation, education level, household size and income level of Chinese outbound tourists – result of two studies (in per cent)[26]

	CTC study 2001	BISU study 2004
Occupation		
Cadre in party and government organization	4	9
State-owned enterprise staff	4	21
Professional/technician	23	7
Enterprise managerial personnel	14	19
Trade/service industry worker	14	2
Industry/manufacture industry worker	7	2
Self-employed	5	n.d.
Unemployed	4	n.d
Student	8	12
Housewife	1	n.d.
Retired	4	8
Peasant	n.d.	1
Military personnel	n.d	1
Teacher	n.d.	11
Others	12	7
Educational level		
No formal schooling/primary school	2	n.d.
Junior high school	7	8
Senior high school/technical school	29	20
College	27	31
University/post-graduate	34	41
Household size		
Average number of persons in household	3.22	
Two persons household		11
Three persons household		57
Two generations		15
Three generations		10
Single		7
Total monthly household income (Yuan RMB)		
1,500 or less	11	34 (below 5,000)
1,500 to 2,999	20	
3,000 to 4,449	22	
4,500 to 5,999	20	31 (5,000 to 10,000)
6,000 to 9,999	14	
10,000 to 20,000	13 (10,000 and more)	17
20,000 to 30,000		8
Above 30,000		10

Note: n.d. = no data given.

For the sociographical features of occupation, education level, household size and income level, the different research results differ too much in their different delimitations to compare their results. Furthermore, answers to questions such as those on income level are not easy to analyse, given the high importance of bonuses and other irregular payments in China and the unavoidable reluctance of interviewed persons to give truthful answers. Given the general fact that especially medium- and long-distance travels are still affordable only for the top decile of the Chinese population, it is not surprising that persons in high occupational positions, with high educational levels, small household sizes and high income are dominating outbound tourism. Other groups are presented mainly by persons travelling on public or company expenses, students or older people invited by their adult children living in China or overseas. It should, however, not be forgotten that in 2004 the 6 million outbound tourists who travelled only across the border from Guangdong Province to Hong Kong and the probably equally large number travelling from Guangdong to Macao, together represent almost half of all outbound tourists. For these travels a very different situation with respect to the costs and efforts involved brings leisure travel into the reach of a much bigger share of the Cantonese Chinese society.

For some more specific information, the results of the Canadian study of 2001 and the BISU study of 2004 are presented here in Table 5.4, as they both included a larger number of participants and were conducted on a nationwide scale.

The average yearly disposable income of the top decile of urban Chinese was officially stated as a little less than 19,000 Yuan RMB.[27] Even taking the higher figure of 35,500 Yuan RMB per year or 2,960 Yuan RMB per month, reported for the national top decile for 2005 (China Daily 2005), 10 per cent of the BISU respondents report a *monthly* income higher than 30,000 Yuan RMB. Two-thirds of the respondents of both studies stated they earn more than the average monthly disposable income[28] of the top decile, again showing that mainly the persons at the very top of the social pyramid of China participate in leisure tourism.

Travel motivations of Chinese outbound travellers

A survey about the dreams of urban Chinese children, conducted in 1996, found world travel in fourth place, surpassed only by becoming *a hero*, *more intelligent* and *taller*. Some 81 per cent of all Chinese city kids expressed their strong desire to see the world (Blok 2002; Li 1998). Law, Cheung and Lo (2004: 355) state 'that the major reasons for Hong Kong residents to travel abroad are to avoid the hectic busy life and the crowded and polluted environment'. It is questionable if this classical motivation explanation of tourism as a flight from unpleasant everyday surroundings (Ren 1988) applies for outbound tourists from Mainland China. The preference for group travel and city tourism does not point into this direction, the idea that crowdedness is synonymous with unpleasantness is a rather western individualistic idea. The Chinese word to describe fun atmosphere is *renao*, composed of *re* 'hot' and *nao* 'noisy'. Chinese long-distance outbound tourists are in most cases not getting a better service or treatment than at home, where most of them are used to having helpers and service personnel available in higher numbers and servility than in foreign countries. As the majority of outbound travellers are beneficiaries of the current Chinese economic and political situation, they are also not very likely to feel especially elated by travelling in democratic societies. Being securely grounded in their own culture, they also do not tend to envy locals for being able to stay in a nicer location or having an assumedly more relaxed lifestyle and better food, as northern Europeans are prone to do with respect to the Mediterranean region and its inhabitants. Unlike the people living in Hong Kong (Breitung 2001), Chinese travellers can choose from an abundance of short-[29] and medium-distance leisure destinations and do not need to travel to outbound destinations for simple relaxation. The case for factors pushing Chinese tourists out of their own country is therefore rather weak; pull factors are more likely to inform their motivational patterns.

The increase in importance of private trips versus 'public' trips can be seen from the growing number and share of trips organized using Private Passports. Since 1999 more Private than Public Passports were used and whereas the number of border-crossings done with Public Passports – for official or leisure trips – did not change much since then, Private Passport travels more than tripled between 1999 and 2003. Three-quarters of all travels are not motivated by the fact that the travellers are sent out in some official capacity with little or no choice where they travel to.

'Little is known about the needs and expectations of Chinese travellers. There are a handful of studies on the mainland Chinese market' (Zhang, Heung 2001: 11). Information on motivations to visit specific destinations

Table 5.5 Outbound travels with Public and Private Passports 1993–2003[30]

Year	Total	Public Passports	1993 = 100	Share (in %)	Private Passports	1993 = 100	Share (in %)
1993	3,740	2,274	100	60.8	1,466	100	39.2
1994	3,734	2,091	92	56.0	1,642	112	44.0
1995	4,521	2,467	108	54.6	2,054	140	45.4
1996	5,061	2,647	116	52.3	2,414	165	47.7
1997	5,324	2,884	127	54.2	2,440	166	45.8
1998	8,426	5,235	230	62.1	3,190	218	37.9
1999	9,232	4,966	218	53.8	4,266	291	46.2
2000	10,473	4,843	213	46.2	5,631	384	53.8
2001	12,133	5,188	228	42.7	6,945	474	57.3
2002	16,602	6,541	288	39.4	10,061	686	60.6
2003	20,222	5,411	238	26.8	14,811	1,010	73.2

Table 5.6 Travel motivations for Chinese outbound
tourists 2001 (in per cent)[31]

Going to a place that is safe and clean	94
Outstanding scenery	92
Increasing my knowledge	85
Experiencing a different culture	85
Chance to see wildlife and unspoilt nature	82
Seeing places important in history	80
Opportunities for playing sport	67
Big modern city	65
Staying at a resort area	62
Outdoor activities	61

is available and will be discussed in Chapter 6, but general studies are again limited in scope and reliability. Compared to the results for the geographical and sociographical characteristics however, the general studies and the destination-specific studies all share a number of common insights.

A Pacific Asia Travel Association (PATA) study puts the existence of 'friendly residents' only on fourth position after 'rich tourism resources', 'good service' and 'reasonable prices' (Travel Daily News 2004). The CTC study asked to rate the importance of 12 travel motivations in selecting a long-haul holiday destination. The top motivating factors turned out to be going to a place that is safe and clean and has outstanding scenery. Opportunities to increase knowledge, experiencing a different culture, chances to see wildlife and unspoilt nature and seeing places important in history were also named as important motivators.

A study conducted by Guo (2004) for long-distance travellers leaving Shanghai airport lists similar pull factors for outbound destinations.

Safety appears here as in national studies as having a high priority for Chinese travellers. Safety here means several things. First of all, safe

Table 5.7 Travel motivations for Shanghainese
outbound tourists in 2003[32]

Safety	5.73
Beautiful scenery	5.69
Well-equipped tourism facilities	5.15
Different cultural and historical resources	5.10
Good climate	5.03
Ease to arrange travel plans	4.98
Good leisure and recreation facilities	4.93
Inexpensive travel cost	4.91
Good place for shopping	4.38
Level of economic development	4.01

transportation infrastructure, stable climatic conditions, protection from dangerous animals, etc., minimizing the danger of being bodily harmed through accidents. Further, safety relates to safety from robbery, mugging and other criminal activities. Chinese tourists are often targeted by thieves, as in most cases they are not very used to having to look after their belongings all the time. They tend also to carry larger sums of cash around, as internationally accepted credit cards are owned only by a minority (Kelemen 2005). Being unfamiliar with the local situation, they are less likely to recognize potentially dangerous situations, or to know how to react in the case of a robbery or theft or to defend themselves. The interest of some travellers in gambling and commercial sex increases the probability of coming into contact with the darker side of a local community, especially in countries where such activities are illegal. However, safety also relates to the danger of being racially abused or ridiculed as Chinese and therefore as a self-perceived representative of *all* Chinese.

The BISU study asked in 2004 nationwide outbound travellers for the most important motivation when deciding on the destination for a long-haul journey.

The issuing of questionnaires can help to gain information about the different choices and preferences of travellers within the given frame of a culture. Unfortunately a lot of research is content to stay at the surface of travel motivation analysis by pretending that the questioned tourists are able to understand and analyse their own motivations and behaviour. However, motivations and behaviour that are taken for granted within a given culture have to be analysed from the *outside* and by comparison to develop a cross-cultural understanding vis-à-vis other cultural sets of motivations and behaviour. The different forms of gaining status from outbound travel across cultures, for instance, are not readily apparent for members of one of the cultures:

> Specifically important to consumption is the Chinese concept of status, which differs in significant ways from Western notions of status. The Chinese consumers are very status conscious, but contrary to the Western admiration of the outsider and the highly individualized James Dean-like rebel, status in China is played out within a consensus and group setting. The Chinese status desire could be called interpersonal-based individuality (Li 1998: 153). Status is attained solely through

Table 5.8 Travel motivations for Chinese outbound tourists in 2004 (in per cent)[33]

Interest in experiencing the local culture	31
Interest in the destinations specific offers	25
Acceptable price	20
Visiting friends and relatives	12
Interest in experiencing nice scenery and clean environment	4
Others	8

the views of one's surrounding group, and criteria can wary between groups. In general, wealth is the most significant determinant of status in Chinese society today, followed by power and knowledge. The stress on wealth makes status oriented purchases important, and many purchases are competitively motivated in order to achieve equal status with one's group. Luxury items in China are generally purchased for status and not because of consumer sophistication and taste, and this is true of international travel as well.

(Blok 2002: 10)

Behavioural background and differences of Chinese outbound travellers

To discuss the influence of Chinese culture on Chinese outbound tourists' past and present behaviour, another full book – if not a shelf full of books – would be needed. This is not the place for instance to discuss the question how 'Confucian' the Chinese society is (van Ess 2003) or how much this concept, originally a Jesuits' invention (Jensen 1997), has been instrumentalized for the support of existing political structures throughout Southeast and East Asia in the form of *ersatz Confucianism* (Bardsley 2003). However, some searchlights can be flashed here on the sources of Chinese tourists' behaviour, including the special role of the Chinese language.

To inform the discussion beyond anecdotal impressions and simple prejudices the tools of Hofstede's cultural dimensions[34] are used for some general observations. Language is the basic instrument to comprehend the world and as the pictographic or ideographic concept of Chinese radically differs from the phonemic concept of western languages, a look at the consequences is necessary. Anderson points out that

> the Middle Kingdom . . . (was) imaginable largely through the medium of a sacred language and a written script . . . Yet such classical communities linked by sacred languages had a character distinct from the imagined communities of modern nations. One crucial difference was the older communities' confidence in the unique sacredness of their languages, and thus their idea about admission to membership.
> (Anderson 1991: 13)

Finally, as reactions to and associations from encounters with new sights and places are shaped by the collective memory canonized in art, some remarks about landscape appreciation and artistic reactions to mobility in classical and modern art are added. 'The cultural experiences offered by tourism are consumed in terms of prior knowledge, expectations, fantasies and mythologies generated in the tourists' origin culture rather than by the cultural offerings of the destination' (Craik 1997: 118).

To exemplify the consequences of the cultural *rucksack* of cultural programming every tourist is bringing to foreign destinations, comparison between Chinese tourists' behaviour and those of their western and Japanese counterparts are presented. Western travellers have dominated international tourism in theory and action since the beginning of mass tourism; Japanese tourists as the forerunners of Asian outbound tourism are often thought to be similar to Chinese tourists, if not outright mistaking one for the other.

Culture, understood as the accumulation of shared meanings, rituals, norms and traditions among members of an organization or society, is the collective programming of the mind that distinguishes members of one group or society from those of another (Solomon 1996). Therefore culture is not a phenomenon in its own right. It is the difference perceived, and only then perceived, by one group when it comes into contact with and observes another group (Jameson 1993). This contact is more often than not connected to spatial mobility, utilizing 'Tourism as Cultural Learning' (Brameld 1977) or, as Clifford puts it, 'Culture *as* travel' (Clifford 1992: 103).

Dann (1993) and others have strongly criticized the misuse of nationality as a dimension of analysis and implied the convergence (Adler, Doktor 1986; Giddens 1990; Weiermair 2000) of tourism cultures as a result of the globalization of tourism (Iverson 1997). Evidence from several decades of history of international mass tourism involving among others Taiwanese and Japanese tourists does rather support the divergence approaches or at least a more complex structure of bidirectional influences, of glocalization (Kaplan 1996; Teo, Chang, Ho 2001b). It is important to point out that the idea of *pure* cultures meeting in *inter*cultural exchanges without much knowledge about the other culture is obsolete, if it ever was useful.[35] For Chinese outbound tourists the problems of clear delimitations between members of different cultural background holding the same passport do not apply. It is rather the opposite way, with many overseas Chinese defining themselves first of all as *Chinese*, even if they are holding a passport from, or were indeed born, in another country.[36] Crotts and Litvin's (2003) study showing that for many studies on cross-cultural tourism behaviour, a category of residency is more useful than a category of nationality, is not limiting the discussion pursued here.

The problem that any statement on *the* Chinese culture or behaviour is of necessity a gross simplification of the differences between different geographical and sociographical segments of the society can however not be denied. Moreover, as cultures are negotiated and present processes (Clifford 1988), the differences between the imagined and the officially supported culture and the lived culture are another dimension neglected here. Another element of distinction has been introduced by Thiem (2001), who points out that in tourism encounters not only is there a difference between the daily life cultures of both the source and the target region but also the difference between the hosts' 'backstage' culture and the 'frontstage' culture (Goffman 1959) shown to the tourist.[37] The fourth dimension relates to the travellers' specific culture away from home, which differs from their daily life culture by showing, for instance, different attitudes towards punctuality and cost considerations in shopping than at home. Crotts (2004: 87) expects accordingly that 'individuals will likely take into account the degree of similarities and differences of the countries they travel to and adapt to the environment in an effort to minimize potential friction during their leisure trip'.[38] For Chinese outbound tourists, a changed shopping behaviour is certainly observable, an adoption of or even immersion into the host culture less so.

The foundation for most cross-cultural research[39] is the work of Hofstede (1980; 2001; Hofstede, Hofstede 2005). He developed, with the help of large-scale samples starting in the 1970s, cultural index scores for five constructs: Power distance, Uncertainty avoidance, Individualism/Collectivism, Masculinity/Feminity and, as a later addition, Long-term/Short-term orientation, which he contends distinguish people from various nations effectively. These dimensions are acquired through 'mental programming', learned patterns of thinking, feeling and potential acting. Three levels of uniqueness are distinguished: the inherited universal human nature, the learned culture specific to certain societies or groups within a society, and the inherited and learned individual personality (Hofstede, Hofstede 2005: 4).

Table 5.9 Hofstede's five cultural dimensions with scores for China, Japan, USA and Switzerland[40]

Dimensions	China	Japan	USA	Switzerland
1 Power distance: the extent to which the less powerful members of a society accept and expect that power is distributed unequally	80	54	40	34
2 Uncertainty avoidance: the extent to which a culture programmes its members to feel either uncomfortable or comfortable in unstructured situations and tries to control the uncontrollable	30	92	46	58
3 Individualism: the degree to which individuals are supposed to look after themselves or remain integrated into groups (Collectivism)	20	46	91	68
4 Masculinity: refers to the distribution of emotional roles between the genders, it opposes 'tough' masculinity to 'tender' feminity	66	95	62	70
5 Long-term orientation: refers to the extent to which a culture programmes its members to accept delayed gratification of their material, social and emotional needs	118	80	29	40

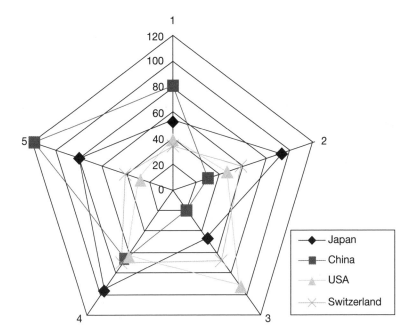

Figure 5.1 Hofstede's dimensions for China, Japan, USA and Switzerland[41]

In four of the five dimensions, China has an extreme position compared to the other countries. For Long-term orientation China scores higher than Japan, USA and Switzerland and indeed has the highest score of all countries that are included in Hofstede's research results. Also Power distance is more pronounced than for the other three countries. For Uncertainty avoidance and Individuality, China scores lower when compared to the other three and also to most other countries in the world. Only for Masculinity are the scores more or less the same for China and for the western countries, with Japan having the highest reading of all countries researched globally.

For Chinese outbound tourists' behaviour, some immediate consequences can be named. A very high Long-term orientation should emphasize the dimension of learning also within leisure behaviour: the spending of time rather for shopping for suitable gifts for important persons at home than for more sightseeing; travel itineraries suitable to the education of the accompanying child. The high Power distance would indicate expectations of being led within a group, of an interest to see the most important and famous sights, places connected to superlatives, of attaching great importance to and imitating travel behaviours of famous Chinese persons. For tour guides as temporary persons of authority, high demands are made with regard to their problem-solving ability and their knowledge. Places should be connected to a clear, somehow authorized and unambiguous story. On the other hand, the high level of Power distance also indicates a clear expectation of behaviour according to social status, of being served and honoured, and of the need to acquire status from ritual actions.

The low level of Individuality, or rather the high level of Collectivism, should result in group-orientated behaviour with a lot of commitment towards the members of the in-group and the harmony within this group, but little commitment to the welfare of the out-group, the general public (Pang, Roberts, Sutton 1998). The size of the group can vary between a family, a travel group or indeed the whole nation versus the others. The importance of strong personal relationships (*guanxi*), personal status and image (*mianzi*, 'face') and human obligation (*ren*) in the Chinese culture are grounded in the importance of in-group relations in large collectivistic societies (Gilbert, Tsao 2000; Li, Wright 2000).

Collectivism supports the wish to go where everybody goes, to do the typical things, buy the typical souvenirs and try the typical local specialities. If the typical things are not known beforehand, it is expected to have them clearly pointed out. If shop assistants do not comply with these expectations, they will be seen as not concerned about the tourists' needs (Reisinger, Waryczak 1994). Memorial group photos in front of sights strengthen and document the collective experience. The low level of Uncertainty avoidance translates into a pronounced flexibility in planning and executing travel arrangements, which is also expected from others. Encounters with strange, unknown situations or persons are not perceived as a threat but as a reason for curiosity and amusement. Tolerance towards Uncertainty also translates into low acceptance barriers for new developments, technology or fashions. The medium level of Masculinity is allowing for a certain level of playfulness, sense of romance and a not very rigid insistence on gender roles.

Ever since the publication of MacCannell's (1999) *The Tourist – A New Theory of the Leisure Class* in 1976, semiotics, the science of signs, has been incorporated in the discussion about tourism, and tourism has been regarded as a consumption of *signs*, made recognizable by *markers*. Dann (1996) moves this approach further by discussing 'Tourism as Language', although China is not mentioned at all.

The same disregard for the substantially different idea of what constitutes a language is also found in semiotics and linguistics, where

> western theories tend to generalize across Indo-European languages. We theorize mainly about languages with phonemic writing systems, inflectional sentential syntax, clear structural distinctions between descriptive and prescriptive forms, and required subject/predicate structure. Chinese theories of language differ in part because the language they explain differs. Their theories account for a pictographic written

language with a grammar that relies almost exclusively on word order for grammatical role-making.

(Hansen 1989: 76)

Western – and in a way also Japanese – commonsense models of language assume that a word is a *sound* and syntax is the major structure of a language. Adding inflections to a word root provides it with syntactic mobility. Alphabets are merely signs that function as carriers of single sounds and therefore are only useful for the spoken language and the historic time in which they are used. Language is supposed to be a spoken system that is written with purely phonetic signs – with Arabic numbers as the only ideographs used in western language-writing.[42] Chinese, on the other hand, is written in pictographs, with spatio-temporal isomorphism explaining the relation between the written word and the parts of the world (in a Wittgensteinian sense) it picks out. Put more simply, Chinese characters as written language are independent of *any particular* spoken language and can therefore transfer meaning across different dialects, languages and centuries. Characters are words[43] and therefore syntactically mobile without any changes to their appearance. As a result, 'Western theory treats the function of language as descriptive or representative. Chinese theory treats the function of languages as socializing, regulating, and co-ordinating behaviour' (Hansen 1989: 77).[44]

Therefore, from a Chinese point of view, Chinese as a language is based on the realities of human life and behaviour, just opposite to the mystifying approach often encountered in the literature: 'The Chinese culture is built on subtlety. For example, the Chinese language is not as precisely structured as the English language. It is based on an abstraction of ideas' (Pitta, Fung, Isberg 1999: 248).

These differences are not just a rather esoteric field of discussion for the followers of Saussure, Lacan or Baudrillard,[45] but constitute the basic mindset of the Chinese civilization and therefore also influence the behaviour of Chinese outbound tourists. The idea that learning to be a tourist is a process of learning a language with its techniques, registers and also its function as a social control as forwarded by Dunn, sounds much less unexpected for a Chinese mind. The visual nature of tourism experiences (Urry 2002) is connected to signs *and* language for Chinese tourists, as language is something to be seen rather than heard.[46] The much stronger 'authority' connected to a Chinese character in comparison to a western alphabetical letter, combined with the high Power distance characteristic as depicted by Hofstede, gives written information precedence over oral information. What is written down has to be clear, unambiguous and should come from a higher authority, not from peer group members.

As Chinese characters are standing for the thing or concept they depict, in a way they *are* the thing or concept. They are 'emanations of reality, not randomly fabricated representations of it' (Anderson 1991: 14). Every

Chinese building, institution, temple hall, museum, etc., must have a name-plate written by a famous artist or ruler in remarkable calligraphy. Calligraphy is a much cherished art form directly connecting the form and the content of written language with the artistic expression. Landscapes based on Chinese characters give the written word direct power over the physical reality, from *ba gua* gardens to the Formula One racecourse in Shanghai based on the shape of the character *shang*. The sign is not only a marker as in MacCannell's usage, but part of the attraction itself, and therefore worth being photographed with and described in information materials.

Behavioural differences between Chinese and western outbound travellers

As was noticed above: 'When Western tourists look at the Yangtze, they see a river; the Chinese see a poem replete with philosophical ideals' (Sofield, Li 1998a: 367). Hashimoto underlines that 'even if a famous attraction of natural landscaping is suffering from environmental degradation, in the Chinese tourists' minds, the place is still as beautiful as described in the poem' (2000a: 132).

> For the Chinese, beauty is the constitutive inter-involvement of many into one, and one with many, until the entire unison becomes both concrete-particular and cosmic-universal, both in scale and in substance. Here is a twofold characteristic – distinction and interchange, even on the level of the subject. As a result, beauty is less of a subject to be independently discussed than a pervasive attitude and atmosphere in which one moves and has one's being.
>
> (Wu 1989: 236)

For Chinese travelling within the Chinese world, the integration into *the atmosphere in which one moves and has one's being*, in a form of 'nostalgia for nature' (Graburn 2001b) or in the context of Chinese poetry and legends, can be the dominant spatial touristical experience.

'We can say that the Chinese mind does not see itself as something apart from nature attempting to understand itself as if it were a foreign "other"' (Allison 1989a: 15). The Chinese world view does not follow the western conception of the self as a bounded, unique, more or less integrated motivational and cognitive universe, set contrastively both against other such wholes and against its social and natural background (Geertz 1984). 'Descartes isolated the *cogito* and split existence into two – thinking and extension, the subjective and the objective' (Wu 1989: 237). The western post-Enlightenment acceptance of the reality of the undivided self as self-evident 'is, however incorrigible it may seem to us, a rather peculiar idea within the context of the world's cultures' (Geertz 1984: 126) or indeed if set against the physical reality. In the Chinese human-centred point of view man cannot exist alone; all action must be in the form of interaction between man and man.[47] In Chinese *shanshui* (mountain and water) land-scape paintings, human figures are shown in tiny scale, in Chinese landscape gardens human *flaneurs* are an integrated, if not very important, part of the landscape.

For Chinese outbound tourists leaving the Chinese world, the disappearance of the context has much greater consequences than for western individualistic travellers. The insistence of Chinese travellers to eat Chinese cuisine daily, reducing the consumption of local food to the sampling of famous dishes perceived as typical for the destination,[48] is often taken as a sign of the missing sophistication or risk-aversion of Chinese outbound tourists. An alternative interpretation of this insistence, which is not even moderated by the mostly mediocre quality of Chinese food outside Asia, could be that with the visit to a Chinese restaurant an outpost of *Chineseness* is reclaimed even within an alien surrounding.[49]

Modern artists in China are responding to these cross-cultural experiences of mobility and resulting disorientation (Urry 2000; Hall 2005). To give but a few examples: Zhang Huan, one of the best-known artists of the Asian-Art scene, used his first personal impressions of arriving in New York City for a performance in 1998 of 'Pilgrimage: Wind and Water in New York'. To put a western metropolis back in line with traditional Chinese geomantic space appreciation of *Fengshui* (wind and water), he lay for hours naked on a Ming Dynasty opium bed covered with slowly melting ice blocks instead of a mattress (Muecke, Sommer 2003).

The Beijing-based artist Cang Xin, transcended the 'tourist's gaze' (Urry 2002) by literarily licking up places with his tongue in 2000/2001, moving horizontally on pavements in slow progression in front of the Vatican or on Trafalgar Square, or licking on monuments and memorials. By first choosing famous sights in China such as the Temple of Heaven or the Confucius Temple and then moving to Rome, London and Stockholm he followed the routes of Chinese outbound tourists (Cang 2002). The 'temporariness' of spaces is an important topic in the work of Chinese artists such as Wang Wei, who, with the help of local brick mongers, built a temporary closed space with 25,000 bricks salvaged from destroyed Beijing *hutong* buildings in the fashionable Dashanzi Art Gallery (Huang 2003) in 2003. Here the gaze is not falling on bricks in their original space but in a changed context: the bricks themselves are moved to create a new non-sensical space of ambiguous authenticity (Wang 2003c).

Taking up again the Hofstede categories described above, the differences between Chinese and western cultures are easy to see. Except for the gender roles, the western countries are always on the other end of the scale from China, in the cases of Power distance and Uncertainty avoidance Central European Switzerland more so than the USA. For Individualism and Short-term orientation the Anglo-American culture prevalent in the USA has the more extreme attitudes. Some differences in the outbound tourism behaviour can be categorized with this insight, again within the limits of such a simple tool. At least descriptions of what is seen as *normal* or preferable behaviour within one of the cultures can however be extracted with greater confidence.

Lower Power distance readings can be translated into an interest in special, unusual destinations and sights within the destination, the search for the insider tip, the 'non-touristical' places and restaurants only known to the locals or other well-acquainted persons. The idea of a higher level of equality between guests and hosts or service personnel will replace the wish for servility and professionalism with the quest for heartfelt, non-professional friendliness, a 'smile from the heart'. This includes tour guides, who will be more cherished for their humour and joviality than for exact information, which can be acquired if so wished individually from other sources. A high level of Uncertainty avoidance, found especially in the Germanic tradition as well as in Iberian and South American cultures, does counteract the wish to think differently as this involves unpredictable outcomes. 'Adventure without risk' offers and guide-book writers that proclaim to convey unusual, but tested, insider tips resolve this contradiction, as do package tours individually prepared out of individually chosen but standardized parts and repeated visits to the same destination. Insistence on itineraries executed as planned and punctuality are other wishes that increase with the uncertainty level readings. Mistrust of foreign, strange persons or outright xenophobia can be the results of such high readings, which are, however, moderated by high levels on Individuality.

Individuality and Long-term orientation are the two categories where Chinese and western cultures differ the most. The disdain for mass tourism, the 'fun today' recklessness as a role model for young beach tourists are some results, but also the medium of common courtesy in encounters with other tourists, for instance, at an airport. Souvenirs will be chosen according to the special characteristics of the prospective person they are supposed to be given to or with post-touristical irony.[50] Business and leisure and different forms of leisure such as sport and culture tourism will be separated more clearly. High levels of Individuality also strengthen the will to make one's own decisions and to make them according to one's self-image. Tafarodi *et al.* (2004) show that in a study of Canadian, Japanese and Chinese students, 88 per cent of the Canadians think that nobody knows themselves better than they know themselves, but only 56 per cent of Chinese share this idea. Two-thirds of the Canadians think that their self-image stays the same in different situations and that this is a good thing, compared to less than one-third of the Chinese. The Chinese 'are the least bothered by self-disidentified conventional behaviour' (Tafarodi *et al.* 2004: 113).

Behavioural differences between Chinese and Japanese outbound travellers

> It is very important for Chinese not to be taken for Japanese. Chinese tourists are increasingly angry about the fact that they are often held to be Japanese. They want to show that not only Japanese can afford long-distance tourism.
>
> (Sun 2004: 29)

When Switzerland was granted ADS in November 2003, a headline in a magazine read 'Sunrise for Swiss Tourism', relating Chinese outbound tourism to the 'Land of the Rising Sun', Japan (Swissinfo 2003). It is a common mistake, especially outside Asia, to assume that all *East Asian* or even *Asian* tourists are more or less the same, a mistake certainly helped by the similarities of travelling in groups and being hard to recognize individually. For those in daily contact with both Japanese and Chinese tourists, this misconception disappears swiftly. For example, Kim and Prideaux (2003) researched the behaviours of airline passengers by asking Korean airline flight attendants to compare Korean, Japanese, Chinese and American customers. They could clearly differentiate between Japanese passengers who scored highest in not expressing dissatisfaction, following the guidance of attendants and not asking for more drinking water or food, whereas Chinese passengers were the least respectful, asking the most for additional beverages and food and were perceived as 'hindering the work of flight attendants due to frequent demands' way ahead of all other passenger groups (Kim, Prideaux 2003: 491).

Looking at the Hofstede's scores the differences become apparent. Only in Long-term orientation do both cultures score high at the same end of the scale (China 118, Japan 80), otherwise there are large and even very large differences. For Power distance Japan's reading is lower (80 vs 54), as Japan's society is based on social control executed by the peer group rather than by higher authorities like in China, which also can explain the otherwise unexpectedly high rating for Individuality for Japan (20 vs 46). The biggest differences are in Masculinity, the dimension in which Japan has the highest score in the world (66 vs 95) and in Uncertainty avoidance, where both countries are at different extremes of the scale (30 vs 92).

Hashimoto (2000b) states:

> There is nothing unique to the Japanese but some characteristics are extreme in degree and in practice, and can be called 'Japanese characteristics', for example, industriousness, collectivism, self-denigration, shyness, politeness and formality in manners. These characteristics are, not surprisingly, based on the Japanese mentality of shame and guilt complexes. The Japanese culture reinforces the sense of 'shame' if a member of the group fails to meet social expectations and this feeling

of shame in turn leads to a sense of 'guilt' in an individual. As it is hard to live with a constant feeling of 'guilt', Japanese individuals tend to work hard to meet or exceed the social expectations.

(Hashimoto 2000b: 39)

Unlike Chinese society, which acknowledges the need of a balance between work and play, or rather tries to avoid such a clear dichotomy, in Japan 'a feeling of guilt in seeking mere pleasure dies hard within a culture of hard work' (Kajiwara 1997: 169). Long absence from work by making full use of granted paid holiday is still frowned upon, less than half of the holiday entitlement is actually used (Arlt 2005d). For Chinese the obligation to bring back souvenirs is mainly connected to the idea to 'pay back' to all those who helped with the preparation of the travel by providing financial or *guanxi* (social connections) support. In the Japanese society, *zoto* (gift-giving) is less connected to the idea of gaining from connections that are reinforced by gifts, but to

maintain social relationships. In other words, the Japanese give gifts when they do not particularly want to, a characteristic of gift-giving that is rarely seen in other cultures, at least to the extent that it takes place in Japan.

(Davies, Ikeno 2002: 234)

Leisure activities and the necessary absence connected to it are seen in Japan as 'letting down the colleagues', only partly redeemed by 'doing something positive on behalf of the whole group' (Graburn 1983: 58) by buying that special kind of gift, *omiyage* (souvenirs).

The strong connection to group orientation and the hard work ethos in tourism behaviour can also be found in the study of Andersen, Prentice and Watanabe (2000) of Japanese independent travellers in Scotland. These tourists present the most independent-minded segment of the Japanese outbound market. Still only 10 per cent agreed to the statement 'I avoid visiting famous places', but around 80 per cent agreed to each of the statements 'I appreciate life at home more when I return', 'It is important for my career to get new inspirations on holiday' and 'It is important for my career to travel widely on holiday'.

Pizam and Jeong (1996), in analysing the cross-cultural tourist behaviour of Japanese tourists in comparison to travellers from South Korea and the USA, found that Japanese travellers strongly prefer to congregate with members of their own nationality, prefer to travel in groups and for short time, strongly prefer local food,[51] are very interested in buying souvenirs for friends and relatives at home, wish to plan trips rigidly and are not at all adventuresome.

Reisinger and Turner (1997) list a number of aspects of Japanese tourists' behaviour, which can be set in juxtaposition to the behaviour of Chinese outbound tourists. Japanese tourists 'are more demanding and have higher service expectations than other international tourists and are driven by a Japanese service philosophy determined by cultural beliefs about how service should be properly performed' (Reisinger, Turner 1997: 1203).

Chinese tourists ask for value for money and respect, but are less annoyed if things are done in a different way than at home:

> In order to maximise the benefits of their holidays, Japanese carefully pre-plan their travel arrangements. They examine all alternative destin-ations, their pros and cons, and consider various pricing policies to save additional funds. As a result, in the prepurchase stage of consumer behaviour Japanese spend a lot of time on decision making.
>
> (Reisinger, Turner 1997: 1209)

Chinese tourists are known for their last-minute decision making, react-ing to special offers or shift of ideas within their group:

> Punctuality is regarded by the Japanese masculine society not only as a sign of good manners and respect to others but also as a measure of professionalism and performance. If service delays are anticipated, providers should make waiting entertaining. Japanese tourists need to be occupied, for example, given menus to look at while waiting in a restaurant or a travel journal to read while waiting in a hotel lobby or at an airport.
>
> (Reisinger, Turner 1997: 1219)

Chinese leisure tourists are normally in a hurry because they want to see as much as possible in as short a time as possible. They do not appreciate wasting time by waiting, but this is for practical reasons rather than object-ing as a matter of principle. If there is some waiting, chatting or napping will help to while away the time without any embarrassment or awkward-ness evolving.

Hashimoto (2000b) provides information about an important segment of Japanese outbound tourism, which seems likely to grow in importance in China as well: small groups of unmarried female outbound tourists between the age of 20 and 29. Such Office Ladies (OLs), living cheaply with their parents, have a high-spending power but fewer obligations towards their company and family. 'They are free to spend their money on themselves and free to travel abroad. These freedoms, however, will be sharply curtailed once they marry' (Hashimoto 2000b: 43). The majority of these young OLs (or *Gals*) are allegedly characterized by a lack of language ability and ignorance of the world, judging the world by the level of *kawaii* (cuteness) alone. They are criticized because they

> enthusiastically take international holidays but do not make any effort to understand the world. The *Gals* are only interested in ethnic gourmet

dishes and handicrafts in developing countries. They pay little respect to the culture of the host countries, and with childish curiosity they do not hesitate to ridicule difficult lives in these countries. In other words, *Gals* travel with their Japanese values without trying to respect and adapt to values in other countries.

(Hashimoto 2000b: 47)

Japanese domestic tourism has been characterized by the promotion of *furosato* (old hometown) tourism (Ivy 1995; Moon 1997; Graburn 1998), travelling to reassert the *Japaneseness* of Japan, even 'to travel within Japan not to rediscover 'old' Japan, but the nostalgia *for* it, the lost feeling of the feeling of loss' (Arlt 2005d). Rea (2000) shows that the *furosato* fiction is taken further to look for 'a furosato away from home' in the symbolic *kawaii* landscapes of the English cottage of Beatrix Potter and in Prince Edward Island, Canada, home to Lucy Maud Montgomery's fictional heroine Anne of Green Gables.

Chinese tourists, as has hopefully become clear, start their career in international travel from a quite different cultural background, unbidden by shame and guilt, not bogged down by masculinity and fear of uncertainty and with a much more straightforward agenda of self-assertion. Even so, some superficial similarities exist as one can find between the Chinese outbound tourists' behaviour and the description of Beer who argues that

in the case of Japanese overseas packaged tourism, modernity seemed less an issue of false or alienated urbanites searching for authenticity. Rather, the Japanese tourists seemed more interested in acquiring status through cultural and material capital, escaping the confines of urban life, cementing social ties, and establishing one's identity as Japanese.

(Beer 1993: 22, in Rea 2000: 639)

Information sources of Chinese outbound travellers

The lack of information sources is often named as one of the main reasons for the concentration of Chinese outbound tourists on a very limited number of destinations and sights (Chinatravelnews 2003b). The People's Republic of China used to be an isolated country with only limited and tightly controlled access to outside information. Since the beginning of the 'Reform and Opening' policy in 1978 and especially with the increased integration of China into the global economy in the 1990s, this has changed significantly. National printed and electronic media are nevertheless still strictly controlled and international media as far as possible censored by the Chinese government. The potential of the Internet as a tool for independent communication, self-determination and civic protest has been recognized and is upset as far as possible by increasingly tight controls. The access to information about foreign countries and international events, not related to 'sensitive' issues, is however possible and rather encouraged by the Chinese government as a way to raise the standard of general knowledge and international outlook. China is the country with the highest number of Internet users outside the USA[52] and TV programmes are accessible to 95 per cent of all households. Books and magazines, including those on international travel, are readily available. With more than 100 million border-crossings by Chinese in the decade after 1990 and over 700,000 Chinese students studying abroad since the 1980s, there is also a lot of first-hand experience as a source of information. Educated Chinese under the age of 40 years are very likely to have some ability to understand information in English and other languages:

> In contrast to an era when all public media were under the tight control of the Communist Party, the contemporary Chinese scene exhibits a bewildering number of new media. One can find all sorts of serious and frivolous newspapers, books, advertisements, radio and television shows, hotlines, and internet sites.
>
> (Blok 2002: 18)

Ever since the days of the reformers Kang Youwei and Liang Qichao at the end of the nineteenth century, the concept of 'Chinese values as base, Western knowledge as tool' has been seen as a way back to the self-image of global supremacy of China, so thoroughly shattered by the defeats in the Opium wars. Learning from the west has also been an important part of the efforts of the People's Republic of China, be it Soviet steam locomotives or German magnetic high-speed trains, the writings of Karl Marx or Lee Iacocca, Stalinist or post-modern European architecture. The self-centredness of European, American or Japanese societies, if projected towards the Chinese society, results in the erroneous expectation that Chinese people know as little about 'us' as we know about 'them'.

In 2001 figures for the relevance of different sources of travel information were given as travel agency 38 per cent, books, newspapers and TV 33 per cent, word of mouth 18 per cent, brochures 7 per cent, Internet and previous visits 2 per cent each (Blok 2002). With the development of tourism and Internet usage, the relevance of the last two items has increased considerably.

Printed and electronic media

Some 25 billion newspapers, 7 billion books and 3 billion magazines are coming off the printing press every year in China;[53] large bookshops reach the size of a whole department store. The size of the travel guide books departments in Chinese bookshops has grown in proportion to the rapid increase in domestic and outbound travels. Guide books to overseas destinations started to appear in greater numbers from the end of the 1990s.[54]

These books were either official translations of western guide books or texts compiled from undisclosed sources, mostly done in black and white only and with many spelling errors in place names, etc. (for instance Jin 2000; Li 2002). Their focus on history and art and the inclusion of many maps, language guides and other features clearly showed their origin as texts for western individual travellers. An *Atlas of Popular Tourist Routes in China*, published in 2001 by the China Cartographic Publishing House (CCPH) included maps and information for Singapore, Malaysia and Thailand as the most important tourist destinations outside China (CCPH 2001).

By 2005, the number, variety and quality of guide books have dramatically improved. Whole series of translated international guide books are available, most of them carefully designed and printed in full colour. The geographical entities covered range from continents, to regions, countries and single cities. Beside them a growing number of professionally written tour guides are published, catering for the more experienced travellers, such as a tour guide to North Europe written by a Chinese author who can boast to have already visited more than 60 countries (Hun 2005). The search for the superlative is reflected in the large number of books about United Nations Educational, Scientific and Cultural Organization (UNESCO) World Heritage Sites, the UNESCO seal being the most prestigious quality label available. In a similar vein, several books now provide a 'shopping list' for must-see places like '66 places to see in a lifetime' (Beijing Gongye University 2005), which shows the confidence of the opportunity for continuous international travelling and lists places as diverse as Rome, Machu Picchu, Disneyland and Antarctica. The decrease of the special status of international travel is documented by the fact that alongside these international destinations, Nanjing, Shaoxing and other destinations in China are also included.

Besides the books published in China, a growing number of travel guides are also published by NTOs and distributed free of charge during marketing activities, fairs or in the aircraft flying to their respective destinations. These booklets are normally financed by advertisements (for instance DZT 2003; DZT 2004). Some companies like Global Refund publish area-cum-shopping guides (Global Refund 2003). Even incoming agencies themselves publish different guide books.[55]

The mixture of national and international destinations can also be found in a number of different glossy travel magazines published in China.[56] The perfect styling, the retail price of about 20 Yuan RMB and the double-paged advertisements for Volvo cars, expensive Japanese cameras and Louis Vuitton luggage clearly show them targeting the upper class of Chinese travellers. Most covers are adorned with European or Southeast Asian female models and the content is not restricted to ADS destinations. Many feature articles are sponsored by NTOs; series include special topics such as 'Small European Cities' 'Island Hopping' or 'Spas', but also 'Travel Etiquette', 'Discover China' or 'My Life as a Student in Holland'. For the October

'Golden Week' 2005, *Voyage* magazine offers travel ideas conveniently sorted according to price: for 10,000 to 20,000 Yuan RMB (approximately 1,000 to 2,000 Euros in 2005) five days in Malaysia, seven days on Sicily or eight days of golf in Australia are offered. For 20,000 to 30,000 Yuan RMB, a visit to Canada or two weeks of safari in Kenya are featured. Even more affluent travellers are directed towards a cruise in the Mediterranean sea, a week in a five-star hotel in Paris or a trip in a luxury train in southern Africa for 65,000 Yuan RMB (approximately 6,500 Euros in 2005). The Chinese edition of *National Geographic* also supplies potential travellers with images of far-away places. Travel sections are also included in many business magazines. Outside China some periodicals also target the incoming visitors, providing information and advice about destinations combined with advertisements of Chinese restaurants and souvenir shops.[57]

Watching television, in many people's view one of the major maladies of modern times, is also holding China in its grip. In 1990, 80 per cent of all Chinese had access to TV programmes and 75 per cent to radio programmes; in 2003 both figures had risen to 95 per cent, including more than 100 million cable TV subscribers. Some 46 per cent of all Chinese households own a colour TV set (Li, Li, Huang 2004). China Central Television (CCTV) provides 12 nationwide channels, each province and municipality has several channels, adding up to 320 TV stations within the whole country.[58] 'Television is by far the most popular medium for information and entertainment among Chinese consumers' (Cui, Lui 2000: 64). Following the increase in domestic and international travel and the recognition of tourism as an industry rather than a wasteful activity by the Chinese government, new formats with travel magazines with titles such as 'Looking around the World', 'Travel Compass' or 'Travelling the Four Seas' were developed in the late 1990s.[59] By the middle of the first decade of the twenty-first century, almost every TV channel provides such programmes, again with little regard to whether the featured destination, often sponsored by the relevant NTO or airline, has already gained ADS or otherwise. For example, ten episodes of ten minutes in length, each about Canada as a tourism destination, were filmed by Beijing Television Station in Canada and aired in the winter 2001/2002, each episode reaching some 200 million viewers (CTC 2002). Shanghai TV aired in eight parts a series about Bavaria in Germany as a holiday destination, which was shot by a Chinese TV team of six persons invited to Bavaria. The series is also sold as a DVD (Hummel 2005). Southern TV presented the German Fairy Tale Route in a ten-part series in its childrens' programme.

Besides the formats that provide direct tourism information, image-enhancing information connected to a destination such as sports events, movies shot at the location, etc., can enhance the perceived value of the country or region as a brand and therefore also increase its attractiveness as a leisure destination. Novels and TV programmes also address the situations and experiences of Chinese in foreign countries directly, with TV series such as 'A Beijing Man in New York' or soap operas such as 'Into Europe' shaping ideas about the foreign countries and their treatment of Chinese (Nyiri 2005b):

> Chinese tourists currently rely on the media (e.g., newspapers and tele-vision) and word of mouth from friends and relatives to provide them with information about potential travel destinations. In the case of first-time or more unsophisticated travellers, travel agents will provide their clients with suggestions and recommend where they should go. How-ever, with the growth of the Internet in China, more and more people are starting to rely on the web to get the information they need.
>
> (CTC 2001: 19)

Travel agencies in China themselves use newspapers and increasingly the Internet as their most important communication media towards their poten-tial customers (Guo 2002), as the Internet has developed into a major source of information in China. From 9 million at the beginning of the year 2000, the number of users increased to 103 million by mid-2005, for the first time surpassing the 100 million mark. Almost half of the users (46 per cent) are older than 24 years, 55 per cent have a college degree or higher education, 20 per cent earn more than 24,000 Yuan RMB (approximately 2,400 Euros in 2005) per year. Email has become a regular feature with 43 per cent of all users having already more than three years experience with it (CNNIC 2005). Destinations around the world, especially outside Asia, have been rather slow to develop websites for Chinese potential visitors. Since 2004, however, a growing number of Chinese language websites are offered by major tourist cities (for example Berlin, Vienna, Tokyo, Wellington), regions (for example Hawaii, Hesse, Western Australia) or countries (for example Australia, Finland, Malaysia, South Africa). Most websites are however just reduced versions of the main website in the Chinese language, disregarding the adaptation needed for successful cross-cultural Internet tourism mar-keting for the Chinese market (Arlt 2005b; Arlt 2006a; Singh, Zhao, Hu 2005). Currently only 7 per cent of Chinese Internet users claim to rely mostly on the Internet to find information about tourism. This number might be related to the information offered, as 11 per cent claim to be especially dissatisfied with the information available online about tourism (CNNIC 2005). For outbound tourism the percentage of Internet usage as part of the preparations is higher (Zheng 2004). A recent study found that tourism marketing websites by Chinese DMOs for domestic tourists lack in quality compared to American DMO websites (Feng, Morrison, Ismail 2003). Another study blames the inbound orientation of China's tourism companies for the fact that China's usage of Information and Communication Tech-nology (ICT) in tourism is still in the development stage of the 1970s:

> However, the demand for eTourism or the driving force of ICT and Internet adoption in tourism lies in domestic and outbound tourist

market. The mismatch between supply and demand markets can partly explain the stagnation of eTourism developments in China.

(Ma, Buhalis, Song 2003: 464)

Other modern electronic media are also used in tourism promotion. The Hong Kong Tourism Board (HKTB) for instance in 2002 held a SMS Quiz Game in selected Mainland cities. Out of the 45 million mobile phone subscribers reached in nine provinces and cities, 1.5 million responses were received (HKTB 2003).

Tourism fairs

Another sources of information for potential Chinese outbound travellers are travel and tourism fairs and exhibitions. As in other commercial fields, there is rather too large a number of competing exhibitions for tourism both at national, provincial and city levels. With the growing tourism market, fairs are seen as an opportunity to earn money from holding a fair in itself. At the same time, many provincial and city governments wish to emphasize their modern, international image by organizing or supporting such activities.

The most established exhibition is the China International Travel Mart (CITM). CITM claims to be the largest travel show in Asia in terms of numbers of participants and stands. The first ever CITM was held in Shanghai in October 1998. Two years later, the second CITM could already attract almost 50,000 visitors to the Shanghai Everbright Exhibition Centre. Reflecting the situation of the outbound market, more than 70 per cent of the exhibitors came from Hong Kong and Macao. For the third CITM, the exhibition was moved to November and held for the first time in Kunming, capital of Yunnan Province in Southwest China. Since 2001 it has been organized each year, with Shanghai and Kunming as alternating venues. After the successful 2002 fair in Shanghai, SARS prevented much interest in the 2003 show in Kunming both from the exhibitors and from the visitors' side. For 2004 the new Pudong International Convention Centre could be used and the renewed interest in travel and tourism resulted in an exhibition space of close to 50,000 square metres, with about a quarter of it dedicated to outbound travel. CITM 2005 again took place in a brand new convention centre erected in Kunming.

CITM has been joined by a number of other exhibitions. The Beijing International Tourism Expo (BITE) was first launched in July 2004 at the old Beijing Exhibition Centre as an alternative to CITM, but could attract hardly any international companies or foreign national tourist boards as exhibitors. In the same year, the Shanghai World Travel Fair (WTF) started as an exhibition mainly for ADS countries. The first two exhibitions were held in January; from 2006 the date was moved to March. In the same month, the Guangzhou International Tourism Fair (GITF) is staged. The GITF goes back to 1993 but is mainly concentrated on domestic and Hong Kong and Macao travel.

In 2005 two new, foreign-produced exhibitions in Beijing were added to the calendar. The Beijing International Travel and Tourism Market (BITTM) opened in April 2005 at the Agricultural Exhibition Hall after intense international marketing but with limited response from the visitors (Lin, Ling 2005). CBITM is another unwieldy acronym, standing for China Business and Incentive Travel Mart. CBITM is a special event focused on the MICE market and was held at the China World Hotel and Convention Centre in July 2005, after a first attempt in Shanghai in 2003 fell victim to the SARS downswing in the market (Graff 2005). Regional tourism fairs also attract more and more international exhibitors. The Hunan Tourism Exposition for example, held in September 2005 in Changsha, capital of Hunan Province, succeeded in bringing tourism organizations from 12 countries to present themselves, including Germany, Japan, South Korea and France (CEN 2005a).

Outside Mainland China, the International Travel Expo (ITE) Hong Kong is also of importance, not for the Mainland end-users but for the travel agencies. China outbound seminars have been held regularly during the ITE Hong Kong since 2003.

Visitors have complained about the lack of information in the Chinese language from international exhibitors. But the exhibitors themselves are often unhappy with the relatively low number of visitors compared to the high costs of participation and the unclear distinction between trade and general visitors even on supposedly 'trade visitors only' days. As in other business relations with China, they find that fairs are more the starting point for business contacts than the place to bring back signed contracts from (Kiefer 2004).

A new approach to provide information about national and international destinations is the World Leisure Expo 2006 in Hangzhou. Between April and October 2006 a number of events to celebrate the diversity of leisure are held in newly constructed gardens and parks, including a recreation of the campanile of St Mark's Square in Venice. National and international destinations are presented in the 'One Hundred Worldwide Cities Folkways' in the Venetian Canal District. Cities such as Sydney, Leeds and Taibei promote their tourism and leisure resources with presentations of arts and crafts and other communications of their image, especially during their assigned 'City Day'.

Peer group members, returnees, travel agencies

The 'Chinese society can be characterized as "low-trust", because trust between in-group members is high while trust is low regarding out-group persons' (Blok 2002: 19). The confidence in the truthfulness of information from official media as well as from commercial sources is not very pronounced in the Chinese society, both are simply seen as 'propaganda'. This can be traced back to the greater tolerance for a manipulation of facts in the Chinese culture, as opposed to the clear imperative 'Thou shalt not lie' in the occidental tradition. In a society where neither the government nor supplier of goods or services have to fear punishment for a distortion of facts by voters or court, this tendency to believe only the members of your in-group becomes even more pronounced.

It is therefore not surprising that a survey by CITS found recommendations by relatives or friends as the most important source of information for the choice of destination (WTO 2003) and that, besides search engines, recommendations by other persons are the most important ways for Internet users to become aware of new websites (CNNIC 2005). With the growth of the number of Chinese who have been travelling to other countries and of Chinese students who returned home from studies abroad increases the probability for having an acquaintance who has been in person to a specific place. In Beijing, Shanghai or Guangzhou more than 7 per cent of the population has been outside Mainland China in 2004 alone (DPS 2005b). Given the extensive personal network especially of persons in higher positions and/or having higher incomes in China's main cities, this almost guarantees them to have at least a friend of a friend available for a peer-group opinion on most destinations.

The concentration on own experiences or those of friends, relatives or colleagues, combined with the wish to visit the most prestigious destinations supports the re-enforcement of the established well-trodden trails: reliable information about places 'nobody' has been to is hard to get and because nobody has been to them they cannot have a lot to offer. Prejudices about destinations are re-enforced in the same way, it becomes 'common knowledge' that destination X is the best place to buy a specific item and that the inhabitants of destination Y have such and such characteristics. Word-of-mouth information will include items that are censored in the official media such as the availability of gambling or prostitution,[60] either for those who seek such pleasures or for those who want to avoid them. Furthermore, compared to guide books or travel magazines, it is also free.

However, for first-time outbound travellers and less well-connected persons, advice from the travel agency is still the major source of information. As mentioned in Chapter 3, in the Chinese tourism industry structure there is no clear distinction between tour operators and travel agencies. The office and the person in the office who is selling the tour, can both be the addressee of complaints if something goes wrong during the trip. Any leisure trip is a ware that has to be bought without the possibility of testing or exchanging it if something goes wrong. In a premature and chaotic market like the Chinese outbound tourism market, a personal long-standing relationship between the customer and the travel agency can provide some trust into the recommendations of the agent, and scale down to a certain degree the uncertainty about the quality of the services offered.

6 Destinations of Chinese outbound tourism

The choice of destinations for Chinese outbound travellers before 2006 was not driven by demand, but by availability. In the 1990s Hong Kong, Macao and some Southeast Asian countries could be visited under the name of VFR, and border area inhabitants could travel to neighbouring areas across the border.

Official leisure destinations came into existence with the establishment of ADS agreements, a relatively small number of mainly Asian countries before 2002. Research in China in 2002 (Guo 2004) found that Germany was similar to Egypt because the official 'ADS world' at that time, outside of Asia and Oceania, consisted only of Egypt, Germany and South Africa.

The state and intensity of economic and political relations has largely determined the destinations for business or study trips with smaller or bigger leisure parts involved. The countries with strong economic relations such as Western Europe, the USA and Japan received more of these visitors as a result.

In the middle of the first decade of the twenty-first century this supply-driven market structure is changing dramatically. ADS has been granted to almost all important destinations with the exception of the USA and Canada. Also other parameters have changed that enable Chinese tourists to make more diverse choices through the availability of more and more detailed information from guidebooks, TV programmes and Internet sources, tourism fairs, travel experiences of friends, and other sources.

Development of destinations

The increase in the number of outbound travellers has been faster than the spread in the diversity of destinations because of the aforementioned political restrictions. The other main restriction is of course the cost of visiting different destinations. Given the limited leisure time, trips tend to be short, with many squeezed into one of the 'Golden Weeks' in May and October.

The vast area of China ensures that all except border trips, including those from Guangdong to Hong Kong and Macao, involve the crossing of distances measured in hundreds, more typically thousands, of kilometres. Therefore transportation costs play an important role as they account for a relatively high part of the total cost of the trip, especially in relation to the short duration of most voyages. This aspect is further aggravated by the absence of low-cost airlines[1] or other substantial discount structures from airlines operating out of China or Hong Kong. With the Chinese currency until mid-2005 pegged to the US dollar, prices in different destinations were also influenced by the currency exchange rates in line with the ups and downs of the American currency.

With status as an important factor, famous destinations such as Paris or Singapore and – unreachable with ADS so far – California or Las Vegas also enjoy a premium over countries unknown in China. Gao Yunchun of the CITS Head Office, as quoted in 2003 about new European ADS destinations, showed remarkably little consideration of the physical realities of the map of Europe: 'Croatia and Hungary have few marketing possibilities if we set a single-stop itinerary. . . . We are considering packaging them with Russia' (Chinatravelnews 2003a).

Recognition of countries and cities can however be changed over time, as examples of the growing interest in Nice/France during the 1990s through word-of-mouth information shows. Major events such as the Soccer World Cup 2002 in South Korea and Japan and especially the Olympic Games 2004 in Athens have also been instrumental in enhancing the profile of these destinations. In addition, the perceived level of safety or danger is an important issue, as all studies agree on, so how 'safe' a destination appears to be is perceived to be a major factor. With the growing recognition of China as an important inbound market for many destinations, in a number of cases visa requirements and fees were lowered and the time needed to process the visa applications cut from several weeks to a few days or even visa-free travel agreements signed.

The diverse leisure tourism destinations can be grouped into seven categories in ascending order of prestige:

1   Border tourism destinations: Russian Far East and Russian Siberia, Vietnam, North Korea, Kazakhstan, Mongolia, Myanmar, Laos
Travels to these destinations are seen as happening mainly in a grey area. Traders cross over daily mostly as 'ant traders', peddling cheap Chinese consumer goods or buying products from these countries, which are sometimes ill-gotten by their trading partners or outright illegal. For some persons, the crossing of these borders might be the first step towards illegal immigration attempts. Others attractions of the border regions are opportunities to engage in activities officially banned in China such as gambling or commercial sex. There are of course also legal and respectable forms of leisure tourism trips going towards these countries, but they alone cannot explain the high number of border-crossings.

2   A former British colony and a Portuguese overseas province turned SARs: Hong Kong and Macao

These are the two destinations first opened to Mainland Chinese in 1983 and by far the destinations with the highest number of outbound tourists. Since the establishment of the SARs in 1997 and 1999 respectively, special regulations to visit the two cities have been applied, making it a 'normal' activity to visit there, especially for the inhabitants from the neighbouring Guangdong Province. For all Chinese, not counting border tourism, Hong Kong and Macao are the easiest and cheapest opportunities to have a first glimpse of 'Elsewhereland', the 'other' world.

3   Southeast Asian pioneer VFR tourism countries: Thailand, Malaysia, Singapore, Philippines

The 'Three Southeast Asian' countries plus the Philippines were accessible since 1990 and 1992 respectively for VFR travels. Thailand especially is the most popular destination for a short trip to a 'real' foreign country. These countries are not perceived as very prestigious destinations as they are connected with sex tourism and cheap 'Zero-Dollar' tourism.[2] In Thailand it is perceived that tourists have to put up with bad service and pressure to buy in shops in exchange for the very low package prices. Singapore differs positively from the other three countries as it is perceived as modern, safe and easy to travel in, especially for southern Chinese because of the background of the majority of the population as former emigrants from Southern China.

4   Less popular ADS countries: lesser known European, African or Latin American countries, Asian ADS countries without large overseas Chinese communities

Eastern European and small western European countries such as Malta and Croatia as well as South American countries are less valued or considered as unsafe and as not offering value for money. For Arab and African countries, little knowledge is often combined with racist prejudices and less interest in 'underdeveloped' regions. Asian countries with a strong and predominant native culture without dominant overseas Chinese influences such as India are also more important as a destination for commercial or official business travellers than for leisure tourists. As a source for gaining prestige, travels to these destinations are useful only for adventurous younger Chinese from intellectual circles or those with a strong western cultural influence such as former overseas students.

5   'Western' destinations perceived as near: Australia, New Zealand

Australia and New Zealand were the first non-Asian countries opened to official leisure tourism through ADS, so a relatively clearer picture of the facts of travelling to these destinations has been communicated through media and word-of-mouth of tourists who have been there already. They are considered as 'near', even though in distance or flight hours they are not nearer to Beijing than Europe or California.[3] Both countries combine *gao bizi*[4] culture with large overseas Chinese communities as a base for visiting famous landmarks, both natural and man-made and 'primitive' aboriginal culture.

6 Popular western or developed Asian ADS destinations:
  EU, Switzerland, Japan, South Korea

The popularity of these destinations is partly based on the high costs of travelling to and in these countries, reflecting on the obvious affluence of those who can afford to go there for leisure. The larger western countries of the EU such as Germany, France and Italy plus Switzerland are well known in China. Greece was added to this list with the help of the Olympic Games in 2004. Europe is perceived as the source of western civilization, with Beethoven, Beckham and BMW playing a bigger role than liberté, égalité and fraternité.

Whereas most EU countries obtained operational ADS only between 2003 and 2005, Japan and South Korea were officially 'opened' earlier. Both restrictions on the possibilities to enter Japan and the lingering misgivings about Japan's role in China's history in the twentieth century influence the popularity of Japan, which is, however, like South Korea, the source of a lot of modern influences on the culture in China.

7 North America: USA, Canada

North America is the last continent to enter the ADS system. In 2005, the remaining superpower, like Canada also the home of many overseas Chinese, was a destination still open only for well-related people who could manage to travel with an official passport or for VFR reasons into the land most envied and idolized in China. Whether the USA and Canada can remain in the top position of the most prestigious destinations for Chinese leisure tourists once they are opened for ADS travellers remains to be seen.

Marketing and research activities of NTOs in China

With the installation of the ADS system the development of representative offices of NTOs in China became both interesting and possible. Switzerland pioneered this movement by opening up an office in Beijing in 1997, even before CNTA issued the 'Provisional Administrative Regulation regarding the Establishment of Residential Representative Office in the People's Republic of China by Foreign Governmental Tourism Department' in June 1998. CNTA approved in the following years NTO representative offices for the European countries Austria, Finland, France and Germany, as well as for Australia, Canada, Japan and Singapore. After 2002 the vision of the growing source market China lured more NTOs to Beijing or Shanghai.

Table 6.1 List of NTOs' representative offices in China[5]

	Country	Office in
NTO office established before 2003	Australia	Beijing
	Austria	Beijing
	Canada	Beijing
	Finland	Beijing, Shanghai
	France	Beijing
	Germany	Beijing
	Japan	Beijing, Shanghai
	Singapore	Shanghai
	Switzerland	Beijing, Shanghai
NTO office established 2003 or later	Greece	Beijing
	Italy	Beijing
	Malaysia	Beijing, Shanghai, Chengdu
	Malta	Beijing
	Netherlands	Beijing
	Nevada, USA	Beijing
	New Zealand	Shanghai
	Philippines	Beijing
	Sweden (together with Norway and Denmark)	Beijing
	Spain	Beijng
	South Africa	Beijing
	Thailand	Beijing
	Turkey	Beijing
	United Kingdom	Beijing
	Zimbabwe	Beijing

Most offices are set up as a part of the embassy or the Chamber of Commerce of the country in question. In line with limitations for other kinds of representative offices in China, they can engage in non-profit activities such as promotion, consultation, liaison and coordination, but are forbidden to carry out any open or covert commercial activities. To promote a country openly as a leisure tourism destination is legal only for countries with ADS, in practice, however, many promotional activities have taken place with countries that had not yet been granted ADS. There is considerable fluidity in terms of the boundaries between official non-leisure study or business trips and leisure tours and this is reflected in promotional work being done often in official cooperation with the big Chinese tour operators or city governments. However, with almost all destinations gaining ADS by 2005, the 'grey' areas of activities for NTOs are disappearing fast.

Many regions and cities also installed their representative in China to promote their destination, often in the form of Chinese students returning home who are familiar with the culture, sights and language of that overseas region. Officially this promotion work takes place within a consulting company or even without any legal framework. Given the size and diversity of the market, one office is not able to cover the whole of China, so some NTOs have set up branch offices in several cities or use their Hong Kong offices to work on the South China market:

> Since the Chinese government's priority is on the inbound and domestic tourism, the government has not put much effort on the study of the outbound tourism. . . . It is therefore important for NTOs representative offices to collect information and follow the market development closely.
>
> (WTO 2003: 95)

Studies and newsletters published by the NTOs, most of which are available to the public, are major sources for information about China's outbound tourism, with all organizations trying to put their limited resources to as good use as possible.

The marketing activities of NTOs in China are manifold. On one level, they include the usual range of destination marketing such as attending tourism fairs, organizing familiarization tours for Chinese journalists and staff of outbound tour operators, producing brochures and websites, trying to place public relations articles into Chinese magazines and newspapers as well as electronic media, building connections to tour operators, government officials, etc.

On another level, most destinations and tourism activities have to be promoted from a very basic starting point as most potential guests are still learning how to become an international traveller. When South Korea started to promote skiing holidays on the peninsula, it had to give information about the basics of skiing alongside the tour offers (WTO 2003: 101). Many countries use the combination of food and tourism promotion to catch the attention of customers in the shopping areas of big cities or bring famous sports teams to China to use their popularity for cross-marketing the home of the teams as tourism destinations. Countries that will not provoke any immediate image in the mind of the Chinese public and have no sights or personalities of current fame in China to offer face a tough job in their marketing activities. For destinations such as Hungary, Tunisia or Peru it is therefore more feasible to promote themselves as part of a bigger region or with a specific theme.

With little information available on the one hand and a strong connection between the general image of a country and the wish to visit it on the other, marketing for tourism destination cannot be divided from the destination image as such. Incidents such as the downing of a Chinese military jet by an American surveillance plane over Hainan in January 2001 (Wang 2005) or the Anti-Japanese riots in 2005 (Giese 2005) will negatively influence the tourism development at least in the short term more strongly than in more individualistic countries. In contrast, positive images such as the Eiffel tower illuminated in red colours for the visit of Chinese president Hu Jintao in 2004 'to honour China' (Xinhua 2004) make the work of the French NTO representative office much easier.

Selected destination experiences

Some destinations have experienced a large number of outbound tourists from China; many others are just starting to become involved. The demand has been influenced by political and economic restrictions as well as by the level of information available from marketing activities of the destinations and other sources. By looking more closely at the experiences of selected destinations, a clearer picture should emerge.

Figures used in the country case studies are taken from national statistics. They tend to be higher for arrivals than the numbers given by the Chinese authorities, which only report the first port of call of citizens leaving the country. As many Chinese tourists travel to several countries in one trip, this systematically under-reports the actual numbers. For instance, according to Chinese statistics, in 2000, 1.057 million Chinese travelled out of China to Thailand, Singapore and Malaysia combined, with only 87,000 going to Malaysia. However, the three countries reported on aggregate 1.563 million arrivals, with Malaysia counting 425,000 visitors. The Chinese departure figures therefore reflect rather the simpler visa procurement and the better flight connections to Singapore and Bangkok compared to Kuala Lumpur than preferences for specific destinations.

Asia

Asia, home continent of China, is by far the most important destination for Chinese outbound travellers. Asian destinations were first opened to official VFR and leisure outbound travel. Asia is the home of at least 50 million overseas Chinese living outside Mainland China, who make visits to their countries easier and who are also the most important source of FDI in China. Asian destinations can be easily and cheaply reached, especially those bordering China or in close proximity. The political relations with several countries or regions in Asia are however not without problems: the island Formosa is regarded by the Beijing government as an integral part of China, relations with Japan and the Russian Far East are tainted by historical problems, tourism is perceived as being connected with prostitution, gambling and drug trafficking in some Southeast Asian countries, India and Vietnam are former enemies in wars with the People's Republic.

Nevertheless, the number of Chinese visitors to Asian destinations increased within the decade from 1995 to 2004 more than sixfold, double the amount of the increase of visitors to transcontinental areas. The lion's share of these visitors however travels – increasingly – to Hong Kong and Macao only. Asia in 2004 was the first destination for about 90 per cent of all travels, compared to about 80 per cent in 1995, but within Asia the share of travels beyond Hong Kong and Macao decreased from roughly one-third to one-fifth in the same period.[6]

Before the ADS process started in 1995, Hong Kong, Macao and the three plus one Southeast Asian countries Thailand, Singapore, Malaysia and the Philippines were opened for VFR travels. Before 2002, all ADS countries apart from Australia and New Zealand were located on the Asian continent. By 2005, more then 20 Asian countries from Turkey to Japan and from Mongolia to Indonesia had been granted ADS, with most of them also receiving tour groups. A number of Asian countries have been operating NTO representative offices in China since the early years of the new century, some opened new ones, such as the Philippines did in August 2004.

Even with the ADS system, questions remain. During 2002 and 2003, 185,000 Chinese visitors to Malaysia are reported to have overstayed their visa. One-third of the foreign prostitutes operating in Malaysia are thought to be Chinese nationals, many officially enrolling as students in private colleges to gain long-stay visa (Sulaiman 2004). Another report cites the case of a person shuttling between Hunan Province and different Malaysian

Table 6.2 Chinese outbound travellers to Asia 1995–2004[7]

	1995	1999	2000	2001	2002	2003	2004
Travellers (in '000s)							
Total visitors to Asia	4,057	7,813	8,846	10,090	14,196	17,940	25,821
Visitors to Hong Kong and Macao	2,791	5,122	5,786	7,120	10,554	14,101	20,492
Total visitors to Asia without Hong Kong and Macao	1,266	2,691	3,060	2,970	3,642	3,839	5,329
Share (in %)							
Visitors to Asia	81.6	84.6	83.1	83.2	85.5	88.7	89.5
Visitors to Hong Kong and Macao	56.1	55.5	54.3	58.7	63.6	69.7	71.0
Visitors to Asia without Hong Kong and Macao	25.5	29.1	28.7	24.5	21.9	19.0	18.5

cities to earn money from begging (Eturbonews 2004). For tourism development in the Asia Pacific region,

> the single biggest obstacle will be visa policies. For years, many Asia-Pacific governments have allowed visa-free access to Europeans, Americans, Australians, Japanese and other citizens of industrialized countries. Now, they are totally confused on how to handle the future travellers from China, India, Russia and other countries within the region. Each of these countries has problems with illegal immigrants, crime syndicates and other undesirables that governments want to keep out. . . . How these visa issues are sorted out, is going to be a major challenge in future.
>
> (Muqbil 2005: 6)

Hong Kong as the destination for almost half of the total Chinese outbound tourism and Thailand, for many years the most important leisure tourism destination, will be treated in more detail below, after some remarks about other destinations in Asia.

Border tourism is the most important form of international tourism for Chinese citizens not lucky enough to belong to the middle and upper strata

1 Hong Kong	1998	8 Japan	2000	15 India	2003
2 Macao	1983	9 Vietnam	2001	16 Sri Lanka	2003
3 Thailand	1988	10 Cambodia	2001	17 Pakistan	2004
4 Singapore	1990	11 Myanmar	2001	18 Jordan	2004
5 Malaysia	1990	12 Brunei	2001	19 Mongolia	2004
6 South Korea	1998	13 Nepal	2002	20 Laos	2005
7 Philippines	1992	14 Indonesia	2002		

Figure 6.1 Map showing ADS countries in Asia[8]

of the big urban centres and coastal regions of China. Of the 15 – or 16 if Sikkim is included – political entities sharing a land border with China, namely North Korea, Russia, Mongolia, Kazakhstan, Kirghizistan, Afghanistan, *Pakistan (Kashmir), India, Myanmar, Nepal,* Bhutan, *Laos* and *Vietnam* plus the *Macao* and *Hong Kong* SARs, only the eight shown in italics are covered by the ADS system. For North Korea, Russia and Mongolia special arrangements are in force; Kazakhstan, Kirghizistan, Afghanistan or Bhutan are not major tourism destinations. In 2004, 123,000 Chinese entered Mongolia, and 91,000 went to Kazakhstan (Du, Dai 2005). India has acquired ADS; however in 2003 less than 50,000 Chinese were among the only 2.8 million international tourists arriving in India in 2003. The number of Chinese visitors to Pakistan-controlled Kashmir[9] and Nepal are also insignificant.

Border tourism started in 1988 (Jenkins, Lin 1997). Agreements with Myanmar and Laos came into force in 1991 (Timothy 2001), soon followed by arrangements with other countries. In 1996, special regulations for border tourism were published by the Chinese government, which allow visits of persons living close to the national borders without the need to go through all the normal procedures of international travel. Even before these regulations Chinese citizens could visit to neighbouring areas, starting with the possibility for one-day travel to North Korea in 1987. An example of border tourism without foreign currencies involving Heihe in Heilongjiang and the city of Blagoveshchensk on the banks of the Amur River has been mentioned in Chapter 3. Maybe this example was remembered when the fear of too much export of foreign currency led to the idea that 'to meet the increasing growth of outbound tourism demand, more forms of border-crossing travel with no or less foreign exchange spending should be worked out, such as frontier tourism' (Zhang, Pine, Zhang 2000: 287).

Entry into Thailand, Laos, Myanmar and Vietnam was further simplified in 1997, partly under a regional tourism development plan of the Association of South East Asian Nations (ASEAN) (Lew 2000; Krongkaew 2004). Border tourism is however actually mainly not about tourism at all. 'The main purpose is the exchange of goods, with tourism in second place' (Dou, Dou 2001: 45). Small traders cross the border, sometimes on an almost daily basis, blowing up the number of outbound travels in the statistics. The *goods* that are exchanged are not always household goods, shoes or textiles. The publication by the China Center for Lottery Studies of the Beijing University of an almost astronomical figure of 600 billion Yuan RMB (approximately 60 billion Euros in 2005) leaving China every year in gambling stakes has provoked heated debates in China. Inside China, gambling is prohibited. In Macao, no more than 2 billion Yuan RMB (approximately 200 million Euros in 2005) was lost by gamblers in the casinos, most of this money returning to Mainland Chinese coffers anyway. The Yunnan police report the existence of 82 casinos in the border regions of Myanmar, Laos and Vietnam, which are open for Chinese citizens only

(Chinanews 2005). Visitors to the frontier town of Mong Lar in Myanmar's autonomous Shan State region count 50 casinos, visited by 200,000 Chinese *special interest* tourists in 2004 alone. Formerly a major drug-dealing centre, gambling and sexual services are now the most important money earners, as they are for Pang Kham, another casino city close by which even offers a nine-hole golf course (McGirk 2004). Another casino luring Chinese gamblers, run by a Hong Kong company, is located in the Rajin-Sonbong Free Economic Zone in the North Korean part of the Tumen River Economic Development Area (Arlt 2001).

Vietnam can claim the highest number of *regular* tourists visiting the bordering countries in the southeast. From 122,000 visitors in 1995, the figure has climbed to 786,000 in 2004. A problem reported here as in other Southeast Asian countries is the over-lording behaviour of Chinese out-bound tourists, which was criticized in the official Chinese media. The *People's Daily* quoted a tourism manager, stating 'Some Chinese look down on our poor neighbours like Vietnam, which offends local people' (People's Daily 2003c).

Beside Hong Kong, by far the most important bordering destination is the former Portuguese province of Macao, since December 1999 a SAR like Hong Kong. Macao is seen as possessing an even greater 'western' touch than Hong Kong and more educational opportunities such as museums, making it a suitable destination for family visits.[10] The Casino Lisboa and other gambling institutions are an attraction more for adults travelling without children. The 'Individual Visitors Scheme' (IVS), which was introduced for Macao in parallel to Hong Kong,[11] has brought an unex-pected surge into Macao. A study giving forecasts for the inbound tourism development of Macao for the coming years expected a visitor number of 6.7 million for 2004. In fact, the 2004 figure turned out to be higher than the number of visitors forecasted for 2006 (Song, Witt 2006), turning Mainland China into the source for more than half of all visitors to Macao for the first time. The seasonality of the Chinese tourism into Macao is not very pronounced, with 660,000 the lowest and 895,000 the highest number of visitors for one month. Standing in the shadow of Hong Kong, Macao is actually *the* success story of the Chinese outbound tourism with an average growth of 100 per cent each year since 2001 in visitors numbers.

A special case is the outbound tourism to Taiwan. Until 1988 the govern-ment that sees itself as ruling the Republic of China allowed no visits of Mainland Chinese to the island. In that year, one year after giving permis-sion to visits of Taiwanese to the Mainland, visits of sick relatives and participation in funerals was allowed, and extended to the visit of first-grade relatives in 1991 and second-grade relatives on year later (Yu 1997). In 1990 the first Taiwan–China tourism conference was held in Shenzhen, followed by further conferences (Zhang 1993) and some further gradual easing of the restrictions. In 2001 more than 100,000 Mainland Chinese visited Taiwan, in 2004 the number moved to 145,000. The entrance of

Table 6.3 Mainland Chinese visitors to Macao 1991–2004[12]

	Total arrivals (in '000)	Mainland Chinese arrivals (in '000)	Share of Mainland Chinese arrivals (in %)
1991	7,489	15	0.2
1992	7,699	27	0.3
1993	7,829	272	3.5
1994	7,834	245	3.1
1995	7,753	543	7.0
1996	8,151	604	7.4
1997	7,000	530	7.6
1998	6,949	817	11.8
1999	7,444	1,645	22.1
2000	9,162	2,275	24.8
2001	10,279	3,006	29.2
2002	11,531	4,240	36.8
2003	11,888	5,742	48.3
2004	16,673	9,530	57.2

Taiwan into the ADS system or a similar agreement was discussed during the visit of a CNTA delegation to Taiwan in November 2005, with direct flights and the possibility of 1,000 visitors per day to travel from selected cities in China to Taiwan envisaged for 2006 (Satish 2005b). Visiting Taiwan, if made possible, would meet strong interest in China, similar to the interest in Hong Kong as a place being Chinese but with a difference.

The Japanese tourism industry has to cope with the change of travel behaviour of the domestic tourists who are moving away from large-group one-time highlights visits to small-group exploratory forms of tourism. Tour groups from China are seen as a possibility to replace the busloads full of less discerning tourists (Arlt 2005c). The big differences between Japanese and Chinese tourism behaviour do not, however, support this hope. Given the very close economic relations between the two countries, Japan is mainly a destination for business or scientific travels, with the unresolved problems of the Second World War still casting a shadow over idea of leisure tourism to Japan. When anti-Japanese protests flared up in April 2005, they resulted in a significant decrease in leisure travels in both directions (Economist 2005b). In terms of modernity, Japan is not seen as the benchmark any more like in the 1980s. On top of that, Japan is perceived as a very expensive destination, with less value for money compared to a visit to Europe. Japan opened the country for inhabitants of Beijing, Shanghai and Guangdong only for ADS travels. In September 2004, Shandong, Zhejiang, Liaoning, Jiangsu and Tianjin were included in the scheme and for the period of the World Expo in Aichi from March to September 2005, all Chinese citizens could apply for tourist visa if they visited the Expo. Visitor numbers increased from 130,000 in 1991 and 207,000 in 1993 (Morris 1997) to 352,000 in 2000 (WTO 2003) and 760,000 in 2002 (TBP 2004). In 2004, the number of 1 million visitors was surpassed for the first time with 1,021 travellers arriving on the Japanese isles (Du, Dai 2005).

South Korea was the first Asian country that received ADS without having been the subject to special border tourism arrangements or a member of the 'three plus one' Southeast Asian countries. Almost 700,000 Chinese visited South Korea in 2004, following approximately 550,000 in the two years before. These results are way beyond the 1997 WTO forecast for 2010, which put the figure at 437,000 only (WTO 2000). The proximity, which results in rather low prices and the opportunity for brief trips including down to just one night in Seoul, could be offset by the lack of perceived clear difference in terms of landscape or culture. South Korea, however, succeeded in injecting some rather special motives for leisure travelling into the Chinese source market. On one hand, the FIFA 2002 World Cup that took place in Japan and Korea brought more than 60,000 Chinese visitors to South Korea in June 2002, creating an echo effect three months later for the Asian Games in Busan, which were watched by another 50,000 Chinese fans. The World Cup enhanced the image of Korea in Mainland China to a considerable degree (Kim, Morrison 2005).

On the other hand, a *Korean fever* swept over China starting in 2001, with pop singers and movies making big inroads into the culture especially of young Chinese, and Korean TV soap operas warming the heart of older generations. Visits that include concerts or even meetings with pop stars or a tour of the soap opera shooting locations pull a growing number of Chinese leisure tourists. The shooting locations were named as the most attractive destination after the capital Seoul in surveys during travel fairs in 2002. Seoul itself combines shopping opportunities with a number of entertainment centres. Finally South Korea has been able to establish outdoor destinations with Jejudo Island, south of the Peninsula, for summer and the ski resorts for winter visitors. As a result, South Korea welcomes almost half of the Chinese visitors as leisure tourists (KNTO 2005).

For the Southeast Asian countries, the example of Thailand will be treated in more detail below. A study of the image of the three countries Singapore, Thailand and Malaysia, conducted with 250 domestic tourists in Hangzhou in 2003, found that Singapore was regarded as the safest destination, with Malaysia in the middle position and Thailand seen at the most risky destination. As reasons the bad experiences communicated with the Zero-Dollar and Minus-Dollar tours and the sex tourism image were cited (Teng 2005).

The Southern and Western Asian countries are trying to get into the Chinese source markets with less success. A taste for beach holidays, especially in long-distance ADS destinations such as Sri Lanka or the Maldives, is only beginning to be acquired by Chinese travellers and was temporarily hindered by the perceived tsunami security risk. Nevertheless a direct air link started in summer 2005 offers the opportunity to stay in Colombo on a four-night package tour for less than 500 US$ (Munzeer 2005). The countries of the Arab peninsula as well as Iran and Turkey can surely offer different landscapes and cultures compared to China but have less developed traditional contacts and are seen as being situated in a less secure part of the world. Finally, Israel was adopted into the circle of ADS countries in June 2005, but not least for the security worries, only about 300 Chinese visited the country per month in 2004 and the first half of 2005, down from three times that figure in 2000 (CBS 2005).

Hong Kong

The British colony of Hong Kong and, after 30 June 1997, the Hong Kong SAR have been the most important outbound destination for the inhabitants of the People's Republic of China and the inhabitants of Mainland or 'Inner'[13] China since the start of the first 'visiting relatives' tours in 1983. Since the return of Hong Kong to China, Mainland China has also been the most important source market for Hong Kong. In the twenty-first century, the share of travels to Hong Kong as part of the overall Chinese outbound tourism has been stable at between 45 to 47 per cent.

The IVS, which started 20 years after the first 'visiting relatives' tours, is gradually opening Hong Kong to individual Mainland Chinese tourists and has brought the Hong Kong incoming tourism market into a strong dependence on the Mainland source market:

> Hong Kong is still perceived as being very different from Mainland Chinese cities in its capitalist economy, cosmopolitan prosperity, colonial history, and East-meets-West culture and lifestyle. . . . While under the shadow of being a shopping paradise, it is these differences that formed the uniqueness and underlying attractiveness of Hong Kong.
>
> (Huang, Hsu 2005: 203)

As mentioned in Chapter 3, the foreseeable end of British rule over the island of Hong Kong, the Kowloon Peninsula and the New Territories in 1997 and the growing demand for outbound travel prompted the organization of 'visiting relatives' tours in November 1983 to Hong Kong. Before that date, tourism between Hong Kong and especially the neighbouring province of Guangdong had been a mostly one-way traffic, with travellers from Hong Kong responsible for approximately 90 per cent of all tourists' arrivals in China after 1978.[14] The price of a one-week bus tour in 1983 was set very high at 370 US$ if the tourist stayed with the group during the whole duration. For those who visited friends or relatives and stayed with them for three nights, the price was reduced to 255 US$. The money was paid by those Hong Kong citizens who acted as invitees for the Chinese visitors. For the Chinese New Year festival of 1985, some 29,000 visitors came to Hong Kong under this scheme, reporting such diverse reasons for travelling as honeymoons, funerals, sightseeing, shopping – and visiting relatives (Chow 1988). Restrictions on the maximum number of visitors per day were set not by the Chinese but by the British colonial government, keeping the total number of Mainland visitors to Hong Kong below 1 million until 1992. Under the SAR government, the restrictions were phased out with 1,500 persons allowed in 1998 and 2,000 in the year 2000. The limitation of the number of visitors was finally abolished in January 2002 (Zhang, Jenkins, Qu 2003).

With the establishment of the ADS system, Hong Kong's position became less clear. In most texts on ADS, Hong Kong is cited as the first ADS country, which is of course only true in retrospective and not taking into account the unique historical relation between China and the 'Incense Harbour'.[15] Only from 1 July 1997 were the restrictions for travel to Hong Kong and Macao as well as Thailand, Singapore, Malaysia and the Philippines to the visits of friends and relatives lifted officially and the three plus one Southeast Asian countries integrated into the ADS system. For SARs of the People's Republic, other rules apply. Multiple entry permits, valid for five years, before only issued for business purposes, became available in 2002 for leisure, business and for visiting relatives in Hong Kong. Following the SARS crisis, which hit the Hong Kong economy especially hard, China and the administration of the Hong Kong SAR signed the so-called Closer Economic Partnership Arrangement (CEPA) in June 2003. Beside abolishing customs duties for 90 per cent of all products made in Hong Kong and opening up distribution and exhibition operations inside China for Hong Kong-based companies (HKTDC 2003), the IVS as part of CEPA gradually opened Hong Kong as a destination for individual travellers from Mainland China. After a pilot phase for the cities Dongguan, Zhongshan, Jiangmen, Foshan and Shunde starting from July 2003 and the extension to most important cities of Guangdong Province Guangzhou, Shenzhen and Zhuhai one month later, Beijing and Shanghai were included in the IVS in September 2003. Six more cities in Guangdong Province, namely, Shantou, Chaozhou, Qingyuan, Meizhou, Zhaoqing and Yunfu entered the IVS in January 2004, in May 2004 IVS access to Hong Kong became available for all inhabitants of Guangdong. In July of the same year, nine major cities in Eastern and Southern China, Nanjing, Suzhou and Wuxi in Jiangsu Province; Hangzhou, Ningbo and Taizhou in Zhejiang Province; Fuzhou, Xiamen and Quanzhou in Fujian[16] were added. The other two municipalities besides Beijing and Shanghai, Tianjin and Chongqing, entered into the IVS in March 2005. From that date, people living in 34 cities with a total population of 172 million were covered by the programme (Greenlees 2005), with a further widening to all inhabitants of the economically more successful provinces in China expected for 2006.

Hong Kong is connected to more then 40 airports in China with almost hourly flights from Beijing and Shanghai. Similarly to the situation in other destinations bordering Mainland China, the majority of the visitors do not arrive by aircraft. In 2004, only 12 per cent used this mode of transportation, whereas 76 per cent used the land border from neighbouring Shenzhen and another 12 per cent arrived by sea, mainly from Macao and the Pearl River Delta harbours (HKTB 2005a). Twelve daily direct trains and numerous ferry services connect Guangzhou, Shenzhen Airport, Zhuhai and about 15 other cities with Hong Kong. The limitations of the transport infrastructure are showing however during the 'Golden Weeks'. Another important aspect of the accessibility of Hong Kong is the function of Hong Kong as a gateway to other destinations. To support the use of Hong Kong as a transit place, from 1993 Mainland Chinese holding a return airline ticket to a third country could get a seven-day 'free' visa without charge. Before 1997, however, less then a quarter of all Mainland visitors to Hong Kong travelled further on, but from 35 per cent in 1997 the figures jumped to 66 per cent in 1998 and 77 per cent in the year 2000.[17] In almost all cases one or several of the 'three countries' Thailand, Singapore and Malaysia were the final destination (Zhang, Pine, Lam 2005). With the IVS, the percentage probably dropped again to pre-1997 levels.

With the exception of a minor decrease in 1997, the number of Mainland Chinese travelling to Hong Kong has been increasing ever since the first official visitors groups in 1983. Before 1997 they were not seen as a major target for Hong Kong tourism marketing efforts, but after the Asian financial crisis of the late 1990s and the health scares of Avian influenza and SARS in the early years of the new century, Mainland Chinese visitors 'rescued' the ailing Hong Kong tourism industry. Whereas the number of non-Mainland Chinese visitors did not experience any substantial growth since 1996, Chinese visitors are almost over-running the SAR with 56 per cent of all visitors from China in 2004. Some 35 per cent of all visitors arrived in Hong Kong under the IVS. The steep increase in the number of arrivals especially since the introduction of the IVS in 2003 is however partly offset by a long-term tendency of the decrease in the length of stay from six nights in 1995 to five nights in 1997 and to only 3.3 nights in 2000 (Huang, Hsu 2005), a trend continuing in 2004 (HKTB 2005a).

The closer economic relations between Hong Kong and Guangdong, especially Shenzhen, also result in the inflation of border-crossing figures by the growing number of Chinese citizens crossing every day into Hong Kong for work, by the drivers of trans-border busses, etc. If the plans for Hong Kong's newly opened Disneyland are realized, 1.5 million border-crossers will be generated by the 4,000 daily visitors being directly transferred by 'Wonderland express' coach from Shenzhen to Disneyland and back (Chow, Lai 2005). But even if the day-trip visitors are excluded, the number of arrivals from Mainland China to Hong Kong in 2004 already vastly exceeds the 1997 forecast of the WTO for 2010 of 5.8 million border-crossings (WTO 2000).

Table 6.4 Chinese outbound visitors to Hong Kong 1984–2004[18]

Year	Visitor arrivals from Mainland China (in '000s)	Visitors from other source markets (in '000s)	Total visitors from all countries (in '000s)	Share of Mainland Chinese visitors of total market (in %)	Development of Mainland Chinese share (in %)
1984	215	3,089	3,304	6.5	+78
1985	309	3,348	3,657	8.4	+30
1986	363	3,690	4,053	9.0	+6
1987	485	4,432	4,917	9.9	+10
1988	684	5,483	6,167	11.1	+12
1989	730	5,255	5,985	12.2	+10
1990	754	5,827	6,581	11.5	–6
1991	875	5,920	6,795	12.9	+12
1992	1,149	6,862	8,011	14.3	+11
1993	1,733	7,205	8,938	19.4	+35
1994	1,944	7,387	9,331	20.8	+7
1995	2,243	7,957	10,200	22.0	+6
1996	2,311	9,392	11,703	19.7	–10
1997	2,297	8,109	10,406	22.1	+12
1998	2,597	6,978	9,575	27.1	+23
1999	3,084	7,594	10,678	28.9	+6
2000	3,786	9,273	13,059	29.0	0
2001	4,449	9,276	13,725	32.4	+12
2002	6,825	9,741	16,566	41.2	+27
2003	8,467	7,070	15,537	54.5	+32
2004	12,246	9,565	21,811	56.1	+3

The number of transit passengers included in the statistics fell in 2004 because the 'SkyPier' facility of Hong Kong Airport makes it possible to go directly from the airport to several cities in the Pearl River Delta without passing Hong Kong immigration.[19] Still more than one-third of all Mainland Chinese visitors, almost 4.5 million people, did not stay overnight during their visit in 2004 (HKTB 2005a), reflecting the ease of going to Hong Kong just for a day of shopping – or working – from Shenzhen.

In 6 of the 12 months of 2004, the number of Mainland Chinese visitors exceeded 1 million. The three 'Golden Weeks' are peak season with border-crossings and transportation infrastructure stretched to the limit. In the last three days of September 2004, almost 100,000 Chinese travellers arrived to use the opportunity to combine the celebration of mid-autumn festival and the start of the October 'Golden Week', followed by 350,000 travellers during the week itself (Lau 2004). In the 'Golden Week' of October 2005, again more than 50,000 persons per day crossed the border from China to Hong Kong (Greenlees 2005). Other reasons for increase visits are festivals such as the Hong Kong shopping festival held in July and August 2004 or the yearly WinterFest.

The pre-1997 image of Hong Kong as the only modern, cosmopolitan city on Chinese soil, albeit without a democratic system, has faded with the development of Shanghai, Guangzhou, Shenzhen and other cities within China's Mainland. Hong Kong is however still for most Chinese the first piece of 'Elsewhereland' to experience. Movies and music from Hong Kong command a strong position within Mainland Chinese culture, not always with a positive effect for Hong Kong's image as tourist destination.[20]

Huang and Hsu (2005) in 2004 conducted focus groups discussions with participants from Beijing and Guangzhou; most of them had not yet travelled to Hong Kong or abroad. Beijing residents saw Hong Kong as a place for shopping and shopping only. They perceived Hong Kong as being not a particularly attractive tourism destination and as unsafe. They were interested to visit Hong Kong, especially to show their children something new and different, but preferred other destinations as first choice for outbound tourism. Guangzhou residents, having daily contact with Hong Kong via television programmes in Cantonese, relatives and economic relations, had a more positive picture about the safety situation in Hong Kong. They also regarded Hong Kong as a 'shopping paradise' and as a destination that was, under the IVS, easier and cheaper to travel to then to northern Chinese cities. For holidays they preferred Southeast Asian destinations, which were also reckoned as being cheaper than a holiday stay in Hong Kong.

Zhang and Lam (1999) conducted a survey with 105 Chinese visitors from Guangdong arriving in Hong Kong in 1997, asking about their motivations for travelling to Hong Kong. Some 42 per cent of them were travelling to Hong Kong for the first time but even in 1997 31 per cent travelled for the second to fourth time and 27 per cent for the fifth or more time. The two main motivations were visits to relatives,[21] especially for the regular travellers, and the attraction of seeing something different, combined with the convenience and quality of the travel infrastructure. Gaining prestige from a visit to Hong Kong was not very pronounced as a motive for travelling. Lam and Hsu (2004) found in another study the experience of the capitalist society, shopping and sightseeing were essential activities.

Hong Kong, lacking major historic monuments like those found in the heartland of Chinese history in northern and central China, is struggling to lose the image with Mainland Chinese visitors that there is 'not much to see'. Beside the efforts to attract Chinese visitors by heritage sites and museums (Henderson 2002), this is done with the addition of new attractions, the biggest one being the Disneyland, which opened in autumn 2005. This impression of Hong Kong as 'a cramped, crowded, expensive and rapid-paced cosmopolitan area where people suffer from high social pressure and work hard' (Huang, Hsu 2005: 199), giving glimpses of modernity and alternative lifestyles but being outside the main Chinese intellectual discourse, is however not without precedent. Wang Tao, intellectual and Hong Kong-based founder of the first Chinese daily newspapers,[22] writes in a letter in 1862:

> Hong Kong is a small out-of-the-way island, where nobody ever goes in for anything but buying and selling. The place is full of traders out to make money. How can you expect to find men of culture in such surroundings? There is nothing to see and nobody to talk to. Apart from going to visit the singing girls, I spend most of my time in my own room.
>
> (Wang Tao[23] in Arlt 1984: 167)

However, Sun Yat-sen, Hong Kong-based revolutionary and first president of the Republic of China, relates to the students of Hong Kong University in a speech in 1923: 'Where and how did I get my revolutionary and modern ideas? I got my ideas in this very place: in the Colony of Hong Kong' (Sun Yat-sen in Cameron 1978).

The different functions of Hong Kong for different Chinese groups are reflected in the reasons for leisure travel to the SAR. For Mainland Chinese living in Guangdong the common language and cuisine, often existing family relations and the geographical proximity make it a rather common and, especially since the introduction of IVS, not very complicated or expensive destination for VFR, shopping and leisure activities. With about 700 scheduled flights reaching Hong Kong from China and another 1,200 leaving Hong Kong Airport to other destinations every week, Hong Kong is also an important starting point for southern Chinese to Southeast Asia and other destinations. For non-Cantonese visitors, Hong Kong's role as the much-cited *shopping paradise* outshines all other reasons to visit. For young families the motivation to 'let the child know and learn something different' (Huang, Hsu 2005: 201) or to visit family-orientated attractions such as Disneyworld can also play a decisive role.

The geographic origin of Mainland Chinese visitors to Hong Kong has come full circle in the decade starting in 1994. From a clear majority of 76 per cent Guangdong-based visitors in 1994, the share of arrivals from the neighbouring province fell below 50 per cent in 1998, while visitors from Beijing, Shanghai, Zhejiang and Jiangsu counted for almost 20 per cent in the same year. In 2000, almost two-thirds of all Mainland visitors to Hong Kong did not originate in Guangdong (Zhang, Pine, Lam 2005). With the easing of travel procedures culminating in the introduction of CEPA, the share of Guangdong visitors increased again, witnessed among other things by the growing number of land arrivals and day-trippers. In 2004, more than 6 million Guangdong inhabitants visited Hong Kong, about half of all Mainland Chinese going to the SAR.

The visitor's demographics did not change much during the 1990s. The growing share of visitors from other provinces is reflected in the rise of single persons and persons in senior white-collar positions. The increase of visitors after 2003 will probably result in a closer mirroring of the data of the general Chinese urban population again.

As shopping is the most important activity of Mainland Chinese visitors to Hong Kong, they are considered as big spenders especially for retailers. For 2004, a total turnover of more than 34 billion HK$ (approximately 3.4 billion Euros in 2005) was reported, a 250 per cent increase

Table 6.5 Visitor demographics of Mainland Chinese visitors to Hong Kong (in per cent)[24]

		1995	*1998*	*2001*
Sex	Male	67	60	62
	Female	33	40	38
Age	Below 15	3	4	4
	16–25	11	10	10
	26–35	33	33	34
	36–45	28	26	30
	46–55	14	16	15
	56+	10	11	8
Civil status	Married	86	83	78
	Female	14	17	22
Occupation	Senior white collar	38	51	54
	Junior white collar	27	27	25
	Blue collar	15	7	5
	Housewife	6	4	3
	Other	13	11	13
Number of visits	First time	58	52	54
to Hong Kong	More than once	42	48	46

against the 2000 figures. The average spending per person per trip in 2004 is given as only 4,355 HK$ (approximately 435 Euros in 2005) by HKTB, a lower figure than those for European or North American visitors. Mainland Chinese statistics report higher expenditures and also another Hong Kong study puts the expenditure higher at 6,250 HK$ (Greenlees 2005). The spending structure differs considerably from those of 'western' visitors. Less money is spent on accommodation but more is used for shopping and visits to leisure parks and other entertainments. The amount spent per day is therefore lower compared to other source markets of Hong Kong and to the average of all visitors, but a greater share of it goes to shops and service providers outside the hospitality business. Given the size of the Chinese tourism into Hong Kong, even with the lower amount per day, Mainland Chinese since 2002 spent more than half of the total tourist monies. In 1993, Mainland visitors to Hong Kong could only exchange 60 US$ per trip. From January 2005, they are officially allowed to take not only up to 5,000 US$ per trip out of the country but also 20,000 Yuan RMB (approximately 2,000 Euros in 2005) instead of only 6,000 Yuan RMB (approximately 600 Euros in 2005) as before.

A major problem for the new surge of Mainland Chinese visitors to Hong Kong under the IVS is the lack of safe, clean and mid-price two- to three-star hotel accommodation (Huang, Hsu 2005). Hong Kong hotels are still positioned mainly at the upscale market focused on business and MICE travellers from mid- and long-distance source markets.[25] As a result, Hong Kong is perceived as too expensive a place to stay for more than a few nights, or as offering cheap accommodation of unacceptable quality for families. This is also reflected in the occupancy rates, which are lower for the most and the least expensive accommodation classes than for the medium-tariff hotels (HKTB 2005b).

Heung (2000) analysed satisfaction with the hotel services offered in a survey of Mainland Chinese visitors using hotels in Hong Kong. Of the

Table 6.6 Spending of Mainland Chinese visitors to Hong Kong 2000–2004 (in HK$)[26]

| Year | HKTB figures | | Chinese figures | | |
	Per person per trip	Total spending	Per person per trip	Per person per day	Total spending
1994	5,469				
1996	6,581				
1998	5,487				
2000	4,868	13,235			
2001	5,169	15,823			
2002	5,487	26,056	5,639	1,250	26,776
2003	5,235	29,800	6,018	1,252	34,258
2004	4,355	33,941			

34 categories presented, only 11 categories were perceived as better then expected, 3 were neutral and 20 were perceived as unsatisfactory.

For European and North American tourists, especially if they are new-comers to East Asia, Hong Kong provides the first taste of China with minor heritage sites such as the Manmo (Wenwu) Temple on Hong Kong Island and local night markets eagerly visited. For leisure tourists from Mainland China, the symbols of modernity and wealth used to be the main attractions with local tour guides providing background stories to the number of Rolls-Royces owned by the inhabitants of Peak villas gazed at from the sightseeing bus. With many Chinese big cities now having their own share of high-rises and expensive private houses, the 'certain degree of glamour' (Huang, Hsu 2005: 203) of Hong Kong is shrinking and the department stores and consumer electronics, camera or jewellery shops remain as the focus of the main activities. 'Hong Kong used to be the number one choice for people here. But now everyone wants to go to Paris', a Beijing-based tour operator is quoted as saying in 2004 (Lau 2004).

This focus on shopping is supported by some group tours offered that follow the Zero-Dollar or even a Minus-Dollar procedure:

> There have even been reports that tour operators in Hong Kong paid money to travel agencies in mainland China before they could secure tour groups from the mainland. . . . Some tour operators appraise tour guides by the commissions or profits they earn for the tour operators. In addition, the basic salary of tour guides is rather low, some even guiding tours without guide-fees, and some also paying to bid for tour groups from tour operators. Therefore, they rely heavily on shopping commissions as their major income.
>
> (Zhang, Chow 2004: 7)

It is therefore not astonishing that in the study of Zhang and Chow out of 20 criteria the bottom 3 were 'My Hong Kong tour guide was able to solve problems', 'My Hong Kong tour guide informed us about customs in Hong Kong that were different from those of Mainland China' and at the very end 'My Hong Kong tour guide introduced us to reliable shops' (Zhang, Chow 2004). In earlier years, the complaints of some members of discounted tour groups from China that in several days of visiting Hong Kong, they did not even have a chance to see the Hong Kong Peak, were reported in the Mainland press (Lau 2004). In 2005, the Hong Kong Consumer Council reported an increase of complaints from individual trav-ellers from the Mainland up to 500 in the months up to August alone. Most complaints were about shopping, especially for mobile phones and similar items and product prices. The Council reacted with the preparation of a pamphlet 'that would give visitors a better understanding of the Hong Kong shopping culture' (Leung 2005).

The expected further opening of the IVS to almost all Chinese affluent enough to travel to Hong Kong will support the integration of Hong Kong into the Chinese quasi-domestic tourism market. With almost one out of six overall Mainland Chinese 'outbound' tourists in 2004 actually just spending a few hours on a day-trip to Hong Kong, the treatment of Hong Kong as a outbound destinations becomes more and more farcical. Within the given statistical definitions, and with IVS open to all Chinese, Hong Kong will remain the top destination for Mainland China – even so the writing is on the wall:

> Hong Kong at the moment can attract Mainland visitors to shop for duty-free items, such as cosmetics, brand-name clothing and electronic products. However, it is uncertain how long such an advantage can be maintained. . . . Some of the coastal Chinese cities may become good shopping destinations in the future when high quality products can be purchased at similar prices.
>
> (Huang, Hsu 2005: 202)

Thailand

For western tourists, Thailand is a major holiday destination in Asia. The government of Thailand is trying to shift the image of the country from *erotic* to *exotic* (Prideaux *et al*. 2004), but the country is still tainted with the image of sex tourism (Hall 1996). 'Probably in no other country in the world has tourist motivation been so obviously linked to sex' (Richter 1989: 86).

For Chinese citizens, Thailand was the first country opened for visiting of relatives. In 1988 the State Council first approved such visits, the 1990 'Provisional Regulations on Management of Organizing Chinese Citizens to Travel to Three Countries in South-East Asia' spelt out in detail the need for the relatives to pay for all expenses and to offer guarantees for their visitors in Thailand as well as in Singapore and Malaysia and from 1992 also in the Philippines.

With the development of ADS after 1995, these four countries became the first benefactors of the new system. Thailand opened a representative office of its National Tourism Authority in Beijing in 2003.

Thailand is an 'easy' destination: the process of visa application is highly streamlined so that visas can be obtained within three days, facilitating last-minute decisions to join a tour. Thailand is also well connected by air to China with many direct flights from 15 different cities in China to Bangkok and Chiang Mai.[27]

The arrival figures for Thailand closely follow the overall development of China's outbound tourism. A strong growth followed the easing of travel restrictions with a doubling of figures in 1993, reflecting the peak of public-funded leisure trips. This was followed by a short-lived stagnation as a result of the campaign against wasting public funds and further growth, but renewed increases, now with a higher percentage of privately paid trips, until 1999.

With the availability of other destinations, the falling demand for Thailand was countered by offers of shorter duration and very low prices, bringing the number of visitors – interrupted by the SARS dent – in 2004 back to the level of 1999. However, the relative importance of Thailand fell with this stagnation from approximately 20 per cent of the non-Hong Kong, non-Macao outbound market to approximately 10 per cent within this period of time.

In the 1990s, many itineraries were based on 15-day journeys and combined Thailand with visits to Singapore and Malaysia, sometimes with stopovers in Hong Kong or Macao as well. Given the bureaucratic problems of outbound travel and the insecurity about whether a second trip would be possible and affordable in the near future, to see as many countries as possible seemed to be a sensible strategy for private tourists. For travellers using public funding it also made sense to make use of the advantage of having secured a place in a travel group as fully as possible.

In the first years of the new century, these kinds of tours were provided mainly for the 'second wave' of first-time outbound tourists from provinces other than Beijing, Shanghai and Guangdong, who could not go to Australia or Japan because of restricted ADS agreements with these countries. Other travellers now travel within the 'Golden Weeks' or even for a prolonged weekend to Thailand only.

Table 6.7 Chinese tourist arrivals to Thailand 1987–2004[28]

Year	Number	Change (in %)
1987	21,464	
1988	33,344	+55.35
1989	52,358	+57.02
1990	60,810	+16.14
1991	75,052	+23.42
1992	128,948	+71.81
1993	261,739	+102.98
1994	257,455	−1.64
1995	375,564	+45.88
1996	456,912	+21.66
1997	439,795	−3.75
1998	571,061	+29.85
1987–1998		**+34.75**
1999	775,626	+35.82
2000	704,080	−9.22
2001	694,886	−1.31
2002	763,139	+9.82
2003	624,214	−18.20
2004	779,070	+24.81
1999–2004		**+0.09**

This is also reflected in the growing number of repeat visitors. In 2000, 85 per cent of all Chinese coming to Thailand went for the first time, in 2004 this figure dropped to 79 per cent (Tourism Authority of Thailand 2005).

The capital Bangkok is the main destination, visited by almost all Chinese visitors to Thailand. In 2002, Bangkok's Chinatown organized for the first time an extravagant celebration for the Chinese New Year,[29] with the prime minister attending the festivities in Chinese attire. Some 75,000 Chinese tourists travelled to Thailand for the occasion (WTO 2003), accounting for the total increase in arrival figures for that year.

Phuket and Pattaya are the next important destinations within Thailand. Phuket is seen as a family destination, whereas the very public commercial sex offers prevalent is denying this status to Pattaya. Domestic Chinese tourism is accustomed to visiting 'minority' people and viewing their 'primitive' folklore and heritage (Oakes 1998; Oakes 2005; Graburn 2001a). This established pattern of behaviour is served by visits to Chiang Mai.

The image of Thailand as a holiday destination has been marred from two sides: by the Zero-Dollar or Minus-Dollar travel arrangements, which have been most rudely used in Thailand, and by the offer of commercial sex and gambling as major tourist attractions.

Chinese tour operators, eager to gain market shares, cut the prices for one-week package tours to Thailand down to 400 US$ or less,[30] clearly below their own cost for visa procurement, transportation, accommodation, food, local transport and guidance. As a result, customer complaints were widely published in China with stories about tourists in Phuket not seeing the ocean once because their trips consisted only of tours from their inland hotel to one shop after another. Commercial sex and gambling are activities many Chinese visitors come to Thailand for, but as these industries are illegal within Mainland China, the 'full weight of the lingering suspicion against the *bourgeois lifestyle form* of leisure tourism crushes down on this combination of sin and waste' (Guo *et al.* 2005b). Stories of groups of Chinese men eating delicacies off the naked bodies of women and of fortunes lost in gambling dens were eagerly snapped up by the public and used as arguments against an 'excessive' development of outbound tourism in general.

Such excesses aside, the low prices, not much above the cost of a domestic trip to another part of China, make Thailand and its neighbours to the south affordable to a much wider group of Chinese tourists, making it the major destination for the first step beyond Hong Kong. Large Chinese communities, the clear image obtained from peer-group information of those who went before, simple visa procedures and the possibility of using Chinese currency in the destination are responsible for an almost home-like feeling but with the added attraction – especially for northern Chinese – of a subtropical climate, beautiful beaches and exotic sights.[31] Thailand is therefore also a foreign destination seen as suitable for elderly tourists.

Nevertheless, within Southeast Asia, the high risks of bad quality tours and sex tourism-related crime are giving Thailand a bad image compared to Singapore and Malaysia with regard to safety. Moreover, this is the risk perceived as most important by Chinese tourists (Teng 2005).[32]

Thailand is still predominantly a holiday destination; even so, business and convention tourism have slightly increased in the first years of the twenty-first century.

The age groups travelling to Thailand are rather tilted towards the older travellers, in line with the 'easy' image and comparatively low costs, without much change between 2000 and 2004. The average length of stay is six days.

The gender ratio is approaching equilibrium. In 2003 approximately 60 per cent of all visitors to Thailand were male, in 2004 this percentage dropped to 54 per cent (Tourism Authority of Thailand 2005).

Chinese visitors spent 104 US$ per day per person on average in 2004, compared to 107 US$ in 2000. The Tourism Authority of Thailand (2005) estimates that 489 million US$ revenue was the total result, an increase compared to the 430 million US$ of 2000 in line with the growth in numbers of visitors.

Thailand illustrates the stormy and unregulated start of Chinese leisure tourism to other countries. Price wars, with quality as the victim, are not unknown in China in consumer goods markets other than leisure and tourism. The fact that 'Zero-Dollar tours' and the resulting cheating and coercing were tolerated by the Thai authorities for a long time seriously

Table 6.8 Purpose of visit of Chinese citizens to Thailand 2000–2004 (in per cent)[33]

Purpose of visit	2000	2004
Holiday	94.3	90.9
Business	3.7	5.5
Convention	0.5	1.0
Official and others	1.5	2.6

Table 6.9 Age structure of Chinese citizens visiting Thailand 2000–2004 (in per cent)[34]

Age group	2000	2004
Below 25	10	10
25–34	27	27
35–44	29	30
45–54	19	20
55+	15	13

damaged the image of Thailand and made it the main target for all the criticism of outbound tourism in the official Chinese media. Thailand's current situation also reflects the long way Chinese outbound tourism has come already from closely regulated, invited VFR visits in 1990 to Shanghai office workers jumping on a plane to spend their weekend at a beach on Phuket.

Oceania

Looking at a Chinese version of the map of the world, which puts Europe and Africa to the left and the Americas to the right-hand corners of the world, Australia and New Zealand appear to be the nearest non-Asian countries to China. Both countries developed good political and strong economic relations with China in the 1990s and could both offer a combination of 'western' culture and a large overseas Chinese community (Chan 2000). They also offered a great variety of landscapes different from those found in China going through the opposite cycle of the seasons, being part of the southern hemisphere.

Australia and then New Zealand were opened to official leisure tourism through ADS in 1997, the scheme becoming operational in 1999. They are the western countries that have the longest-standing experience with Chinese visitors, which besides tourists also include large numbers of business, VFR travellers and students.

Using Chinese figures, Oceania's share of outbound travellers from China in 2004 was 1.3 per cent, down from 1.5 per cent in 2002. For travellers travelling further than Hong Kong and Macao it was 4.3 per cent, a slight improvement on 4.2 per cent in 2002 but still a decrease from the 4.9 per

Figure 6.2 Map showing ADS countries in Oceania[35]

Table 6.10 Chinese visitors to Oceania 2000
and 2004 (in '000s)[36]

	2000	*2004*
Total Oceania	150	361
Australia	127	270
New Zealand	18	62
Other Oceania	5	29

cent reading of 2000 (CNTA 2003a; WTO 2003; Du, Dai 2005). More than 90 per cent of all Chinese visitors to Oceania travel either to Australia or New Zealand. The other ADS countries in Oceania, which include Fiji, the CNMI, Tonga, the Cook Islands and Vanuatu, can only attract small numbers of Chinese visitors. The CNMI has been most active in working on the Chinese market since 2002, with the National Visitors Organization conducting surveys in China and starting a Chinese-language tourism website in early 2005. For all island states in the Pacific, the lack of trained personnel, of Chinese restaurants and most of all of a recognized 'Pacific' brand identity are identified by experts in the region as the major problems for development (Magick 2005). In the Chinese group-orientated collective mind, there is no tradition of a yearning for a South Pacific paradise, neither of the Robinson Crusoe nor of the Gauguin variety.

Australia

Australia has been a destination for Chinese ever since the 1850s goldrush era in Australia, which brought at its highest point almost 40,000 Chinese to the goldfields of Victoria in a single year in 1861. The majority of the Chinese gold seekers returned to China after the goldrush, but about 2,000 stayed permanently (Guo 2002), the nucleus of the overseas Chinese community of today. Souvereign Hill in Melbourne as a memorial for the strong presence of Chinese workers in the goldfields in 1850s today serves as a major attraction for visitors from China.

Abul Rizvi of the Australian Department of Immigration reported to the Senate in June 2005 about the ADS:

The scheme started in 1999, and prior to that there were extensive discussions between us and the Chinese government on this particular scheme, closely involving the minister at that time. I really could not say why the Chinese government decided to grant Australia this status ahead of others, other than perhaps to reflect the close relationship we have developed with the Chinese authorities. Since it started in 1999, we have expanded from the three regions that were covered at the time, which were Beijing, Shanghai and Guangdong. In July 2004 we expanded it to an additional six provinces.

(Australian Senate 2005)[37]

The first ADS travel group arrived in Australia in August 1999. Australia opened its tourism representative office in Beijing soon after. The experimental form of the ADS agreement for both the Chinese and the Australian government as the first time ADS was granted to a 'western' country, is reflected in the fact that only Chinese citizens living in the three most affluent provinces of China were included initially, making it easier for the Chinese government to monitor the scheme and minimizing the risk of illegal immigration for economic reasons for the Australian government. Only 39 travel agencies in China were granted the right to organize groups to Australia. 'Groups' however could consist of no more than two participants plus a tour guide.

Australian cities are connected to China with direct flights between Beijing, Shanghai and Guangzhou and Sydney and Melbourne. In the early years of the twenty-first century, shortage problems in the number of available seats occurred following the bankruptcy of Ansett, but these would be resolved later on.

The figures available for the arrival numbers of Chinese to Australia vary, partly because of the parallel accounting in fiscal years and calendar years and because of the inclusion or exclusion of the considerable number of students. But even sources relying on the same basic tables of the Australian Bureau of Statistics quote different numbers. The trend, however, is quite clear with figures hovering around the 20,000 mark before and after 1990 and doubling from 1994 to 1996 and again to 2000 with continued if slowed growth afterwards.

For some time, the forecasted growth figures envisaged an even more pronounced development with the Tourism Forecast Council (TFC) prediction in 1999 for 2005 of a 21 per cent market share of Chinese visitors to Australia (Pan, Laws 2001) overshooting the real development by a wide margin.

In 2000, the Mainland Chinese market was the tenth largest origin market for Australian tourism. The TFC expected a growth of 25 per cent for the following ten years, which would have led by the year 2010 to China sending 1.1 million visitors, overtaking New Zealand as Australia's largest source of international visitors (Guo 2002). The 2001 forecast of no less than 1.4 million Chinese visitors in 2012 was revised downwards in 2004 to 850,000 visitors in 2012, predicting China to be in fourth place after New Zealand, the UK and Japan (People's Daily 2001) by that year. Even so the Australian Tourism Export Council (ATEC) is still quoted in 2004 as upkeeping the 1.4 million prediction for 2012 (Open China 2004).

Table 6.11 Chinese visitors to Australia 1982–2004[38]

Year	Fiscal year	Calendar year
1982		2,180
1985		5,400
1990		24,000
1992		19,000
1993		22,000
1994		30,000
1994–1995	34,275	
1995		43,000
1995–1996	54,952	
1996		54,000
1996–1997	63,804	
1997		62,180/66,000
1997–1998	66,101	
1998		73,000/76,540
1998–1999	71,713	
1999		87,000/92,600/97,995
1999–2000	94,092	
2000		115,000/120,300/124,354
2000–2001	125,435	
2001		148,000/158,000/172, 000
2001–2002	162,000	
2002		180,000/190,000
2003		176,100
2003–2004	171,500	
2004		270,500

Until 1999, the year of the effective start of the ADS system, work, not leisure trips were the most important reason for Chinese visitors to travel to Australia.[39] In 2002, however, almost half of all trips were not for work, education or VFR, but for holidays, with education overtaking VFR as the third important reason for the first time (Australian Bureau of Statistics 2003).

The spatial patterns of Chinese visitors to Australia follow the common theme of concentration in a few places only, with big cities taking the first position. Four million nights are spent in Sydney and 1.4 million in Melbourne. The Gold Coast and Cairns in Queensland attract 0.9 million overnight stays each.[40]

Some 82 per cent of all Chinese tourists who visit Australia want to see Sydney with the landmarks Opera House and Harbour Bridge. In 1999/2000, 80,500 out of 94,100 Chinese visitors to Australia visited New South Wales (People's Daily 2001).

A typical tour to Australia will last one week, many taking place during the 'Golden Weeks'. One or two nights will be spent in Victoria, two to three nights in Sydney and at the Gold Coast each (Junek, Binney, Deery 2004).

Cairns has become more popular, even so it is a detour on most itineraries, following the visit to Cairns of a string of Chinese leaders after 2000 including Li Peng, Zhu Rongji and Jiang Zemin (Open China 2004). As leisure tourists until 2004 came only from Beijing, Shanghai and Guangdong, the number of first-time visitors is still very high even though Australia has been open for ADS tourists since 1999:

> The selling points of Australia in the first place are the beautiful and vast natural scenery, the Sydney Opera House, Golden Coast, the Movie World, the Ocean Park, Aboriginal culture . . . In Australia, there are a lot of tourist activities in which travellers can participate, such as sheep shearing and horse riding on a farm and to watch animals unique to Australia such as koala bears, kangaroos and platypuses. Shopping is also a selling point. Sheepskin and woollen products are much cheaper than in China. It is attractive to buy these, especially as gifts to friends. Most of the passengers have a good impression of Australia.
>
> (WTO 2003: 76)

Compared to western tourists, the level of intensity of the participation is however low and mostly prefer a photo opportunity rather than immersion. This can be also deduced from a study in which Wilks, Pendergast and Wood (2003) analysed the accidental deaths of overseas visitors in Australia between 1997 and 2000. The odds of being among the more than 300 accidental deaths that occurred among them during that period, were several times higher for European or American visitors than for Japanese or Singaporean. Chinese are not among the ten main groups of victims.[41]

In his survey of 210 Chinese tour operators, Guo (2002: 271) names the following five attributes as the most important ones for Australia's image as a tourist destination for the Chinese: clean environment, world-famous landscape, attractive wildlife, world-famous buildings, and being different. The least important ones are shopping opportunities and nightlife and entertainment. There is an obvious gap between the image and the reality of the Chinese travelling in Australia, with shopping and entertainment in the form of visits to casinos being major elements of the tour whereas landscape and wildlife are experienced in very small doses compared to most other visitors to Australia.

This is also reflected in the not too successful efforts of Western Australia to attract more Chinese tourists. In a 2004 promotion the Western Australia government tried to convince Shanghainese travellers that this part of the country was the 'real Australia', pointing out the unique wildlife and the especially bright stars in the night sky (Shanghai Daily 2004). Being different from the rest of Australia is a preposition that might work in other markets (Crockett, Wood 2004), but is not an argument for tourists who travel to the country because it is different from their own and for whom individualistic encounters with Mother Nature are not among the most sought-after experiences.

A more important motivation for travel to Australia is the high number of young Chinese going to Australia for their studies. Parents, sometimes together with their future overseas student, visit different cities to select the most suitable university or visit their child during his or her studies there.

Being located on the southern hemisphere adds the possibility for Australia to attract Chinese visitors escaping the cold during the spring festival 'Golden Week' or the heat of the northern summer. Furthermore being an English-speaking country increases the possibilities for direct communication for many visitors from China compared to destinations with their own native language.

The Olympic Games of 2000 further helped to upgrade the prestige connected to visits to Australia for leisure as well as for incentive tours.[42] The higher price and the higher 'exoticism' compared to Southeast Asian destinations position Australia in the middle between the high-end destinations America and Europe, especially France, and the nearer Asian destinations. This can be seen from the findings of the Australian Tourism Commission (ATC) 'Brand Health and Communications Tracking Study' of June 2003 among potential travellers[43] to Australia from Beijing and Shanghai, which checked the awareness of the possibility to travel to the destination, the wish to travel, if time and money would be available and the realistic intention to travel in the next few years outside China. Thailand and Singapore were the realistic, if not so much coveted destinations, whereas Australia after France, but before the USA and Europe were named as other possible and wished for, if harder to achieve, destinations. Japan was more important for travellers from Shanghai than from Beijing.

Junek, Binney and Deery start their analysis of the needs of Chinese visitors in Australia by saying: 'Little is known about the Chinese market in terms of what attractions they expect to see whilst in Australia and what expectations they have of these attractions' (Junek, Binney, Deery 2004: 149).

Pan and Laws state categorically:

> Chinese tourists know little about Australia – at most they have some general ideas about koalas, kangaroos and the Sydney Opera House. They think of Australia as an agricultural country with flocks of sheep but otherwise a wilderness. They know nothing about other Australian tourism products.
>
> (Pan, Laws 2001: 46)

The higher price and more pronounced cultural difference between China and Australia compared to Southeast Asian destinations is also reflected in the background of the visitors. Furthermore, for reasons of restrictions of ADS to Beijing, Shanghai and Guangdong, the majority of visitors stem from these provinces, with Guangdong being the most important source market. For example, during the 'Golden Week' period of October 2003, the first after SARS, about 100,000 Chinese tourists visited Australia, half of them originating from Guangzhou, 30 per cent from Beijing and 20 per cent from Shanghai (People's Daily 2003a).

The majority of them had tourism experience with other destinations before coming to Australia and belonged to the higher income and education sectors of the Chinese outbound tourists. Out of 250 tourists interviewed by Yu and Weiler (2001), 44 per cent recorded an income higher than 30,000 Yuan RMB and 70 per cent had a university degree. The repeat visitors rate is however not very high, staying in the single-digit region according to several studies (Yu, Weiler 2001; WTO 2003; Guo 2002).

In terms of spending, visitors from Mainland China spend per person almost double the average of all visitors to Australia (Yu, Weiler 2001). Besides education,[44] food and accommodation, shopping and entertainment and gambling feature most prominently in the spending structure. Package tours mostly use three-star hotels, with more expensive accommodation used by business travellers. As Chinese tourists are much less likely to visit the outback or go on walking tours, they use other forms of accommodation such as camping grounds, backpackers hostels, etc., much less.

Accordingly, the mostly city-orientated travelling of Chinese tourists in Australia leads to preferred activities in line with this behaviour. A survey of the ATC (2004) finds shopping, market visits, beach visits, visit to gardens and casinos as the most popular pastimes.

Yu and Weiler (2001), concentrating on package tours only, offer a different result, however: the five activities that are seen as most important according to their survey are beach life, visits to national parks, visits to historic sites, visits to aboriginal villages and visits to city gardens and parks. Least important are participating in sports and dining in western restaurants.[45]

Australia is also considered to be a typical 'Golden Week' destination for *6-adults-1-child* family groups. These consist of an affluent, well-educated

Table 6.12 Leisure activities of Chinese tourists in
Australia (in per cent)[46]

Activities	
Go shopping for pleasure	72
Go to markets	47
Go to beach, swimming, surfing, diving	44
Visit botanical and city gardens	41
Visit casinos	36
Visit national parks	32
Experience wildlife, zoos, aquariums	31
Visit historical buildings	27

couple with high income, inviting their respective parents to travel together as a sign of appreciation for their support in earlier years. The main person of the group, however, is the single child of the couple. 'As there is only one child in the typical Chinese family, these young Chinese tourists might have great influence on their parents' decision making' (Yu, Weiler 2001: 89).

Junek, Binney and Deery (2004) analyse the readiness of the tourism industry in Melbourne to serve Chinese visitors, showing positive results: at most attractions Chinese food is offered, Chinese-speaking staff and guides are available, and printed or at least photocopied brochures are provided both in simplified and non-simplified characters. The shops also try to cater for Chinese tastes but get little business, as the guides try to steer the tourists away for lack of commission agreements. Audio-guides in Chinese language are also offered but not often used.[47]

Prideaux (1998) is less optimistic, warning the Australian tourism industry, 'not to treat all Asian visitors as if they were Japanese'.[48]

In Australia, Chinese tourists are sometimes taken advantage of by tour operators and guides. Because the prices for the package tours are too low compared to the real cost, incoming companies have to earn money from shopping commissions and sometimes outright fraud. In 2005 a scandal developed when the media in Sydney reported on the fact that some guides pretend that a 'council fee' has to be paid to walk on beaches or to take photos of famous sights such as the Sydney Opera House or Harbour Bridge. The tourists are, according to these reports, not brought into the city centre because of the centre allegedly being unsafe for Chinese visitors, when in fact this is a measure to prevent the visitors seeing for themselves that shops in Sydney's Chinatown charge only half for favoured items such as sheepskin boots compared to the duty-free shops outside the city centre that the tour groups are brought to (Connolly, D'Costa 2005).

Australia will probably enjoy a steady flow of Chinese tourists from the growing number of provinces included in the ADS agreement in the foreseeable future. Its positive image and well-known images such the Sydney Opera House and the kangaroo, its strong Chinese community reinforced by a growing number of Chinese students with an Australian University as their Alma Mater, and the perception as the nearest truly 'western' destination will form a sustainable competitive advantage. The expectations of early ADS times, that China will become the number one source market by 2010 may however be over-optimistic.

New Zealand

New Zealand became an ADS destination together with Australia in 1999. However, the development of Chinese outbound tourism to this country differs from the Australian experience. Similar to Australia's original arrangement, not all Chinese can get ADS visas, but only citizens from

Beijing, Shanghai and Guangdong provinces.[49] Leisure trips, however, play a smaller role as a reason for Chinese to travel to New Zealand. The NTO of New Zealand established a Shanghai representative office, but the marketing efforts and perception of market development for Chinese tourism to New Zealand have been much less enthusiastic when compared to Australia (Becken 2003).

It is questionable if the absence of a direct air link is the result or the cause of the limited development of Chinese outbound tourism towards New Zealand. In any case, the necessity to change planes in Hong Kong or Sydney results in higher costs and longer travel durations than for most other major ADS destinations.

The statistics for the number of Chinese visitors again need careful interpretation, with Chinese statistics not counting visitors who travel to New Zealand via Australia and national statistics following fiscal years and including students. Before 1996 the number of visitors from China to New Zealand was well below 10,000, rising above the 20,000 mark for the first time in 1999 and continuing to grow faster after the ADS agreement with a tripling in numbers between 1999 and 2002. After the SARS year 2003 without any growth, the New Zealand statistics report a significant rise again for 2004/2005 whereas the Chinese statistics report stagnation since 2002.

The seasonality of Chinese visits to New Zealand is less pronounced than from other source markets. However, if differentiated according to purpose of visit, the example of 2004/2005 shows the effect of the spring festival in February with strong increases in the number of holiday travellers partially offset by a decrease in business travellers. In November and December, however, both holidaymakers and business travellers manage to schedule their visits to enjoy the mild temperatures of the southern summer.

Table 6.13 Chinese travellers to New Zealand 1998–2005[50]

New Zealand		China	
Period	Statistics	Year	Statistics
June 1998– May 1999	17,059	1999	12,591
June 1999– May 2000	28,212	2000	18,288
June 2000– May 2001	42,060		
June 2001– May 2002	63,563	2002	58,700
June 2002– May 2003	73,641	2003	59,400
June 2003– May 2004	71,249	2004	61,800
June 2004– May 2005	83,860		

Table 6.14 Number of visitors from China to New Zealand in 2004/2005 by purpose of visit (in '000s)[51]

	05/2004	06/2004	07/2004	08/2004	09/2004	10/2004	11/2004	12/2004	01/2005	02/2005	03/2005	04/2005
Total	5,724	4,752	6,400	5,742	6,400	6,867	9,471	10,425	6,786	10,098	4,925	6,864
Holiday/ vacation	2,808	2,208	2,620	2,992	2,480	3,549	4,536	4,825	3,103	6,600	2,175	3,528
VFR	756	768	940	660	1,100	798	945	1,550	957	1,386	800	1,104
Business	1,620	1,264	1,400	1,232	2,280	1,785	3,066	3,025	1,566	1,122	1,350	1,728
Other	540	512	1,440	858	540	735	924	1,025	1,160	990	600	504

The spatial consumption pattern of Chinese visitors to New Zealand shows a pronounced 3:1 preference for the North Island. In 2002, they stayed approximately 2.3 million nights on the North Island and only approximately 800,000 nights on the South Island.[52] In comparison, German tourists for instance spend almost the same number of nights on both islands. More precisely, New Zealand for the Chinese is mostly composed of two places only: Auckland plus Rotorua. For 91 per cent of all Chinese visitors the main city Auckland is the most preferred destination, with the attractions connected to Maori culture and geothermal activities in the Rotorua region having a pull factor of 52 per cent.

> Visitors from China tend to spend the majority of their time in the major centres. In the year ending December 2003, 70 per cent of all Chinese visitor nights were spent in the Auckland region, showing the domination of Auckland as the main destination of Chinese visitors.
>
> (TRCNZ 2005)[53]

Auckland is also the home of more than half of the overseas Chinese population of New Zealand, itself the largest and fastest-growing Asian minority population in New Zealand (Feng, Page 2000). From Hong Kong, tours that cover just Auckland and Rotorua are offered, the shortest lasting only four days including eight hours flying each way.

Becken (2003) distinguishes between the spatial consumption of six different types of international travellers to New Zealand: coach, auto, camper, backpacker, VFR and comfort travellers plus a special category of 'gateway-only' visitors for tourists not leaving their gateway of arrival at all. The style of travel is closely linked to travel purpose. Most Chinese visitors, especially holiday tourists, travel as coach tourists in an organized coach tour. About one-quarter of Chinese visitors are gateway-only tourists, staying only in Auckland in most cases. For non-holiday and non-business visitors, the ratio of visitors not moving around in New Zealand is even close to 50 per cent. VFR and auto tourists, both using private cars, are responsible for the remaining number of travellers. Camper, backpacker and comfort travel styles have no significant role.

According to Becken (2005), one-quarter of all inbound tourists to New Zealand participate in coach tours. In comparison to Australian, European and North American visitors, tourists from East, Southeast and South Asia

Table 6.15 Dominant travel style among Chinese tourists in 2001 for different travel purposes (in per cent)[54]

Holiday		Business		VFR		Conference		Education	
Coach	70	Coach	65	Gateway	45	Coach	50	Gateway	54
Gateway	13	Gateway	23	VFR	45	Gateway	45	VFR	39

are the most limited in spatial variety. 'Only few places other than the four hubs (Auckland, Rotorua, Queenstown, Christchurch) were visited by Asian coach tours' (Becken 2005: 31). As 70 per cent, more than the other groups, of all Asian coach tourists also fly by domestic aircraft; they travel the highest distance per day in kilometres, but the shortest total distance in overall terms. Trains and ferry are used only by 3 per cent of Asian coach tourists.

The image of New Zealand is similar to Australia connected mainly to clean nature and difference from home. With Auckland's Sky Tower providing the famous building, the Maori culture is perceived as an important attraction. From their survey Ryan and Mo (2001) developed four clusters for reasons to visit New Zealand: social investigative reasons such as VFR, looking for business and study opportunities; New Zealand-specific reasons connected to the image and the scenery; the general reason to look for relaxation; and finally the reason to visit a place not been to before. However, one of the major reasons to visit New Zealand, or rather Auckland and Rotorua, is the simple fact that these places are part of an Australia–New Zealand package tour. This is supported by the short duration of stay in New Zealand with 42 per cent in 2002 staying less than five days and another 19 per cent five to seven days only.

According to the New Zealand statistics, the increase in holiday visitors has been faster than in the other categories, bringing them in 2004/2005 up to 50 per cent of the total visitors.

In terms of age and gender, the group of 45 to 54 year olds has taken over as the most important age group after 2002. Together with the 35 to 44 years old they represent 60 per cent of all visitors. The proportion of female visitors grew from 33 per cent in 1998/1999 to 39 per cent in 2001/2002 to stay at this level in the following years.

Nine out of ten Chinese visitors to New Zealand surveyed by Ryan and Mo (2001) travelled there for the first time. In this survey, all participants had travelled to another place outside Mainland China before travelling to New Zealand, one-quarter of those asked even to four or more different countries. Some 65 per cent reported to hold a university degree. ADS tourists by definition could only originate from Beijing, Shanghai and

Table 6.16 Reasons to visit New Zealand 1998–2005[55]

	1998/ 1999	1999/ 2000	2000/ 2001	2001/ 2002	2002/ 2003	2003/ 2004	2004/ 2005
Total	17,059	28,212	42,060	63,563	73,641	71,249	83,860
Holiday/ vacation	6,002	10,266	18,377	25,045	29,074	31,538	41,428
VFR	2,674	4,609	6,358	8,595	10,013	11,415	11,673
Business	6,413	8,746	10,813	15,922	17,339	18,765	21,091
Other	1,970	4,591	6,512	14,001	17,215	9,531	9,668

Table 6.17 Age structure of Chinese visitors to New Zealand (in visitor numbers with percentages in parentheses)[56]

	1998/9	1999/2000	2000/1	2001/2	2002/3	2003/4	2004/5
Total	17,059	28,212	42,060	63,563	73,641	71,249	83,860
Under 15	452 (2.6)	1,056	1,532	2,943 (4.6)	3,126	2,563	2,371 (2.8)
15 to 24	799 (4.7)	1,890	3,866	9,815 (15.4)	11,577	5,298	5,194 (6.2)
25 to 34	3,333 (19.5)	5,417	7,812	10,441 (16.4)	10,649	10,245	11,759 (14.0)
35 to 44	5,181 (30.4)	7,488	11,712	16,976 (26.7)	18,768	20,148	24,889 (29.7)
45 to 54	4,198 (24.6)	7,314	10,625	14,960 (23.5)	19,807	21,537	26,361 (31.4)
55 to 64	2,310 (13.5)	3,809	5,189	6,428 (10.1)	7,499	8,468	10,045 (12.0)
65+	786 (4.6)	1,238	1,324	2,000 (3.1)	2,215	2,990	3,241 (3.9)

Guangdong until 2004, since then also inhabitants from Chongqing, Hebei, Jiangsu, Shandong, Tianjin and Zhejiang are included.

It is reported that 68 per cent of all Chinese visitors stay in hotels. This preference is more pronounced for holiday and business visitors, with VFR guests more often staying in private homes and 'educational tourists', i.e. students, staying in student flats.

In terms of preferred activities, Chinese tourists wish to experience Maori culture, visit a National Park, see gardens and farms, do sightseeing in cities and enjoy boat cruises. Older visitors find tours and heritage sightseeing more important, with younger visitors preferring being active (Ryan, Mo 2001). These results correspond with official New Zealand figures about the actual activities of Chinese tourists. Such activities are concentrated on relatively few places and do include a visit to the casino but do not include activities such as trekking, glacier walks or visits to museums and galleries, which are important items for European or American visitors. The special interest for Maori culture may be explained by the fact that visiting a 'minority cultures' preferably of archaic forms in a touristified way for easy consumption is an activity well established within domestic tourism in China.

Ryan and Mo (2001) report that only 15 per cent of the Chinese visitors to New Zealand interviewed expressed the wish to return. For those who did wish to return, the 'clean and unspoiled environment' provided the major reason, but not the 'scenery', again reflecting the difference between the image of the big open countryside and the actual stay centred on the city of Auckland. Besides whale-watching, shopping and language problems were quoted as less satisfactory experiences. Interestingly criticisms

Table 6.18 Leisure activities of Chinese
tourists in New Zealand
(in per cent)[57]

Activities	
General sightseeing	48
Watching Maori performance	30
Visit to Sky Tower	29
Visit to geothermal sites	25
Sightseeing tour	24
Farm show	20
Visit beaches	20
Visit casino	20

concentrated on the fact that Japanese language signs are available but no signs were in Chinese. This might be interpreted as a symbol of being held in less esteem than the Japanese, creating a very unpleasant impression for most Chinese.

Becken provides some background information that might help to understand the low rate of returnees:

> There is some belief that New Zealand is not sold at its optimum to the Chinese market, because some large companies took possession of a very specific segment (low-quality coach tours), which does not provide customer satisfaction and creates negative word-of-mouth. A similar situation is described for Australia, where Chinese travel agencies negotiate very low prices with Australian suppliers, which results in tours of a poor quality and service. Moreover, little variety is given in current products, limiting potential Chinese tourists' choices.
>
> (Becken 2003: 11)

Overall, Chinese outbound tourism to New Zealand will remain closely connected to the developments in Australia, especially as long as there is no direct air link. Following Becken's remarks, after gaining ADS status it seems that a anomaly has developed between the sophisticated, high-income leisure visitors from China and the not very sophisticated or varied tourism products offered to them. This in return may have influenced some critical views of Chinese tourists being not the kind of '"interactive travellers", who are experience-focused, interested in the local culture and environment, and aware of sustainability issues' (Becken 2003: 11), preferred as customers by the New Zealand tourism industry.

A differentiation between the VFR- and education-orientated visitors and the high-end leisure travellers will probably help to regain the momentum of Chinese tourism to New Zealand.

The Americas

North and South America represent two opposite ends of the development of China's outbound tourism. The USA is among the top ten destinations for Chinese travellers, but in North America neither the USA nor Canada had received ADS at the time of writing in early 2006, even so these countries are among the most preferred destinations for leisure travellers from China. In Central and South America, Cuba, Argentine, Brazil, Chile and Peru as well as Mexico, Barbados and Antigua and Barbuda all received ADS but can register only limited numbers of Chinese tourists, making South America the least important global region for China's outbound tourism.

The share of America as a destination has been stable at 8 per cent. In 1997, 241,000 out of 3.03 million outbound travellers from China travelling beyond Hong Kong and Macao went to the Americas. In 2000 both figures

1 Cuba	2003
2 Brazil	2005
3 Chile	2005
4 Peru	2005

5 Mexico	2005
6 Barbados	2005
7 Antigua and Barbuda	2005
8 Argentine (ADS Agreement)	2004

Figure 6.3 Map showing ADS countries in North and South America[58]

had risen in parallel to 362,000 and 4.87 million and by 2004 to 680,000 out of 8.360 million respectively. More than 90 per cent of these travels had the USA and Canada as their destination. In 2004, South America still could barely register 50,000 visitors from China.

South America

Despite efforts by various airlines, by 2005 South America was still not connected by any direct flight to China, making visits expensive and tiresome with travel times of 20 hours or more. Accordingly, the annual number of visitors for instance to Brazil has stagnated at around 14,000 since 1996. Big expectations have been nevertheless connected with the granting of the ADS. The Brazilian Tourism Administration predicted 200,000 Chinese visitors per year after ADS (Chinaview 2005). Peru's head of the National Tourism Chamber estimated the number of Chinese visitors for 2006 at 100,000[59] after just 4,800 in 2003 (Muzi 2004). Chile recorded only 705 visitors in 2002 but reacted to the award of ADS by deliberating the inclusion of teaching the Chinese language into the school syllabus (APEC 2004).

As attractions Brazil sees the rainforest but also soccer stars as pull factors. Peru is banking on Machu Picchu, its natural features including snow-capped glaciers and rainforests and the Titicaca Lake. The Peruvian consulate general in Shanghai anticipated the granting of ADS with a study about 'Strategies to Promote Peruvian Touristical Offers in China'. It identified the new middle classes with an income of 1,200 to 2,400 US$ per month as the main target group for Peru. The authors recognize the problem of marketing Peru alone for Chinese tourists especially in the light of the long distance between the two countries. They recommend regional cooperation to serve the interest of the Chinese visitors to see as many countries and attractions during one trip as possible and to facilitate better marketing (Belaunde, Rios 2004).

Politically, China's foreign policy has tried to strengthen the ties with South American countries in the early years of the twenty-first century with substantial economic cooperation between Brazil and China for instance. In the field of leisure tourism it is hard to see similar developments, even so there exists sizeable overseas Chinese communities in several South American countries. The physical distance will remain a factor, the fascination of Americans and Europeans radiating from the Inca and Maya cultures is much less pronounced in China with its own millennia of history. The image of Central and South America in China includes the element of widespread crime, dangerous wildlife and insecurity, anathema to Chinese tourists, with security as their highest priority.

North America

The United States of America, the ultimate travel destination for many Chinese, home of the strongest global economy, Walt Disney and Las Vegas, remains an elusive destination for leisure travellers from the People's Republic of China. In December 2004, an MOU about the granting of ADS to the United States was signed, but by early 2006, despite the intensive lobbying of the states of Nevada and Hawaii, there was no sign of an implementation. The hurdles to enter the USA were high before 2001 and have been even higher since. Those Chinese visitors who manage to obtain a visa have to travel either for business, study or VFR reasons. Group tourism is not officially acknowledged to exist by the immigration administration. Both the USA and Canada have large and growing overseas Chinese communities in their big cities and are a highly coveted place to study, with approximately 60,000 students studying in the USA alone.

The USA has not been very keen to enter into the ADS system with its basic ideas of state-controlled group tourism, visa issuing to persons who did not visit the consulate in person and the principle of reciprocity, sometimes to the disdain of the tourism industry.[60] Even after the signing of the MOU, the visiting US Assistant Secretary of State for Consular Affairs Maura Harty told a press conference in Beijing on 2 March 2005 when asked about ADS:

> The ADS is certainly a conversation that we can have with the government of China if they want to. But I've got to tell you right now, we encourage Chinese – individual Chinese travellers to travel to the United States. [ADS] is a Chinese concept. It's not an American concept.
>
> (Harty 2005)

Chinese media carried the rumour that in the case of group visa being established, each tourist would have to put down a deposit of 100,000 Yuan RMB (approximately 12,000 US$), refundable after return (China View 2004). The USA does not have a national tourism representative office in China. The only state that somehow managed to get the permission to open a tourism office in Beijing in 2004 has been Nevada:

> Delay in granting ADS to the USA appears to be the result of stalling tactics on the part of the US government, which sees the ADS negotiation as part of a wider trade issue. Canada, in turn, appears unable – or unwilling – to conclude an ADS agreement with China since there are concerns that Chinese tourists, once in Canada, might try to cross the border illegally into the USA.
>
> (TBP 2004: 14)

Unlike the United States, the CTC has tried actively to gain ADS since 1999. In January 2005, the intention of the Chinese government to recognize Canada as an ADS travel destination was officially proclaimed in Beijing during a visit of Canada's prime minister. CTC officials speculated that the refusal of Canada to bargain for ADS in exchange for the deportation of a fugitive, living since 1999 in Canada but sought for alleged crimes in China, slowed down the process. Nevertheless, like their southern neighbour, Canada was still waiting in early 2006 for the actual start of ADS travels. CTC has had a representative in Beijing from 2001 but an official CTC office was opened only in January 2005, a few days after the announcement of the pending ADS for Canada (CanadaTourism 2005).

For the United States, in 2004 China could only achieve the twentieth rank[61] in the list of source markets, behind small countries such as Switzerland or Sweden. The number of Japanese visitors is almost 20 times higher than the number of Mainland Chinese. With regard to Canada's incoming tourism, China ranked twelfth in terms in trips in 2004, but seventh in terms of overnight stays (CanadaTourism 2005).[62]

Accessibility with regard to air connections is not a problem. Major cities in North America can be reached by direct air links from Beijing and Shanghai, with flights to the west coast (Vancouver, San Francisco) more frequent and also originating in cities such as Guangzhou, Shenzhen or Shenyang. Beside several Chinese airlines, United, Northwest and since 2005 also Continental Airlines offer services between China and the USA (Alcantara 2005). The air seat capacity from China to Canada grew between 2000 and 2004 from 136,000 to 212,000, sufficient to represent no constraint for bilateral travels. Since 2005 direct flights also connect China to eastern Canada with a Beijing–Toronto connection.

Unlike other destinations, after initial growth in the 1990s the number of visitors to both countries in the new millennium did not increase substantially or even decrease. In the case of the USA this development, not counting SARS, has more to do with the visa-issuing policy than with the potential demand. For Canada, the competing offers of other countries with ADS are blamed by CTC for the lacklustre performance of a figure staying shy of 100,000.

The USA is a year-round destination for Chinese travellers. The differences between the arrival numbers in the four quarters of the year 2004 are small: 24 per cent in spring, 25 per cent in summer, 29 per cent in autumn and 22 per cent in winter are recorded. The smallest number of

Table 6.19 Visitors to North America from China 1994–2004 (in '000s)[63]

	1994	1995	1996	1997	1998	1999	2000	2001	2002	2003	2004
USA	158	167	199	210	209	191	249	232	226	157	203
Canada	42	52	64	80	56	63	77	86	95	77	87

visitors arrived in February (21,000), and the largest number in August (34,000) (OTTI 2005). For Canada, climatic realities determine a rather strong seasonality of Chinese visits with summer seeing approximately one-third of all visitors, spring and autumn one-quarter each and winter just one-sixth (CTC 2005). For leisure travellers the seasonality will be even more pronounced.

Half of all Chinese travellers to the USA visit California, about a quarter include Los Angeles, San Francisco or New York in their itinerary. Other important destinations are Las Vegas, Orlando (Disneyworld) and Washington DC. The Grand Canyon and the Hawaiian Islands are the main nature-based attractions. Beside the general attraction of Chinese outbound tourists towards big cities, historic sites and famous sightseeing spots the proximity of the Pacific Coast to China and the big overseas Chinese communities in Californian cities are influencing the spatial consumption pattern of Chinese visits to the USA.

In a similar way, most of the interest in Canada is centred in the western part of the country because of its proximity, milder weather and scenic beauty. Places visited include Vancouver, Victoria, the Rocky Mountains, Banff and Jasper. The big cities in central Canada, particularly Toronto, Ottawa, Montreal and Quebec City but also the Niagara Falls, attract visitors from China. 'Very rarely are people interested in other parts of the country' (CTC 2001: 18). Accordingly, Ontario (37 per cent) and British Columbia (36 per cent) are the most visited provinces of Canada with Quebec (17 per cent) already a distant third (CTC 2005).

USA, the country seen in China as the main competitor in the effort to regain the global 'No. 1 country' position, which China could claim to occupy on many counts in terms of scientific, economic and cultural achievements until the sixteenth century, is the obvious choice to visit as a place to learn some of the tricks of the modern western world. Even so most 'western' cultural influences actually reach China after going through the filter of the Japanese or South Korean culture, icons such as the original White House[64] in Washington, Hollywood, Disneyland or the Golden

Table 6.20 Chinese visitation estimates for 2004 (People's Republic of China and SAR Hong Kong combined)[65]

	Market share (in %)	*Volume (in '000s)*
Pacific	52	170
Middle Atlantic	34	112
Mountain	20	66
South Atlantic	18	60
California	49	160
Los Angeles	29	94
New York City	27	89
San Francisco	24	79

Gate Bridge in San Francisco and the big casino-hotels of Las Vegas have enormous *brand value* to transfer to the prestige and cultural capital of Chinese visitors. The contacts, especially of southern Chinese, with ABC,[66] descendants of many decades of 'coolie' emigration to the USA, or with more recently arrived well-educated relatives as well as the popularity of Hollywood movie stars and NBA players are other pull factors:

> They don't want to see spectacular scenery (in the USA), they've got plenty of that at home. What they do want is to see how America measures up to the American Dream. They're familiar with the stereotypes of the United States as the richest and most advanced nation in the world, its lifestyle as the holy grail of development. And they want to see it in all its brilliant modernity, to understand how far China has to go to catch up, and whether the struggle will be worth it.
>
> (Dunlop 2004)[67]

More than half of all visitors to the USA claim business reasons for their travel, reflecting – beside the lack of ADS visa – the intense economic relations between the two countries.

Canada does not have a very clear image as a tourism destination separating it from the USA. It is perceived, however, as having a tradition of friendly relations with China, starting with Norman Bethune, a Canadian doctor who fought on the Communist side in the 1930s and whose merits are still taught in history lessons in Chinese schools.[68] Canada is also the home of more than a million overseas Chinese, with Vancouver and Toronto each having around 400,000 inhabitants regarding themselves as ethnic Chinese:

> Canada is close to the US and felt to be similar to it in many ways. So some Chinese view Canada as the next best thing, given that visas to the US are usually much more difficult to obtain . . . Virtually everyone who visits Canada wants to see the US.
>
> (CTC 2001: 19)

Canada's overall touristic image in China consists of being a beautiful, wide-open country that is safe and clean, but lacking famous historical sites and offering 'nothing special' with regard to cultural attractions. Another drawback in the eyes of Chinese travellers is the 'inconvenience' of Canada. Winter is considered as much too severe to travel and the major points of interest are perceived to be scattered across a great distance so that it is expensive as well as difficult to get to all of the key spots in a one-week trip. This is also true for distance between the three large cities that can be visited – Vancouver, Toronto and Montreal – which are perceived as being too far apart.

With North America being a white spot on the ADS map, most visitors do not cite holidays as their major reason to go to the USA or Canada.

Table 6.21 Purpose of trip of Chinese visitors
to USA in 2004 (multiple response,
in per cent)[69]

	2004
Business	56
VFR	41
Vacation/holiday	39
Attend convention	14

Table 6.22 Primary purpose of trip of Chinese
visitors to Canada and to Ontario in
2003 (in per cent)[70]

	Canada	Ontario
Business	25	31
VFR	45	52
Leisure	15	12
Other	15	5

For the USA, business trips – with bigger or smaller leisure component –
are responsible for more than half of the successful visa applications, with
39 per cent also naming leisure as a reason. For Canada, VFR is the most
important reason to visit with only 15 per cent stating leisure as the main
travel motivation. The example of Ontario shows the same characteristics
but even more pronounced.

The geo- and sociographical characteristics of travellers to North America
are determined by the costs of travel and the difficulties of getting a visa.
Also keeping in mind the strong business and VFR components, big city
and Southern China dwellers are most likely to provide the majority of
travellers. For 1997 CNTA stated that 70 per cent of visitors to the USA
are between 31 and 50 years old, while 22 per cent are over 50 years, 80
per cent of them have at least college education (Chen 1998). Figures looking specifically at Chinese visitors to Hawaii in 2003 find 45 per cent in the
35–44 years age group and 40 per cent in the 45–54 years cohort and 38
per cent having college and 46 per cent university education (HTA 2003).
For Canada in 2003, the number of younger visitors was higher with 41 per
cent of all visitors below 35 years old, 43 per cent between 35 and 54 years
old and 16 per cent 55 years or older (CTC 2005). A surprising figure is
the claim of 29 per cent of Chinese visitors to the USA in 2004 to be on
their first international trip outside China.[71] Chinese visitors to North
America follow the pattern of spending large amounts of money especially
while shopping. Due to the fact that official figures include students, the

amounts of 99 US$ for the USA in 2004 (OTTI 2005) and 61 CDN$ for Canada in 2003 (CTC 2005) probably underestimate the spending behavior of Chinese tourists in North America. Cai, You and O'Leary (2001), distinguishing between the three groups of Chinese business only, business and leisure, and leisure only travellers to the USA, found that business visitors spent less on entertainment, but all three groups spent a substantial amount of money on gifts. Business and leisure travellers participated more in destination activities than the business only travellers. The most popular destination activities and attractions are shopping, dining, city sightseeing, visiting historical places, amusement and theme parks, national parks, and casinos visits. For leisure travellers or for VFR tourists who tour North America with their relatives, these activities occupy most of the available time. The existence of Chinatowns in many cities, Chinese signage at many airports and not least the large number of Chinese restaurants serving rather authentic cuisine from different parts of China in North America ease many problems Chinese outbound tourists find in other parts of the western world.

The wide spaces, offering opportunities for outdoor sports and intense experiences of the relation between man and nature, which occupy a large part of the imagery of American cowboy country and the wildernesses of Canada, are experienced by most Chinese tourists to an even lesser degree than by the city dwellers that visit in Los Angeles or Toronto. The China Office of the Nevada Tourism Commission at least started to offer a ten-days tour, which beside San Francisco, Los Angeles, the Grand Canyon, Disneyland and of course Las Vegas also offers a visit to Reno, the 'cowboy hometown' (People's Daily 2005b). For elderly tourists, it is hoped that the slower pace in Canada's relatively uncrowded cities compared to the pace of life in Chinese cities, may make even the cities a more relaxing place. Younger Chinese tourists interested in ecotourism, the fall colours of the forests in autumn or winter festivals and skiing, will fit more easily in the tourism products Canada offers. The CTC admits, however, that this is a niche market, a relatively small segment (CTC 2001).

The former kingdom of Hawaii, situated halfway between Mainland USA and Mainland China, enjoys a special image as the home of the most famous beaches and the destination of choice for Japanese pleasure-seekers. The Hawaii Tourism Authority (HTA) studied its Chinese visitors in 2003. Ninety per cent of the interviewed travellers were stopping either en route to, or more frequently en route from, the US Mainland, most staying for two nights. Seventy-one per cent visited Hawaii for the first time, none of them alone but in groups with a number of six to ten persons as the most common size. Only 30 per cent of the tourists travelled with a Private Passport, even so 43 per cent named vacationing as purpose of the trip. Of all visitors, 78 per cent were male and 75 per cent fell within the age group of 35–54 years. The six most popular activities reflect both what Hawaii offers and Chinese tourists' behaviour: dining in Chinese restaurants

(95 per cent), Swimming, beach, pool (91 per cent), Visiting historical sites (84 per cent), Tour bus excursions (75 per cent), Duty-free shopping (68 per cent) and Polynesian shows, *luau* (58 per cent).

Some other common tourists' activities in Hawaii found little echo with Chinese tourists: Helicopter tours or Windsurfing participation was below 1 per cent and for Sun-bathing only 15 per cent spared some time, even though almost everybody went for a swim.

Of all tourists, just 60 per cent were satisfied or extremely satisfied with their trip to Hawaii, for all parts of the offer no category received more than 50 per cent positive responses or a satisfaction mark higher than the importance mark. Accomodation, restaurants and airport each got bad marks from around 15 per cent of the visitors. 'Many felt discriminated against, or as if the staff were not "user friendly" or knowledgeable in Chinese cultural expectations. In addition, meals were a big disappoint-ment [to] many tourists' (HTA 2003: 21). The unsatisfied travellers could be identified as mainly younger, more experienced travellers with higher income, whereas the older first-time visitors with mid-level incomes were easier to satisfy. The older group was more interested in being able to communicate in Chinese, getting value for money but also prestige upon return home from their trip. For the younger, more affluent Chinese visi-tors to Hawaii information about the business environment but also Hawaii as a romantic destination were more important. Significantly they com-plained that they did nothing different from their last trip to the Hawaiian Island not only because of the lack of alternative package tours but also because of the absence of casinos.

From the HTA study a clear differentiation is possible. On the one hand are the savvy younger English-speaking Chinese travellers who are 'selecting destinations on their attractions as a destination rather than the fact that they are allowed to travel to that destination' (Bailey 2004: 5) and are not satisfied with run-of-the-mill package tour offers. On the other hand exist the more traditional first-time visitors who still can't believe their luck that the political changes in China and the money of their company or their children brought them to Hawaii, giving them 'bragging rights' after their return.

The Japanese government promoted outbound tourism to the USA in the 1970s and 1980s to show the willingness to decrease the American merchandise trade deficit of that time through tourism spending by Japanese tourists in the USA. In the early years of the twenty-first century, Japan-bashing has been replaced by China-bashing in Washington based on the merchandise trade deficit with China, which since 2002 exceeds 100 billion US$ annually and is for 2005 approaching 200 billion US$ (Economist 2005a). With Chinese visitors currently spending about one billion US$ each year in the USA,[72] a wider opening of the door in the form of a working ADS agreement or some other form would help at least partially to ease the deficit. The pent-up demand would be certainly there.

For Canada, the possibility to combine both North American countries in one trip would also give a boost to its incoming numbers from China. For now, CTC is concentrating its marketing efforts on business and frequent travellers and on young, western-influenced Chinese between 20 and 40 years of age, who hopefully respond to individual trips and niche offers such as ecotourism and winter sports. With the help of ADS, TourismVancouver optimistically forecasts 500,000 Chinese visitors to its city by 2015 (TourismVancouver 2005).

Europe

Many Chinese visitors will select the EU as their destination in the coming years for Europe's rich cultural heritage and very different lifestyles and customs from Asian countries, and there are dozens of convenient non-stop flights between China and the EU.

(China Daily 2004e)

Liu Wuxiong, managing director of the CITS Outbound Department named some of the reasons for Chinese outbound tourists visiting the EU after the Final MOU for an almost EU-wide ADS agreement was signed in February 2004 in Beijing.

However, Europe has been the most important destination outside Asia for Chinese outbound travel before that. The share of Europe without Russia has been stable at 10–12 per cent of all travellers going beyond Hong Kong and Macao since 1995. In this year the number of Chinese travellers reached 600,000. In 2000 the figure went up to 1 million for the first time, representing 10 per cent of all Chinese crossing the border or 22 per cent of all travellers going beyond Hong Kong and Macao. In 2004 the number had risen to 1.8 million, representing 6 per cent of all Chinese crossing the border but still 22 per cent of all travellers going beyond Hong Kong and Macao. Even when travellers to Russia are disregarded, the Americas were 50,000 visitors ahead compared to Europe in 2000, but by 2004 Europe could welcome 300,000 visitors in excess of the Americas, representing 11.9 per cent of all travellers going beyond Hong Kong and Macao compared to 5.3 per cent for the Americas.

Europe is the most important trading partner for China in the world (Hu, Watkins 1999). Politically, Europe is an important element, a counter-weight, in the foreign policy of China, which is supporting a multi-polar world as opposed to the dominance of the USA. Before 1960, the People's Republic had cooperative relations with the Soviet Union and Eastern Europe. Diplomatic relations with Scandinavian countries and the United Kingdom started in the 1950s, with France following in 1964 and most other European countries in the early 1970s. The European Economic Community, the predecessor to the EU, officially established ties with China in 1975, fashioning a defined China strategy since 1994.

Table 6.23 Europe's share of China's outbound tourism 1995–2004[73]

Year	Country/region	Total outbound	Total outbound without Hong Kong/Macao	Outbound to Europe	Percentage of total outbound	Percentage of total outbound without Hong Kong/Macao
1995	Europe incl. Russia	4,972,391	2,181,619	608,535	12.2	27.9
1995	Europe without Russia			218,065	4.4	10.0
1996	Europe incl. Russia	5,598,986	2,747,883	598,968	10.7	21.8
1996	Europe without Russia			249,519	5.0	11.4
1999	Europe incl. Russia	9,232,365	4,110,365	823,553	8.9	20.0
1999	Europe without Russia			385,813	4.2	9.4
2000	Europe incl. Russia	10,649,455	4,862,843	1,079,089	10.1	22.2
2000	Europe without Russia			472,987	9.5	9.7
2001	Europe incl. Russia	12,133,097	5,012,219	1,177,110	9.7	23.5
2001	Europe without Russia			570,129	4.7	11.4
2002	Europe incl. Russia	16,602,347	6,048,218	1,397,661	8.4	23.1
2002	Europe without Russia			706,533	4.3	11.7
2003	Europe incl. Russia	20,221,939	6,121,247	1,351,109	6.7	22.1
2003	Europe without Russia			689,878	3.4	11.3
2004	Europe incl. Russia	28,852,850	8,360,724	1,807,375	6.3	21.6
2004	Europe without Russia			997,769	3.5	11.9

The total stock of European FDI in China amounts to more than 35 billion Euros, the total trade volume to 135 billion Euros.[74] About 100,000 Chinese students are enrolled in European universities, half of them in the United Kingdom. 'The breadth and depth of Europe-China relations are impressive, and the global importance of the relationship ranks it as an emerging axis in world affairs' (Shambaugh 2004: 243).

All these intense commercial, political and social relations, including the experiences until very recently with European rule in Macao and Hong Kong, have supported interest in and knowledge about Europe in China. The perception of Europe as the major source of western civilization including culture, philosophy and technology enhances the attention directed towards Europe. 'Europa fascinates most Chinese because of its long history, magnificient buildings, its many countries with different cultures and its beautiful nature' (Sun 2004: 27). Europe, especially Western Europe, is quoted as the most coveted destination in several studies (Wang 2003a; Australia Tourism Commission 2004; Guo *et al.* 2005b).

For leisure tourism, these facts translate into large numbers of business-induced travel opportunities especially to Western Europe in the 1980s and 1990s, well-known attractions as material evidence of the cultural heritage and the opportunity to visit many different cultures within a short time because of the close proximity and easy accessibility of the European countries.

As Duan Jidong, CTS Outbound Department marketing manager, is quoted as saying:

> Europe is not strange to Chinese people since we have a lot of business contacts. And personally I feel that the main body of Chinese tourists seem to be much more interested in countries that are more developed than China.
>
> (Wen 2004)

In speaking of *Europe* as destination, this term has to be differentiated in several ways. Firstly, Russia is treated as a part of Europe in the tourism statistics, being the most important destination by far. In reality, the vast majority of these visits to *Europe* are border-crossings from Manchuria and Inner Mongolia into the Russian Far East and Siberia, taking place thousands of kilometres away from Europe. Eliminating Russia from the statistic deflates the figures for visitors to Europe for 2004 from 1.8 to 1 million travellers.

Second, *Europe* is dominated by the EU. As a result of the enlargement in 2004, the EU now comprises of 25 countries[75] with more than 450 million inhabitants. By 2005 all EU countries had acquired ADS.

Third, a number of European countries have entered the so-called Schengen Agreement, which started in 1995 and abolished internal border controls. For Chinese tourists this has the positive effect that a visa issued by one of the Schengen countries automatically entitles to visits to all of them. As of 2005, most western and southern EU countries are part of it,[76] with the new EU members in Eastern Europe expected to join soon.

ADS agreements in Europe were first concluded with Malta in 2001 and Germany in 2002, becoming effective in 2003. CNTA was interested in an ADS agreement with the EU or, rather, with the Schengen area, in 2001. However, since a European initiative did not materialize, CNTA changed strategy and decided to negotiate on a bilateral basis first (EUCCC 2001). The awkward situation with Germany as a Schengen country issuing tourist visas only for Germany under the ADS agreement speeded up the process of granting ADS to almost all relevant tourism destinations in Europe as a group, several of whom had been in intensive discussions about the individual granting of ADS for some years already.

The agreement was initially started during the sixth EU–China Summit in October 2003, the MOU was signed in February 2004 and the ADS became effective in September 2004. Holders of ADS tourist visas are allowed a maximum stay of 30 days, and each tour group must have a minimum of five people (Tiplady 2004; European Commission 2004). The three EU countries outside this agreement concluded their own agreements in 2005, as did Norway and Iceland, which had been mentioned in the EU MOU by virtue of their inclusion in the Schengen Agreement. Denmark signed the MOU in February 2005 (People's Daily 2005c), Ireland in May 2005 and Norway in June 2005 (Norway 2005), sending the first groups within a few months. The first ADS groups to Britain arrived in July 2005. Switzerland and Croatia also became ADS destinations in 2004 and 2005.

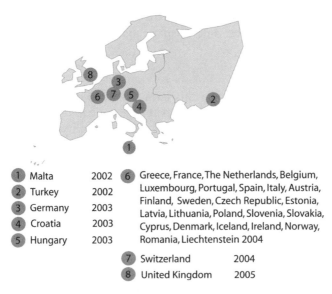

1	Malta	2002	6	Greece, France, The Netherlands, Belgium, Luxembourg, Portugal, Spain, Italy, Austria, Finland, Sweden, Czech Republic, Estonia, Latvia, Lithuania, Poland, Slovenia, Slovakia, Cyprus, Denmark, Iceland, Ireland, Norway, Romania, Liechtenstein 2004	
2	Turkey	2002			
3	Germany	2003			
4	Croatia	2003			
5	Hungary	2003			
			7	Switzerland	2004
			8	United Kingdom	2005

Figure 6.4 Map showing ADS countries in Europe[77]

Russia had already signed the MOU for ADS in 2002 but chose another way of offering visa-free travel instead of ADS groups. Since August 2005, selected travel agencies in China have the right to arrange package tours for groups of at least five persons, which can enter Russia without visas (CEN 2005b). The remaining European countries without ADS, situated in the Balkans and Eastern Europe, are not in high demand at the moment with Chinese tourists.

Before looking in some detail at Germany as the country most visited by Chinese in Europe – except Russia – and the Northern European countries as future repeat-visit destinations, information about the destination Europe is provided here.

Most destinations are visited by Chinese leisure tourists as single countries because of their size or geographical isolation or as a combination of just two countries. This is the case for the USA, sometimes in tandem with Canada, for Australia, sometimes in tandem with New Zealand, for Japan and for South Korea. Egypt's or South Africa's African neighbours are perceived to be of less interest for Chinese visitors. For Hong Kong, Macao and also Thailand the brevity of the visit and/or the low price prevent combinations with other countries. However, in opposition to this, Europe is seen as *one* destination that offers the opportunity to visit many different countries in close proximity but with varied cultures and attractions and to visit several times for different parts and aspects. The existence of the EU and its representations and offices in China, the common ADS MOU, the Schengen Agreement and the existence of a common currency in most major European destinations support the perception of *one* Europe within the tourism industry. The European Union Chamber of Commerce in China (EUCCC) has a Tourism Industry Working Group, with airlines, hotels and incoming companies as members, the European Commission and CNTA started in 2005 a joint ADS committee (ETOA 2005).

The first NTO in China was opened by Switzerland in 1998 (Droz 2004), Germany, France, Austria, Finland, the Scandinavian countries and other European countries followed with representative offices mostly located in Beijing or Shanghai. Big cities such as Vienna or London have their own representative offices.

The Trans-Siberian Railway is of no importance anymore for passenger transport from China to Europe even for the European part of Russia. From the early air connections to Beijing from Moscow in the 1950s and from Paris since 1973 to about 100 weekly connections in 1998, China is today connected to Europe by several hundred flights every week, with several flights a day from major hubs such as Paris or Frankfurt and daily direct connections between more than a dozen European airports from Madrid to Moscow and Beijing, Shanghai or Guangzhou. Many relations are organized in code-sharing between European and Chinese airlines.

The development of the visitor numbers to Europe, according to Chinese outbound statistics, shows the relative decline of Russia but otherwise a rather stable relation between the shares of Germany, France, the United Kingdom and the other European countries since 1999. The fact that travel to Russia did not participate in the overall increase is another sign that these border-crossings are mainly happening in the Asian part of Russia and mainly not for leisure purposes. With the start of visa-free 'quasi-ADS' travels to Moscow and St Petersburg from autumn 2005, this may change in the future. The big caveat to the outbound statistics is, however, the fact that these figures only reflect the first port of call. On the one hand, some Southern Chinese travellers taking a plane from Hong Kong to Europe do not show up in the statistics as travellers to Europe but to Hong Kong. On the other hand, as 80 per cent (ETOA 2005) of all Chinese visitors to Europe travel to more than one, sometimes up to ten, countries, the figures are also influenced by the development of the air connections available and by the consulate favoured for the issuing of Schengen visa.[78] To check the figures against national inbound statistics is also not always successful as in *Schengenland* no border controls are executed, national statistics did not or still do not give separate figures for visitors from the People's Republic. The Swiss official publication 'Swiss Tourism in Figures' for 2004 (STV 2004) still showed for Asia only figures for Japan, India and Israel, with Chinese visitors lumped into 'Other Asia'. In Austria, China also was a part of 'Other Asia' in official statistics until April 2003 (Wien Tourismus 2004). Germany and some other countries' statistical services mix Mainland and Hong Kong Chinese. Using accommodation statistics leaves out the VFR traveller section, which is an important segment in some European countries. As a result the city tourism board of Paris reports 400,000 Chinese visitors, double the numbers of Chinese travelling according to the Chinese statistics to the whole of France in 2004. Britain reports 65,000, 68,000 and 95,000 visitors for 2002 to 2004 (Visit Britain 2003; Visit London 2005; Yuk 2005) while the Chinese figures are almost double that numbers. In this case the explanations connected to Schengen do not even apply as the United Kingdom is not part of the Schengen Agreement. The ETC figure for 2000 of 323,475 Chinese visitors to Italy (WTO 2003) is widely quoted but obviously not realistic.

The seasonality of business travels to Europe is more structured by the European holiday seasons in summer and during Christmas time than by any climatic influences. For leisure travels summer is the busiest season for climatic reasons, with spring and autumn more relevant for Southern Europe. The 'Golden Week' peak seasons are less relevant for Europe as many, especially first-time, visitors want to see a number of countries and therefore stay longer than one week.

The destination of Europe is perceived by Chinese visitors rather as a network of cities than a geographic puzzle of countries. The first and foremost destination is Paris, coming out on top of every survey about the

Table 6.24 Visitors to European countries 1995–2004[79]

	1995	1996	1999	2000	2001	2002	2003	2004
1 Europe	608,535	598,968	823,553	1,079,089	1,177,110	1,397,661	1,351,109	1,807,375
2 Russia	390,470	349,449	437,740	606,102	606,981	691,128	661,231	809,606
Russia %	64.2	58.3	53.2	56.2	51.6	49.4	48.9	44.8
3 Europe without Russia (1) – (2)	218,065	249,519	385,813	472,987	570,129	706,533	689,878	997,769
Germany	117,069	132,950	93,617	11,2824	138,371	165,687	165,168	222,878
Germany % of (3)	53.7	53.3	24.3	23.9	24.3	23.5	23.9	22.3
France			87,826	96,485	114,435	136,692	135,407	210,533
France % of (3)			22.8	20.4	20.1	19.3	19.6	21.1
UK	31,000	31,000	42,807	61,129	88,440	128,000	134,088	177,601
UK % of (3)	14.2	12.4	11.1	12.9	15.5	18.1	19.4	17.8
Other Europe	69,996	85,569	161,563	202,549	228,883	276,154	255,215	386,757
Other Europe % of (3)	32.1	34.3	41.9	42.8	40.1	39.1	37.0	38.8
of which:								
Belgium	17,415	30,866						
Belgium % of (3)	8.0	12.4						
Spain	14,477	15,010						
Spain % of (3)	6.6	6.0						
Malta		320	669	882	995	1,446		
Malta % of (3)		0.0	0.0	0.0	0.0	0.1		
EU			260,000	312,000		645,000		900,000
EU % of (3)			67.4	66.0		91.3		90.2

most coveted city in Europe, followed by Rome, Venice and Vienna (Vienna Tourist Board 2005). 'I am looking forward to having romantic pictures taken in front of the Eiffel tower' (China Daily 2004e) is the way Chinese media quoted a potential traveller to Europe to cement the ADS agreement with the EU. The capital city is in most countries also the most important destination for Chinese leisure visitors, with other cities chosen sometimes for their easy accessibility along the coach routes, the attraction of factory outlets or the opportunity to sample yet another country, such as the Vatican or Luxembourg.[80]

The general image of Europe as the picturesque continent full of culture, art, old cities, architecture and inhabitants with strange but interesting customs, the place of heritage and shopping opportunities as 'head brand' as supported by clichés and typecasts as 'supporting brands' for different countries. Images include Germany as the home of scientists, artists and engineers, orderly and safe; France with Paris as the romantic place of art and sophisticated culture; Italy as the sunny country with monuments and Roman art; Greece as the place well-known from the Olympic Games; Austria as home of Queen Sissy and famous composers; Switzerland as beautiful and rich, etc. Specific cities have also acquired clearly distinguishable functions within the destination of Europe. Paris as the 'romantic' place for Eiffel tower and Crazy Horse shows; Amsterdam for buying high-value low-volume diamonds and for watching women in shop windows in the red-light district; Nice and Monaco for the millionaire-for-a-day feeling; Rome to show interest in culture; Munich to drink beer; Berlin to look at the remains of the wall. In a survey of Chinese tourism students about their image of Europe (Schwandner, Gu 2004) the difference from China was stressed and Europe envisaged as being romantic, small enough to allow travel to several countries in a short time, safer than the USA, but less modern.

The expectations of finding modernity in Europe of the early visitors from China have mainly been substituted by the anticipation of a visit to a big open-air museum with shops attached. This is not a sign of a lack of maturity, as some tourism planners think, but based on the comparison of earlier travellers of European cities with places such as Beijing or Shanghai: 'As a comparatively immature long haul travel market, perceptions of the UK are very old fashioned. Famous sights and the Royal Family may be recognised, but London is not associated heavily with creativity or modernity' (Visit London 2005: 4).

The reasons for a visit to Europe vary with the different kind of travellers. WTO (2003) distinguishes between intellectuals, who are interested to see the original places of European history and art they know a lot about, affluent travellers who have been to Asian and Oceanic destinations already, mass tourists who realize a lifelong dream by travelling to Europe, and young white-collar workers, who are influenced by technology and fashion. To these groups can be added the business travellers, who have been to Europe before as a member of a government or company group and now want to travel with their family or friends, parents looking for the right university for their offspring or visiting them there, especially in the United Kingdom, and sponsored travellers, who get part or full funding of their trip to Europe as a gratification from their state-owned company, their ministry or the Communist Party.

The sociodemographic profile of Chinese travellers to Europe shows that two-thirds of all visitors are male, with almost half of all visitors in the age group of 25 to 34 and therefore younger than the average Chinese outbound traveller. High income and education levels reflect the higher costs of travelling to Europe.

Unlike Australia, New Zealand or Japan, ADS agreements for European countries do not put restrictions on the place of abode within China for potential visitors. The relatively high cost of travelling to Europe does, however, support the focus on the high end of the source market China, favouring the inhabitants of the big cities and the coastal area. The repeat visitor rates differ within Europe, connected among other things to the level of commercial relations between China and the country. In 2005, most countries estimated that about 20 per cent of their visitors came at least for the second time, a rate bound to increase over time.

Chinese visitors to Europe are high-spenders. For tax-free shopping in Germany, Chinese customers are the second-most important group of customers behind visitors from Russia (Global Refund 2004). In some big department stores, such as the Galeries Lafayettes in Paris, Chinese customers spend more than any other group of foreign customers. Even though most Chinese tourists stay in hotels, spending is rather directed towards souvenirs, gifts and high-quality branded consumer goods than

Table 6.25 Sociodemographic profile of Chinese travellers to Europe (in per cent)[81]

Sex	Male	67
	Female	33
Age	15–24	12
	25–34	46
	35–44	19
	45–54	15
	55+	8
Family income	Low/middle	4
	High	32
	Very high	64
Education level	Low	1
	Middle	12
	High	87

towards accommodation. At airports, this might even include Chinese premium cigarettes such as Zhonghua, which are less expensive in European
duty-free shops than in China (Min 2005). For some visitors, shopping for
reselling is a way of financing the travel.

The main activities of Chinese tourists differ considerably from those
of European tourists as beach holidays play almost no role for Chinese
visitors to Europe. The typical activity for first-time visitors is a coach tour
through many different European countries, staying nearly every night in
another city or in fact another country. 'The Chinese outbound tourism to
Europe is pure city tourism. In many cases, the long-haul flight and the
strenuous bus tour from one destination to the next leaves passengers after
a few days tired, bored and irritated' (Roth 1998: 17).

Within the cities the main attractions are visited in hurried sightseeing,
taking photos often taking precedence over detailed examination, and shopping tours. 'At the present time most of the Chinese tourists do not have
much interest to see antiques with a completely different cultural background' (Huang, Fang 2003). The 'beautiful landscapes' often cited as a
major attraction of Europe are mostly seen rushing by alongside highways.
Repeat visitors may spend more time in one country or at one destination.
Festivals such as the Oktoberfest Beer Festival in Munich or Christmas
Marts are also fitting into the expectations of Chinese visitors to Europe.
Themed tours are only starting to become part of the market, with the self-
driven luxury cars racing along the German *Autobahnen* a much-publicized
example. Chinese tour operators promote themed seven- to ten-day trips
to Europe such as a European Football Tour, or an Ancient Castle Tour.

Typical itineraries accordingly consist mainly of coach tours, with an
intra-European flight sometimes added to connect for instance Paris
and Rome. Most tours stay within *Schengenland* to avoid the problem
of acquiring a second visa. This fact puts non-Schengen countries such as
the United Kingdom and – until 2005 – Switzerland at a disadvantage.
First-visitors' Central European tours would often follow a route such as
Frankfurt–Munich–Berlin–Hamburg–Amsterdam–Brussels–Luxembourg–
Paris–Cologne–Frankfurt for ten-day tours with the addition of Nice–
Venice–Florence–Roma or of Salzburg and Vienna for longer tours. A long
Southern European tour offered for 16-day travels to Frankfurt–Berlin–
Hamburg–Hanover–Amsterdam–Brussels–Luxembourg–Paris–Nice–
Monaco–Barcelona–Madrid and Lisbon (Weyhreter, Yang 2005). Many
Chinese visitors find out during such whirlwind trips, that Europe is actually
bigger than it looks on a Chinese map of the world and complain about
the many hours spent in the bus.

The fierce competition on the Chinese outbound market has driven the
prices for travels to Europe down to a level bemoaned by the European
Tour Operators Association, which cites payments dropping from 70 Euros
to 40 Euros only per person per day for the land operators (ETOA 2005).
In China, this enables travel companies to offer retail prices as low as

9,999 Yuan RMB (approximately 1,000 Euros in 2005) for five-day trips. Average tours of 10 to 15 days and themed tours cost between 15,000 to 20,000 Yuan RMB (approximately 1,500 to 2,000 Euros in 2005) (China Daily 2004b; Tiplady 2004). Such prices leave almost no room for any profit margin and invite activities to earn extra money from special agreements with shops etc. Zero-Dollar groups have not developed in Europe but systems of kick-back payments, meals paid by shops and other measures, not common in European tourism and illegal in many countries, present a threat to the quality of the tourism product.

In mid-2005 the euphoria connected to the widening of the ADS system to almost all countries in Europe lessened with the realization that the increase in the number of visitors could not keep up with the level of the post-SARS and new-ADS year 2004. The problems of 'overstaying' Chinese tourists disappearing into the overseas Chinese communities in large numbers[82] resulted in changes in the visa-granting procedures towards stricter controls and random check interviews at the consulate, contradicting the basic idea of ADS. Chinese tourists as a result felt discriminated against by being seen as potential criminals[83] and tour operators threatened to stop offering tours to EU countries.

The visa problems will be solved and Europe will stay a favoured destination for Chinese tourists. With the increased competition within Europe and with the rest of the world, improvements in service quality and product adaptation will be necessary for the European destinations.

Germany

Germany, the country whose inhabitants themselves spent the highest amount of all nations for international travelling in 2004, has for many years been the most important destination in Europe for Chinese visitors. From half of all travels in the 1990s to Europe (without Russia), the share has gone down to about a quarter in the early twenty-first century, but is still still ahead of France and Britain, the other main gateways to Europe for Chinese travellers. Germany is the most important European investor in China and the most important trading partner for China in Europe. It is not surprising, therefore, that about half of all visits to Germany are business-induced.

The DZT was among the first NTOs to open a successful representative office in Beijing at the beginning of 2001. Since then, several cities, provinces and airports located in Germany have also appointed representatives in China to promote their services. Following a visit by Chancellor Schröder, Chinese media reported at the end of 2001 that Germany had been included in the list of ADS countries. The MOU was signed in July 2002 and the first Chinese ADS groups to Germany arrived in February 2003 (Finck 2004) to large and enthusiastic media coverage, as the German inbound travel industry was looking forward to very good business as the only Western European ADS destination so far, only to be hit especially hard by the SARS crisis, which reduced the number of leisure travellers in 2003 considerably.

China is connected to Germany with daily direct air connections from Beijing and Shanghai and less frequent direct flights from Guangzhou and some other airports mainly to Frankfurt and Munich. As most visitors do not only travel to Germany, connections via other European airports and following land transport are also utilized. The national carriers Lufthansa and Air China have a long-standing cooperation including maintenance arrangements and are both members of Star Alliance.

Following German national statistics, which do not distinguish between Mainland Chinese and Hong Kong Chinese arrivals, the number of Chinese visitors to Germany cleared the 100,000 hurdle for the first time in 1993. Until 2004 this figure increased almost fourfold, somewhat less than the overall fivefold increase of Chinese travelling beyond Hong Kong and Macao. The average length of stay continuously decreased from 2.9 to 2.1 nights per stay.

The seasonality of arrivals of Chinese visitors to Germany is rather strong with 30 per cent of all overnight stays in the three months between August and October and 20 per cent during December to February. For leisure visitors the seasonality is even more pronounced than for business visitors.

Chinese visitors to Germany do not visit the whole country but follow well-trodden paths. Eighty-two per cent of all overnight stays are concentrated on 5 of the 16 German provinces, all in the western and southern parts. Ten per cent are covered by the city provinces of Berlin and Hamburg, leaving the southwest and the north and east of Germany as *terra incognita* for Chinese visitors.

However, the accommodation statistics prove that in fact it is not even certain provinces that are favoured by the Chinese visitors. Tourism to Germany is almost exclusively city tourism, so almost 40 per cent of all overnight stays in 2004 took place in the six biggest cities organized in the marketing association 'Magic Cities'. At the same time they show that half of all touristic regions in Germany had less than 500 overnight stays of Mainland or Hong Kong Chinese within the year. These figures support the results of a survey conducted in 2002 in China. Among the German cities Berlin, Hamburg, Frankfurt and Munich were known by more than 75 per cent of all surveyed Chinese, but other cities did not reach scores higher than 30 per cent (BAT 2003).

Germany is mainly known as a destination for business or MICE (Davidson, Hertrich, Schwandner 2004) travel, as home of famous inventors such as Siemens or Benz and products of good quality from cars to machines and beer. Artists such as Beethoven or Bach, romantic landscapes and medieval city centres are also a part of the image of Germany for most Chinese visitors. Safety and cleanliness, important attributes for Chinese outbound tourists, are attributed to a high degree to Germany (Xu, S. 2005;

Table 6.26 Arrivals and overnight stays in Germany by Mainland and Hong Kong Chinese 1994–2004[84]

	Arrivals (persons)	Overnight stays[a] (nights)	Average length of stay (days)
1994	100,396	288,649	2.9
1995	115,460	333,266	2.9
1996	130,764	336,044	2.6
1997	141,878	362,101	2.6
1998	161,454	388,380	2.4
1999	177,467	397,309	2.2
2000	213,897	467,654	2.2
2001	236,443	512,866	2.2
2002	270,308	572,594	2.1
2003	267,803	577,646	2.2
2004	387,166	794,021	2.1

a Only commercial accomodation with more than eight beds are included.

Table 6.27 Percentage of overnight stays in Germany by Mainland and Hong Kong Chinese in 2001 and 2002 per month according to share of visitors[85]

2001		2002	
Percentage of overnight stays	Month	Percentage of overnight stays	Month
10.5	September	10.6	September
9.6	August	9.2	October
9.5	October	9.0	August
8.7	July	8.9	March
8.4	April	8.2	July
8.0	March	7.7	February
7.9	June	7.6	June
7.4	November	7.5	May
7.4	January	7.2	April
7.3	May	6.5	January
6.9	February	6.5	November
6.5	December	6.0	December

Arlt 2005a). Specific images connected to Germany are also related to such diverse figures as Karl Marx, the Grimm brothers, and – unfortunately – Adolf Hitler (Huang, Fang 2003). The number of Chinese living in Germany is recorded as about 50,000 Chinese from Mainland China and Taiwan plus another 50,000–100,000 ethnic Chinese from Hong Kong and Southeast Asia. The vast majority of them live in the western part of the country,[86] running among other things around 15,000 Chinese restaurants and takeaways (Leung 2002).

Germany is a major economic power, an important commercial partner of China and the home of some of the biggest fairs and exhibitions in the

Table 6.28 Percentage of overnight stays in Germany by Mainland and Hong Kong Chinese in 2001–2004 per month according to provinces[87]

Province	2001 Overnight stays	2001 Market share (%)	2002 Overnight stays	2002 Market share (%)	2003 Overnight stays	2003 Market share (%)	2004 Overnight stays	2004 Market share (%)
Bavaria	114,545	20.01	113,636	22.16	115,090	19.92	154,734	19.63
North Rhine-Westphalia	98,038	17.13	81,495	15.89	106,445	18.42	151,398	19.21
Hesse	121,435	21.22	104,685	20.42	106,747	18.48	151,347	19.20
Baden Wurttemberg	60,505	10.57	54,855	10.70	68,390	11.84	113,871	14.45
Rhineland Palatinate	45,108	7.88	41,270	8.05	59,089	10.23	79,732	9.35
Berlin	49,489	8.65	42,590	8.31	47,236	8.18	56,082	7.11
Hamburg	23,867	4.17	21,560	4.21	22,852	3.96	27,346	3.47
Lower Saxony	21,021	3.67	18,955	3.70	20,715	3.59	23,106	2.93
Saxony	8,753	1.53	7,774	1.52	10,005	1.73	9,972	1.27
Bremen	3,684	0.64	3,818	0.74	3,492	0.60	6,329	0.80
Saarland	5,273	0.92	8,730	1.70	3,785	0.66	4,547	0.58
Brandenburg	7,538	1.32	3,280	0.64	4,921	0.85	3,926	0.50
Schleswig Holstein	3,199	0.56	2,737	0.53	3,445	0.60	3,588	0.49
Thuringia	2,997	0.52	2,311	0.45	2,630	0.46	3,867	0.49
Saxony Anhalt	5,230	0.91	3,986	0.78	1,992	0.34	2,412	0.31
Mecklenburg Western Pomerania	1,676	0.29	1,039	0.20	953	0.16	1,764	0.22

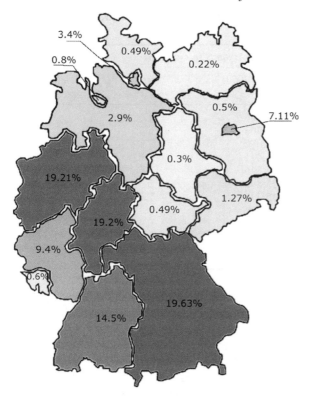

Figure 6.5 Percentage of overnight stays in Germany by Mainland and Hong
      Kong Chinese in 2004 by provinces (in per cent)[88]

world. Together with an underdeveloped image as leisure and fun destination this led to an unusually high percentage of business visitors from
China, which is responsible for 47 per cent of all arrivals compared to 22
per cent for all Chinese outbound travels. Leisure-orientated travellers
make up only 17 per cent of all visitors, opposed to 55 per cent as average.
The rest is reported as 'Other private visits' (Xu, S. 2005), including for
instance visits of relatives to the 10,000 Chinese students enrolled at
German universities. Visitors to fairs and exhibitions from China amounted
in 2002 to approximately 17,000 persons (Finck 2004). Many of the fair
visitors are of course also embarked on leisure trips as a part of their
journey, but not necessarily within Germany. With ADS granted to all
Schengenland countries, the number of leisure tourists visiting Germany
within a multi-country trip will however probably increase, given the active
marketing of the DZT and the central position of Germany within the
enlarged EU.

    The socio-demographic profile of Chinese travellers to Europe shows
that two-thirds of all visitors are male, with almost half of all visitors in

Table 6.29 Overnight stays in Germany by Mainland and Hong Kong Chinese in 2004 in the 'Magic Cities'[89]

City	Overnight stays in Germany		
	Total (2004)	*Percent*	*Cumulated percent*
Frankfurt	90,796	11.5	
Munich	65,363	8.3	
Berlin	56,082	7.1	26.9
Cologne	36,441	4.6	
Hamburg	27,346	3.5	
Duesseldorf	25,896	3.3	38.2
Stuttgart	15,796	2.0	
Hanover	5,476	0.7	
Dresden	3,027	0.4	
Total 'Magic Cities'	326,223	41.3	
Total Germany	789,429		

the age group of 25 to 34 and therefore younger then the average Chinese outbound traveller. High income and education levels reflect the higher costs of travelling to Europe.

Slightly more then half of the Mainland Chinese visitors to Germany are reported to arrive from Shanghai (24 per cent), Beijing (16 per cent) and Guangzhou (12 per cent), with 48 per cent originating in other regions of China. A survey of group leisure travellers to Germany, as part of a Europe coach trip, found that 77 per cent of all Mainland Chinese tourists visited for the first time; only 3 per cent had been more than once before to Europe (Weyhreter, Yang 2005).

The total expenditure for a trip to Germany has been estimated as high as 2,900 Euros or 195 Euros per day (Xu, S. 2005). A large part of this

Table 6.30 Sociodemographic profile of Chinese travellers to Germany (in per cent)[90]

Sex	Male	66
	Female	34
Age	15–24	18
	25–34	39
	35–44	28
	45–54	11
	55+	3
Family income	Low/middle	6
	High	29
	Very high	65
Education level	Low	0
	Middle	10
	High	90

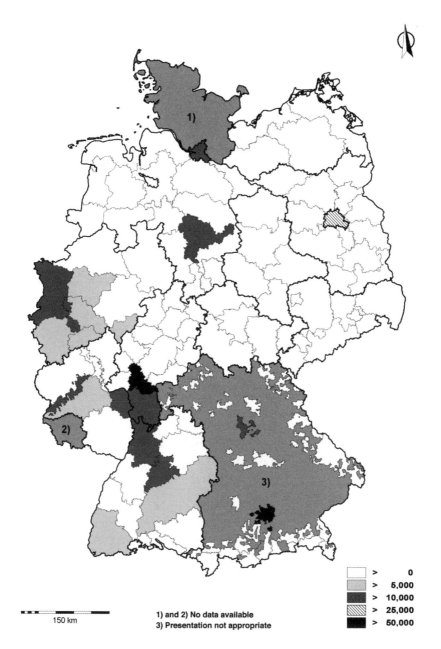

	>	0
	>	5,000
	>	10,000
	>	25,000
	>	50,000

150 km

1) and 2) No data available
3) Presentation not appropriate

Figure 6.6 Overnight stays in Germany by Mainland and Hong Kong Chinese in 2004 by tourist regions[91]

is not spent on transport, accommodation, food or entrance fees but on shopping. The turnover in shops in Germany that are members of the Global Refund tax-free structure surpassed 5 million Euros per month in 2004. Per year the turnover increased from 12 million Euros in 1999 to 45 million Euros in 2004. For other European destinations, Japanese and American as well as Russian customers are more important, but in Germany only Russian citizens, who buy goods of a value of 196 Euros per visit on average with apparel and watches and jewellery and 'authentic' items such as cuckoo clocks as the main items (Kelemen 2005), are still ahead of Chinese in tax-free spending.

The short time of the visit for leisure tourists and the importance of business travellers results in the dominance of hotels as the main form of accommodation for 90 per cent of the 87 per cent of all travellers who stay in commercial hospitality. The other 13 per cent make private arrangements with friends or relatives (Xu, S. 2005).

The method of travel of most Chinese leisure visitors to Germany influences their main activities. A considerable amount of time is spent on the highways sitting in a coach between big cities.[92] In the city centres the icons of Germany such as the Brandenburg Gate in Berlin or the cathedral in Cologne are visited and shops are raided for branded goods. For officials on the leisure part of their visit, the quote from the head of the German representative office in Beijing is still true that it is 'an absolute Must is the visit of a red-light district, the birthplace of Karl Marx, and a casino and music program in the evening' (Xu 2002). Some outdoor destinations have succeeded in becoming part of the trail, such as the Titisee in the Black Forest, and the Munich Oktoberfest, which with 6 million visitors each year is considered to be the biggest festival in the world, and also Christmas Markets, especially the one in Nuremberg. Other activities that are important elements of visits to Germany for Japanese tourists such as the visit to Neuschwanstein Castle, a cruise on the Rhine River and visits to small cities such as Heidelberg and Ruedesheim, are only slowly entering Chinese itineraries as well. A special visit leads many Chinese visitors to the small city of Metzingen, which is completely unknown to the majority of Germans. Starting with an outlet for Boss suits, this small city in the northern part of the Black Forest is now offering about 20 factory outlets for branded goods that produce a higher turnover of tax-free purchases than all tax-free shops in Berlin taken together. A combination of two other specialities of Germany, the production of famous limousines and the absence of speed limits on the highways, has produced an unusual thematic tour. For about 2,000 Euros for a one-week trip, Chinese visitors can drive a Mercedes-Benz or BMW, daringly speeding up and down the *Autobahn* at up to 250 km/h (Sussebach 2004; Klawitter 2004).

Weyhreter and Yang (2005) found in their survey that the five items seen as most important for the coach tourists were to see how other people live, see beautiful landscapes and new places, as well as the friendliness and knowledge of the guide accompanying the tours. Visiting casinos and nightclubs was declared to be the most unimportant item. For the important items, a high level of satisfaction was reported. However, the level of satisfaction was much lower for the time available for free use as a result of the packed itinerary, especially the average 400 km travelled every day. The quality of the Chinese food, the hotel and the coach were also criticized. Travellers who had already been to Europe before were less satisfied than those travelling for the first time. Similar criticisms have also been voiced in several articles in the Chinese press (China Business Weekly 2004).

German shops, restaurants and hotels are not used to and are uncomfortable working with *unofficial* incentives and payments as an important part of their business. However, the pressure to reduce the prices of leisure tours to Europe has in Germany also been translated into a growing pressure for inbound operators to supplement their direct income from the tour operator with other sources and to diminish costs. Out-of-town hotels, badly serviced coaches and little choice at lunch or dinner are sometimes the result, particularly annoying the more sophisticated and/or self-paying visitors.

'Germany has made good use of the one year advantage in the implementation of the ADS agreement', argues Horst Lommatzsch, the 'Mr China' of the DZT (Ebner 2004). The advantage of being the most important destination for business travellers in Europe, who could use a smaller or bigger part of their publicly paid travel for leisure purposes, is disappearing with the growing percentage of self-paid leisure tours under the ADS system, where no excuses are needed anymore. The advantage of starting ADS – after the false start because of SARS in spring 2003 – one year earlier then the other European countries resulted in a higher awareness of Germany as a leisure destination. To attract repeat visitors, further efforts and quality drives will be needed.

Northern Europe

With the first leisure tourism visit to Europe in most cases having Paris, Rome or Munich in focus and few overseas Chinese living in the Scandinavian countries or Finland, *Northern Europe* is rather a destination for repeat visitors.

The Chinese term for Northern Europe, *Bei Ou*, includes the three Scandinavian countries Denmark, Norway and Sweden as well as Finland and sometimes also Iceland. The political relations between Northern Europe and China have been traditionally very good. The Scandinavian countries already recognized the People's Republic of China in 1950, Sweden and Denmark being the first two western countries to establish diplomatic ties with China.

All five countries are members of the Schengen Agreement, so one visa is enough to visit all of them or the other *Schengenland* countries. However, Norway and Iceland are not members of the EU. Whereas the other countries received ADS together with most European countries in September 2004, Denmark opted for a special agreement. In tourism, Sweden, Norway and Denmark are represented by the Scandinavian Tourist Board (STB), founded in 1993 by the tourism authorities of the three countries and representated with an office in Beijing since 2004. Finland is represented by its own office. With ADS, the possibilities to organize legitimate leisure tourism has improved the reliability of travel arrangements.[93]

Accessibility is one of the strengths of the destination. Helsinki is only 6,300 km or less than eight hours flight-time away from Beijing, the shortest distance to any major EU airport. Together with Copenhagen, several daily flights are offered, often for transit passengers to other parts of Europe.

The number of Chinese travellers increased sharply – albeit from a very small base – around the turn of the century. The increase slowed down, partly because of SARS, from 2003. Exact figures are not easy to find, even the Norwegian Embassy in China has to rely on the guestimate of the STB of 60–80,000 visitors to Scandinavia in 2004 (Norway 2004).

The seasonality of arrivals is quite pronounced.[94] In Finland 60–70 per cent of all visits took place between May and October in the years 1999 to 2004 (MEK 2005); the number of overnight stays of Chinese visitors to Denmark in May 2005 was three times higher than in February 2005 (Statistikbanken 2005).

In Finland, 83 per cent of all Chinese visitors visit only Helsinki and its surroundings (Finnish Tourist Board 2005). Accordingly two-thirds of all overnight stays, 51,000 in 2003 and 58,000 in 2004, took place in the capital (Artman 2005), compared to only a few thousand in other overall important tourism regions such as Lapland or the Lake District.

In Scandinavia, the same concentration on the capitals of the three countries is very pronounced. In Sweden two-thirds of all Chinese visitors have

Table 6.31 Chinese outbound tourism to Northern Europe (in '000 arrivals)[95]

	1998	1999	2000	2001	2002	2003	2004
Denmark							
Norway							
Sweden				65	55		
Scandinavia total	27[a]		5–6[c]	24[a] 7–8[b]	27[a]		60– 80[c]
Finland	12	14	15			54	71
Finland (incl. Hong Kong)		7	12	25	68	53	68
Iceland	0.3	0.4	0.6				

a   Number of visas issued by Scandinavian consulates in China (STB 2004).
b   Estimate STB 2002.
c   Estimate of land operators (STB 2004).

Stockholm as their main destination and only one-third have destinations in the rest of the country, just opposite to the average figures for foreign visitors to Sweden (STA 2005).

The image of Northern Europe is not very well developed with Chinese potential visitors. A study of the STB in 2002 found that

> quite a few of our interview persons said that they did not know very much about the Scandinavian countries and that this was the major reason for not being interested in going there. Some persons did not even answer which country they would prefer to go to, because they knew too little to decide between the three countries.
>
> (STB 2002: 41)

However, most Chinese connect the perception of the Scandinavian countries as having a relatively high price level and high standard of living and, therefore, they can be a prestigious destination to which to travel. Other features connected with Northern Europe were beautiful scenery, safety, peacefulness, high levels of freedom, social welfare and economic development and history (STB 2004: 16). The last point is not connected to any in-depth knowledge though. The Vikings or indeed any other historical periods or points of interest were not mentioned by the people interviewed in the STB study. Rather better connected to the destination seems to be the Danish writer H.C. Andersen, who is well known to many Chinese as an author of fairy tales that have been translated into Chinese (STB 2002: 42).[96] Nightlife, excitement and ethnic heritage are, however, missing in the Northern Europe image (Blok 2002).

Many images of Scandinavia as an area with rather relaxed attitudes towards sexuality, or the imagery of reindeer-rearing, Santa Claus and aurora borealis, have not entered into the collective consciousness of China, as can also be seen from a guide book on Northern Europe. For it's target group, the younger, more independent travellers from China, a list of the 12 best things to do in Northern Europe includes: a Fjord boat trip, a visit to Icelandic volcanos, a visit to the Ice Hotel, a romantic trip on a big ferry, bathing in the hot-water Blue Lagoon in winter, a visit to a small fishing harbour town, a visit to Copenhagen inner city, the experience of the seafood festival in August, a visit to the Little Mermaid, visiting old city centres and the experience of modern Scandinavian design in museums and shops (Hun 2005). As a visual icon probably only the Little Mermaid will be recognized by a larger number of Chinese, even so those who actually travel to Copenhagen will find it smaller – and harder to be photographed in front of – than expected.

Given the underdeveloped image of Northern Europe and the small number of overseas Chinese and Chinese students living there, it is not surprising that only a quarter of all visitors to Northern Europe come as leisure tourists. For Sweden, two-thirds of all Chinese visitors cite business as the main reason to travel there and one-quarter leisure, with VFR, studies and other reasons only responsible for a few per cent of travel motivations (STA 2004).[97] For Finland, 39 per cent of all Chinese visitors in 2004 claim business as the main reason, 25 per cent are on a leisure trip, 24 per cent are just in transit, and only 2 per cent visit friends or relatives (Finnish Tourist Board 2005).

Supporting these figures are the average length of stay. Some 30 per cent of all Chinese entering Finland in 2004 did not stay overnight at all, 40 per cent spent only one or two nights, another 15 per cent three to five nights. Just 7 per cent of all visitors spent more than seven nights in the country (Finnish Tourist Board 2005). For Sweden the figure is higher with 30 per cent staying more then seven nights (STA 2005).

For logistical and probably also for image reasons Northern Europe is rather a destination for Northern Chinese. In 2003, around 42,000 passengers arrived at Finnish airports from Beijing, compared to 12,000 from Shanghai (Civil Aviation Administration Finland 2004). According to Swedish statistics, 75 per cent of all Chinese arrivals were men, no less then 92 per cent claimed to have an education level of college or higher. Seventy-nine per cent were between 25 and 44 years old, with 11 per cent in the 45 to 64 bracket (STA 2005). In Finland 65 per cent of all arrivals were male travellers, with the average age of men and women of 39 years and 36 years respectively.

Not much information is available about the number of repeat visitors or the levels of income of visitors to Northern Europe, especially those who travel mainly to this destination opposed to a stop-over break. The hoped-for customers are identified as 'a small but growing number of travellers, especially the young professionals, ... looking for relaxing, quality holidays in a single destination. Quality and an image of tranquillity fit well into the image of Scandinavia' (STB 2002: 30).

As in other destinations, Chinese visitors are spending above average. In Sweden, the average amount spent daily by Chinese is one-third higher than the average amount of all foreign visitors with 827 SEK compared to 623 SEK (approximately 88 and 67 Euros in 2005) (STA 2005). In

Table 6.32 Age and sex of Chinese residents visiting Finland in 2004 (in per cent)[98]

	Male	*Female*	*Total*
Under 15	3	5	4
15–24	5	14	8
25–34	29	34	31
35–44	31	27	30
45–54	22	10	18
55–64	8	7	7
65 and more	2	4	3

Finland the difference is even bigger despite the large number of short-stay visitors. With 86 Euros per day the Chinese are second in spending only to the Japanese visitors (98 Euros per day) but reach almost double the average foreigner who spends 47 Euros per day (Finnish Tourist Board 2005). Part of the reason for this is the fact that more than 80 per cent of all nights are spent in hotels, double the average figure for all visitors to Scandinavia.

The main activities of Chinese visitors to Scandinavia are sightseeing tours and shopping in the big cities. Only one out of seven Chinese visitors in Finland participates in summer or winter outdoor activities. In winter, 7 per cent of the Chinese went skiing or a snowboarding, compared to, for instance, 17 per cent of all Japanese visitors.

Travel itineraries to Northern Europe accordingly tend to follow the focus on the capitals of the Scandinavian states and Finland, i.e. Copenhagen, Stockholm, Oslo and Helsinki. Typical elements of other visitor groups to Northern Europe such as the visit to Santa Claus in Rovaniemi in Lapland or a trip to the Polar Circle are not included. However most leisure itineraries are not only centred on Northern Europe but are rather stop-over programmes at the port of entry before or after going to other parts of Europe. Following Japanese travel patterns, combinations of Russia and Finland will probably increase in importance.

It is interesting to note that the level of satisfaction of Chinese travellers to Sweden in 2001 and 2002 was not very high, especially when compared to the average for all international visitors.

A reason for the lack of enthusiasm, which is also shown in the less than average positive answers by Chinese visitors to the question whether they would recommend Sweden as a destination to others, might be the limited number of famous attractions and the large distances between them. Blok (2002) in his study of Norway cites the example of Australian caves as a danger for Scandinavia as well: 'When visiting caves in China, colored lights and emperor-style robes make for photo opportunities deep within grottoes. According to Australian experience, Chinese visitors to Australian caves find them too boring and educational, sometimes leaving after a few minutes' (Blok 2002: 13). As in other countries, there is also a pronounced gap between the image of Northern Europe as a beautiful countryside and the reality of visits to only the capitals' inner cities.

Table 6.33 Level of satisfaction with stay in Sweden in 2001 and 2002[99]

	2001 Chinese visitors	*2001 all international visitors*	*2002 Chinese visitors*	*2002 all international visitors*
Very satisfied	35	60	32	59
Mostly satisfied	63	37	66	38
Dissatisfied/no comment	2	3	2	3

Northern European destinations expressed high hopes for leisure tourism development from China after the establishment of ADS. Finland was reported to aim for 260,000 tourists from China by 2010 (People's Daily 2003b). Whether these hopes for more and more sophisticated visitors can be fulfilled remains to be seen but for the first four months of 2005 the visitor numbers to Finland were below the figures of the three previous years.

Africa

Africa, the forgotten continent of globalization, has traditionally attracted a rather large share of attention by Chinese foreign policy. As an important part of the Third World, of which the People's Republic of China claimed some kind of leadership since the 1955 Bandung Conference,[100] China supported left-wing governments politically and economically with support for infrastructure projects such as the Tansam Railway between Tanzania and Zambia in the 1970s. With little competition around, at the beginning of the twenty-first century Chinese companies are engaged in many different fields in Africa and are one of the major investors. Chinese construction companies have erected bridges and sport stadia in almost all African countries. Chinese businessmen, government officials and construction workers have therefore already travelled to Africa for many years, but outbound tourism is a new phenomenon.

The share of Africa as a destination is tiny. It amounted in 1997 to just 35,000 out of 3.03 million outbound travellers from China travelling beyond Hong Kong and Macao. In 2000 both figures had risen in parallel to 55,000 and 4.87 million respectively, leaving the share at 1.1 per cent. For 2004 the share increased slightly to 1.4 per cent, or 115,000 out of 8.36 million (Du, Dai 2005). Africa will, however, remain together with South America the least important continent for Chinese outbound tourism.

Among the first non-Asian ADS countries, Egypt got ADS in 2002, followed by South Africa in 2003. In December 2003, Premier Minister Wen Jiabao announced at the China–Africa Cooperation Forum in Addis Ababa the inclusion of another eight African countries from 2004, namely Ethiopia, Kenya, Mauritius, Seychelles, Tanzania, Tunisia, Zambia and Zimbabwe.

The following text will look more closely at the case study of South Africa. South Africa is the most important of the ADS countries in Africa, with only Egypt in competition because of its early start and the famous sight of the Pyramids as an alternative lure for Chinese visitors. Egypt is also a country that can be included in round-trip centred on the eastern Mediterranean sea. Some 6,300 Chinese visited Egypt in 2003, a figure similar to the level of Chinese visits to Egypt in 1996, with leisure tours concentrating almost exclusively on Cairo, Luxor, the Aswan Dam, Alexandria, the Red Sea and the Suez Canal.

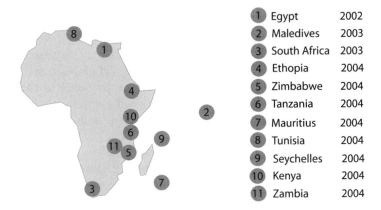

1	Egypt	2002
2	Maledives	2003
3	South Africa	2003
4	Ethopia	2004
5	Zimbabwe	2004
6	Tanzania	2004
7	Mauritius	2004
8	Tunisia	2004
9	Seychelles	2004
10	Kenya	2004
11	Zambia	2004

Figure 6.7 Map showing ADS countries in Africa[101]

In 2005, 500 Chinese tourists visited Egypt for a special Chinese New Year ceremony with Chinese red lanterns, music and dance in front of the Pyramids (People's Daily 2005a; Travelwirenews 2005a). The tourism minister of Egypt, Al-Maghrabi, visiting China in July 2005, gave a total figure of 100,000 Chinese tourists as the expected number for the end of 2005. The main difficulty in attracting more Chinese tourists, according to him, is food, with the Chinese visitors not fancying the local cuisine: 'I welcome more Chinese to come and invest in Chinese restaurants' (Xinhua 2005).

The countries who were awarded ADS in 2004 are, however, also trying to get their share of the Chinese outbound market. Tanzania for instance launched a tourism marketing campaign in late 2004 after sending two teams of tourism officials to China. In mid-2005 the Chinese-language version of the promotional website of the Tanzania Tourist Board was put online, giving information on wildlife parks and beaches as well as historical and cultural sites. Before ADS, the number of Chinese visitors to Tanzania averaged 3,000 persons (Tairo 2005).

South Africa

Inbound tourism is an important economic factor for the Republic of South Africa. In 2003, 6.5 million inbound visitors[102] spent 54 billion Rand.[103] Only 31,000, or half a per cent, of these came from the People's Republic of China (SAT 2004a).

South Africa received ADS in 2003 and established its own South Africa Tourism (SAT) office within the embassy in Beijing. Consulates also exist in Shanghai and Hong Kong.

Air transport is the only important form of accessibility.[104] With no direct flights existing, most travel routes are via Hong Kong, Singapore or Kuala Lumpur, from where daily connections exist.

The number of Chinese inbound tourists to South Africa increased strongly, if from a small base, since 1998, especially after ADS was approved. However, the growth was not more than proportional to the overall growth of the Chinese outbound tourism. Of all Chinese leaving Mainland China, the 2002–2004 numbers each represent approximately 0.2 per cent of the total.

South Africa enjoys a relatively low seasonality with the lowest proportion of arrivals in June with 7 per cent and the highest in December with 10 per cent. For Chinese visitors the first quarter of the year with 6–7 per cent of all arrivals per month is the low season and September to November with 10 to 12 per cent the high season (SAT 2004b).

Chinese visitors to South Africa are also visiting only specific places in a rather uniform way. Gauteng and Western Cape are visited by the majority of all Chinese visitors and the North West Province by approximately one-third. All other provinces are at or below a 10 per cent share. This translates mainly into visits to Johannesburg, Cape Town and the Kruger National Park. Such a spatial distribution is much more focused than the average foreign tourists' spatial distribution patterns. In terms of bednights, Gauteng and Western Cape are even more prominent, accounting for 90 per cent of all nights stayed (SAT 2004a).

The image of South Africa as a holiday destination is mixed. Only 14 per cent surveyed by SAT thought of South Africa as 'Very positive' as a holiday destination, but 18 per cent 'Very negative', with 27 per cent as majority answering 'Neutral'.

Beside concerns about personal safety and lack of knowledge, almost half of Chinese surveyed answered 'Not comfortable with African blacks', making racism the openly admitted top reason for not going to the country, clearly ahead of questions on price or distance.

Table 6.34
Chinese tourists to South Africa 1998–2004[105]

1998	12,790
1999	19,337
2000	19,722
2001	20,577
2002	25,849
2003	30,753
2004	43,000

Table 6.35
Provinces visited (with and without overnight stays) by Chinese visitors in South Africa in 2003 (in per cent)[106]

Gauteng	77.9
Western Cape	61.8
North West	37.3
Kwazulu Natal	10.4
Mpumalanga	8.8
Eastern Cape	7.6
Free State	2.8
Northern Cape	1.5
Limpopo	1.4

Approximately half of the Chinese visitors to South Africa went for business reasons. For those who went to South Africa for leisure, the natural attractions and the hospitable people got the highest marks, but safety and public transport including domestic flights were seen as very unsatisfactory.

Visitors to South Africa originate mostly in the three main outbound tourism source provinces, with Beijing and Guangdong being more important than Shanghai. Given the fact that package tours to South Africa are among the most expensive of all leisure tours on offer, it is to be expected that mainly the high-end income groups are travelling to South Africa. Figures from surveys which put almost half of the visitors into the less than 1,000 Euros per month bracket probably reflect the relatively high number of sponsored trips, paid by the government or companies.

Seventy per cent of all Chinese visitors are first-time visitors, 20 per cent are coming for the second or third time and 10 per cent have been to South Africa even more often, a fact corresponding to the high proportion of business travellers (SAT 2004c).

The gender ratio for Chinese visitors to South Africa is 70:30, a clear predominance of male visitors. Almost two-thirds of all visitors are between 25 and 44 years old, with 7 per cent aged 55 years or older.

Half of all visitors are travelling with some kind of package tour, half are individual travellers. Both kinds make very interesting customers for the South African tourism industry. Of all source market visitors, only the Chinese spend more than half of the total cost of their trip inside South Africa. On average Chinese travellers spend 16,761 Rand, with holiday travellers spending 11,351 and business travellers 28,412, which translates into daily expenditures of 1,494 Rand (holiday travellers 1,348, business travellers 2,140). These figures are considerably higher than the comparative figures. The daily expenditures of holiday travellers for all other non-African visitors for instance are below the 1,000 Rand limit (SAT 2004a). The money is spent not only on shopping, but also on accommodation, where 53 per cent of all Chinese visitors prefer four- or five-star hotels (SAT 2004b).

The main activities of Chinese visitors are not concentrated on the natural attractions of the country but on shopping, nightlife and casinos. One-third

Table 6.36 Reasons for Chinese not to travel to South Africa (in per cent)[107]		*Table 6.37* Age distribution of Chinese visitors to South Africa in 2003 (in per cent)[108]	
Not comfortable with African blacks	44	18–24	11.5
Personal safety	42	25–34	28.3
Don't know enough about country	41	35–44	34.2
Concerned about health	30	45–54	18.9
Language problem	27	55–64	5.9
Other better destinations	25	65+	1.2

Table 6.38 Main activities of Chinese visitors to South Africa (in per cent)[109]

Shopping (malls, flea markets)	83
Nightlife (theatre, shows, nightclubs, bars)	65
Visiting a casino	58
Wildlife, e.g. viewing game in nature reserve	55
Visiting natural attractions, e.g. scenic drives, parks	50

are partaking in beach life, but other sport or adventure activities are not important.

A typical itinerary for a one-week travel to South Africa would look something like this (SAT 2004b):

Day 1: Shanghai–Hong Kong–Johannesburg.
Day 2: Johannesburg. Gold Mine City, watch gold mining.
Day 3: Sun City, wildlife at Pilanesburg, casino in night.
Day 4: Lion Park, museum, watch diamond mining.
Day 5: Cape Town, ostrich farm.
Day 6: Seal Island, Cape of Good Hope, penguin beach.
Day 7: Shopping, airport after lunch.
Day 8: Hong Kong–Shanghai.

In conclusion, South Africa is a destination at the high end of the market. Half of all Chinese visitors to South Africa have been to Southeast Asia, a quarter of the visitors to Europe, Japan, or Australia, before they travelled to Johannesburg (SAT 2004b). The sophisticated travellers, half of them travelling not in package tours, are not kept back by racist biases against African people and spend a lot of money on shopping, gambling and accommodation. South Africa is less plagued by Zero-Dollar arrangements but the security problems and the long travel time will probably prevent it becoming a major destination for the developing Chinese outbound mass tourism. Even though neighbouring African countries have been granted ADS, they will not easily fit into a package aimed at the high-price, high-quality-orientated Chinese clientele visiting South Africa.

7 Product adaptation and marketing of tourism destinations and products for Chinese outbound tourists

From the discussion of the cultural, political and economic background of Chinese outbound tourists in Chapters 2 to 4 and their behaviour and experiences in Chapters 5 and 6, it should have become apparent that Chinese visitors differ substantially not only from western, but also from Japanese and other Asian tourists. They require not only to be informed about the offers of specific destinations and tourism companies but also to be offered specific products and communicated with in a specific way. Furthermore, the sharp divisions between the different levels of income and sources of funding for different groups of outbound travellers call for an even greater differentiation. Finally, some 20 years of leisure tourism experience have already produced some seasoned travellers who differ from the majority of first-time visitors to the world beyond the borders of the People's Republic of China.

The considerations of the following chapter are based on the analysis presented here as well as on practical experiences of the author as organizer of a large number of Chinese business, study, culture and tourism groups in Europe in the 1990s and the results of the work for the ECTW Award.[1] They relate less to the 70 per cent of all outbound tourists who visit the quasi-domestic destinations of Hong Kong and Macao than to those travellers visiting other destinations, especially when leaving the realms of the Chinese world by travelling to countries with small overseas Chinese communities.

Chinese outbound tourism is taking place in the framework of the consumerism and the nationalism prevalent in today's Chinese society. From the viewpoint of consumerism, destinations and tourism products are perceived and evaluated as *branded* products, the construction of the *imagined community* of Chineseness results in a definition process involving the difference between *us* and *them*.

The consumption of a tourism product is taking place in three phases: anticipation and preparation, the actual journey or activity, and finally the remembering and communicating to others of the experience:[2] 'Travel for leisure is increasingly permeating the discourse of everyday life, but remains a highly involving experience, extensively planned, excitedly anticipated and fondly remembered' (Morgan, Pritchard, Pride 2004b: 4).

With the perception of the destination as a brand, the consumption is clearly first of all symbolic in the sense of Baudrillard, leading to the anticipation of consumption as being frequently more enjoyable than the act of consumption itself and to the fact that 'there are no limits to consumption' (Baudrillard 1988: 24). The value of an outbound trip for the enhancement of the prestige of the traveller in a group-orientated and hierarchical society, the gain in cultural capital (Bourdieu 1979), underlines the importance of the time after the return. If bringing back 'romantic pictures taken in front of the Eiffel tower' of oneself[3] and branded goods of the touchable variety are the main goals of a leisure journey, the importance of the 'middle part' of the vacation, the trip itself, diminishes in the circumstances of the Chinese outbound tourism to an even greater degree than in other cases.

Chinese tourists abroad see themselves as *Chinese* tourists: every real or imagined slight against one of them is therefore also seen as an attack on the *face* of China as a nation. Every honour given to a symbol of China such as flags, language or culture are honours given also to the individual traveller encountering it.

Product adaptation and marketing for Chinese outbound tourists therefore need to reflect the holistic approach of brand development rather than the concentration on single events during a visit to a destination or the usage of the services of a company. Face-giving activities beyond the purely practical needs of adaptation, translation, etc. are main instruments to enhance the satisfaction of the visitors from China.

This satisfaction is currently not very pronounced and the need for adaptation is clearly seen on the receiving side:

> As China is a new generating country of international tourists, its destination countries and regions have only preliminary knowledge of its tourists. As a result, the completeness of service facilities, the creation of service items, the adaptive renovation of relevant environments and the convenience of language and signage of the destinations are far from the extent that Chinese tourists would feel comfortable at. Therefore, those countries and regions that target China's outbound tourists as their major market should make an effort to improve their environment in order to better attract Chinese tourists and increase their satisfaction.
>
> (Du 2003: 125)

For Europe, the contrast between the products and services demanded and those often received by groups hosted by European companies or organizations, has been summarized by Arlt (2004) in Table 7.1.

Table 7.1 What Chinese visitors want and what they get[4]

What Chinese visitors want	What Chinese visitors get
Being greeted with respect, given face and prestige	Being greeted with paternalistic attitudes and treated as Third World visitors
Contact with the 'exotic' locals and their festivals, customs and daily life	Cathedrals, monuments, museums
Photo opportunities for their group in front of destination icons	Lectures into the finer points of European art and history
Chinese-language information material, relating local situation to Chinese culture	1:1 translations or no Chinese-language material
Shopping opportunities as part of the itinerary	Shopping in between the 'real' sightseeing
Information about business opportunities	'Leisure-only' itineraries
As many highlights as possible in as many countries as possible in rapid succession	Marketing trying to convince them to spend prolonged periods at each location and hence fewer locations visited
Differentiation of products for different regional and sociocultural target groups	One product for all Asian visitors, modelled on the needs of Japanese tourists

The role of the Chinese language

As discussed in Chapter 5, the Chinese language is first of all a written communication system, with each character representing not a mere sound but a complex meaning. With written information carrying more importance and authority than in alphabet-based languages, a picture taken in front of a sign stating the name of a sight instead of the sight itself is of equal value:

> Nor are they [Chinese tourists] disappointed when, arriving at a renowned tourist site, they find nothing but a stele marking the spot. If they know their Chinese history or legend, the place, while offering nothing to look at, will assume a significance in proportion to the associations it will evoke in their minds.
>
> (Pan 1988: 134)

Different spoken Chinese dialects are incomprehensible for Chinese from outside the respective part of the country, but written Chinese does not differ throughout Mainland China,[5] making it possible to communicate in writing with all Chinese regardless of their origin from different parts of China. The usage of Chinese is necessary in all offers and products for Chinese outbound tourists. From a functional point of view, the ability of most travellers to understand English is limited especially among those born before 1980. Other languages are known only to experts, foreign-language students and members of the national minorities. Besides the only partial knowledge of vocabulary and grammar, communication difficulties arise also from the very different usage of language in the way of using irony or questions, levels of modesty and frankness, etc. However, most outbound tourism especially for less sophisticated travellers is organized as group travel, with Chinese-speaking guides in attendance. The more important part is, therefore, the non-functional relevance of the usage of Chinese in the communication with Chinese outbound tourists. Providing Chinese-language written information and signage is not only helpful to the individual Chinese traveller, but shows respect to *all* Chinese and to China in general by acknowledging its importance in the world.

The opposite effect is reached when Chinese visitors detect Japanese-language information without equally being provided with Chinese-language brochures, explanations or signs. Even though one of the three Japanese writing systems is based on Chinese characters, such signs are not perceived as an aid to understanding but as an offence. To convey *face* and to make visitors feeling welcomed is the major task of communication between host and guest. 'PRC [People's Republic of China] outbound travel executives are suggesting that . . . simple actions such as improving signage at tourist spots and boosting the number of qualified Chinese-speaking guides would help PRC visitors feel welcome and boost the popularity of that destination' (Travel Daily News 2004).

More endearing than just using Chinese is the correct usage of it. The fact that Chinese names are written with the family name first, the small number of common family names (made worse by the common usage of a title instead of the first name), the difficulty in knowing the gender of a Chinese person from his or her first name: all these problems almost guarantee mistakes and misunderstandings.[6] Foreign place names are transcribed with Chinese characters,[7] resulting in different, but codified pronunciations of names for cities, mountains, etc. To know these names for the local destination will ease communication in many situations.

Information in Chinese has not only to be translated in the linguistic sense. Taking into account the limited familiarity but also the limited interest in every detail of the architectural or historical features of a destination, information has to be given on a general level without assuming previous knowledge and for greater geographical units than just a city or a province. Europeans are used to impressing North American visitors with buildings that are several centuries old, Americans are proud to show off their modern cities, Japanese hosts expect guests to pay reference to their art, New Zealanders want to convey a deep relation to nature. Chinese tourists often react to all this with curiosity rather than with awe, as they compare it to their own heritage and their own modernity and have no time or interest in close, active involvement with nature. Combining Chinese language with local customs is a sure way of satisfying Chinese outbound visitors.[8]

Consequences of differences in motivations and behaviours of Chinese outbound tourists

The Chinese outbound market is rapidly compartmentalizing into very different parts. The differences between first-time visitors from the countryside, young urban white-collar workers, retired cadres, etc. are quite distinctive. In many – domestic and overseas – destinations, repeat visitors complain that only one kind of package tour is offered and high-end market offers or packages for grandparents/parents/child groups are seldom organized. Geographically, there are also pronounced differences among the outbound travellers.

A matrix of customer groups, with each group requiring different offers, can be developed from the roughly sketched – necessarily overlapping – different market segments according to travel experience, age, geographical location and special interests:

- *First-timers*: given the short history of outbound travel and the only recent lowering of bureaucratic hurdles for Chinese not living in Beijing, Shanghai or Guangdong, the majority of Chinese outbound travellers going beyond Hong Kong and Macao will continue to be first-time visitors, among them many who will not expect to travel regularly to overseas destinations. A dense programme with many highlights in short succession, a focus on the prestige coming with outbound travel, a low level of active participation and ample opportunities to shop for regionally typical and branded goods will fulfil their expectations. Price sensitivity is very high in this group, as the necessary amount of money will be greater compared to the income of most of these tourists and the lack of experience makes the travellers underestimate the cost involved in reaching destinations outside China.
- *Repeat visitors:* the more popular destinations report on average a quarter of visitors from China who spent time in the country before. The majority of these are not leisure, but business and VFR travellers. Leisure travellers are, however, also returning to places to show them to their parents, spouses or child, or to visit more thoroughly a region where they spent only a short time before. Thematic travels, along thematic routes or visiting thematic highlights are sought-after products. For this segment a higher percentage of individual travellers and a greater price elasticity can be expected.
- *Westernized travellers*: the growing number of young urban professionals,[9] many of whom studied abroad, have a much higher level of knowledge and interest in international destinations. Offers that follow western patterns of individual or small-group travel and show a clear distinction from the 'loud, rude and culturally naïve Chinese tourist' (Satish 2005a) are looked for.

- *MICE travellers*: even in times of ADS, Chinese travelling on official or business missions are a major force and also very likely to be involved in touristic activities during a smaller or larger part of their journey. Offers for MICE customers are easier to place in destinations that are strongly connected with China in economic and political terms. However, this does not prevent destinations organizing stop-over programmes as Finland or South Korea are doing for Chinese business travellers en route to Europe or Hawaii for travellers to the USA.
- *Teenagers*: education for the single offspring is the most important investment for Chinese families. Language summer camps, soccer schools, educational round-trips – the possibilities to develop products for the 'little emperor' generation are much larger than the current touristic offers.
- *Young office girls*: like their Japanese counterparts, Chinese 'OLs', unmarried young women living with their parents and being able to spend most of their income from office jobs for leisure and shopping, start to integrate – mainly regional – shopping and pop culture travelling into their spending pattern. Asian shopping malls and Korean pop and movie stars draw their attention.
- *Honeymooners*: unknown a few decades ago, one way to gain prestige in early adult life is a luxurious honeymoon trip using all the kitsch elements connected to Hollywood-inspired romance. Many destinations try to gain a foothold in this market.
- *One mouth, six pockets*: '*One mouth six pockets* is the description for a typical family in China today: a single, precious offspring doted on by its parents and two sets of grandparents' (Economist 2005c). As described for Australia, the particular result of the one-child family, the economic upswing of the last decades, the high value of filial piety and the view of travelling as a way to learn new things, means young urban affluent couples invite their parents to go abroad together with their child. The itinerary will, however, be more or less completely modelled according to the wishes of the offspring.
- *Empty nesters*: couples between the age of 45 and 65, whose children or child has left the household are identified as 'empty nesters', an important target group for several analysts and also for the Chinese outbound market (TBP 2004; Visit London 2005). However, the persons belonging to this group today spent their youth or at least their childhood in the Cultural Revolution. The majority of them were neither able to acquire large personal savings nor will they embrace consumeristic ideas of personal indulgence easily. Only in the coming decades will this group – the section of Chinese society with the greatest purchasing power and keen to enjoy their wealth – also become international travellers.
- *Older people*: this group, who suffered war and cultural revolution, mostly see consumerism as evil and frugality as a virtue. 'The self-

lessness of older people is a real marketing challenge' is the insight of a public relations agency manager in China (Economist 2005c). They also do not have the financial means to travel abroad and will rely on government, the Party, companies or children to sponsor trips to other countries. Foreign-language abilities could not be acquired during their schooling and working time except some Russian for those born before 1940. Few older people will travel outside groups and most of them will be first-time visitors if they travel beyond the border at all.

- *Coastal and urban citizens*: as the wealth is very unevenly distributed within China, the vast majority of outbound tourists travelling beyond Hong Kong and Macao and not engaging in border tourism are living in and around the big cities of Eastern and Southern China. Among these, the Cantonese have the longest experience in outbound travel and also the largest number of overseas relatives. Simple generalizations such as the Cantonese like entertainment, the Shanghainese like shopping and people from Beijing like culture (WTO 2003) are to be taken with a rather large pinch of salt. Coastal and urban dwellers are clearly more likely to have more money, travel experience and language abilities than the Chinese living in the hinterland and in rural areas.

Consequences for human resource development and training

Companies and institutions involved in receiving Chinese outbound tourists need to prepare their members and personnel for the specific needs of – in most cases – the new group of customers. Junek, Binney and Deery (2004) found that of the six major Melbourne attractions visited by Chinese tourists, four have Chinese-speaking staff or specially trained staff and two have generally interculturally trained staff. In destinations with smaller overseas Chinese communities in Europe or Africa, the training situation is in most cases much less impressive.

Like all customers, Chinese outbound tourists visiting foreign destinations have to be taken serious by their hosts. Differences in behaviour or misunderstandings are not necessarily rooted in the inexperience of local ways, the Chinese guests may well follow their own way on purpose. Many local persons working in tourism are fast to conclude for instance in cases of complaints in hotels that the source of the problem is the unfamiliarity of Chinese with modern facilities, falling back on the old image of China as a Third World country. The preference of shopping to visiting yet another local museum is often interpreted as a lack of education. Similarly, for many people outside Asia, 'East Asia' appears as one monolithic cultural area. Signage in Japanese will be expected to be understandable also to the Chinese, the image of shy group tourists, always exactly following the planned schedule, is expected to be true for Japanese and Chinese alike. Few non-Asians working in tourism have a clear idea about the difficult history of the relations between Japan and China.

For non-Chinese service personnel it is not always easy to accept the behaviour of the Chinese guest as the boss, regarding the staff member as a servant who can be loudly summoned and who is expected to always blame him- or herself even for mistakes that were made by the guest.[10] Within the framework of a pragmatic acceptance of the respective role distribution between host and guest, no idea of a 'heartfelt' or authentic friendliness versus a 'false' friendliness can be developed. The smile and the service are paid for and have to be delivered as part of the role played by the personnel. In a destination, all local persons are part of the product. Human resource development should therefore try to involve not only the persons working in tourism but also the general public in a better understanding of Chinese visitors and their behaviour. Some industry associations have published brochures as a information base for human resource training, for instance, the Bavarian Hotel and Restaurant Association (BHG) (2005) with the title 'Different Countries, Different Customs' or the *hotelleriesuisse* together with Switzerland Tourism (2004) welcoming 'Chinese as Guests in Switzerland' in German, French and English.

The majority of leisure tourists from China use hotels as their form of accommodation. Some hotel chains have introduced special programmes to adapt their offer to Chinese tourists, like the 'Chinese Optimum Service Standards' programme of Accor Hotels, including Chinese-speaking front-desk staff, Chinese breakfast, information materials in Chinese, etc. Other hotels are still struggling to fulfil the minimum requirements of Chinese travellers for a regular supply of hot-water thermos flasks and single-use toothbrushes. Human resource training here more often than not uses the concept of 'the customer as employee' (Johnston 1989) by trying to convince the Chinese guests that the whole hotel is not a private sphere, therefore, frowing upon walking in the lobby in slippers and speaking loudly from room to room.

Probably the most important employee in Chinese outbound tourism is the tour guide. Apart from practical problems such as lack of language skills, for tourists from group-orientated societies tour guides are of greater importance than for tourists from individual-orientated societies. They are less likely to oppose guidance but instead will expect to be guided and having the guide as the major source not only of 'mediation' (Ooi 2002) between the destination and themselves, being less used to supplement official interpretation with own 'research' with the help of guide books etc., but also as trouble-shooter.

The study by Zhang and Chow (2004) about the perceived performance of Hong Kong tour guides versus the expectations as seen by a sample of 500 Mainland Chinese tourists, reports the worst result scored for 'ability to solve problems', the second highest on the list of expected quality attributes.

At the bottom of the list is 'a sense of humour', a quality perceived as important in most individuality-orientated societies in line with the peer-group behaviour, but not seen as important in a person one has to rely on.

Complaints about the quality of tour guides are often voiced by Chinese outbound tourists (Yu, Weiler, Ham 2001). In the study of Zhang and Chow not a single attribute is found where the performance is perceived as level or even higher than the expectation (2004: 88). Training and, if possible, certification of tour guides is a demand high on the agenda of Chinese tour operators. Examples from Australia and Germany show an improvement in this sphere. To put tour guides in a position where they have to cheat the tourists to realize Zero-Dollar tours or to earn their own money through commissions is therefore especially bad for Chinese groups who rely more on the guides. Of the 700 customer complaints from outbound travellers that the National Holiday Affairs Office in Beijing received in 2003, most reflected problems with guides cutting sightseeing to add more shopping stops to earn more commissions. 'A mainland tourist to Hong Kong complained that he was incited by a local guide to buy an expensive but low-quality home electric appliance' (China Daily 2004d).

Consequences for the Internet, media usage and marketing

To enhance the brand value of and the perceived prestige connected with a visit to a destination, the establishment and maintenance of a Chinese-language tourism website is necessary. The use of Chinese helps the successful location by Chinese Internet search engines. A 'Welcome' in a language other than their own is not very convincing to any foreign visitor (Arlt 2005b). The content needs be adapted to the special interests and needs of Chinese visitors within the general rules of cross-cultural Internet tourism marketing. As has been pointed out in several studies (Arlt 2002a; Arlt 2002b; Arlt, 2002c; Arlt 2006a; Singh, Zhao, Hu 2005), simple measures can help with the success of a targeted website. For instance, photos showing other Chinese tourists together with the main attractions, or if possible a Chinese celebrity, will transport the information that this is a place seen as worthwhile and safe by other Chinese. Information on local Chinese infrastructure or twin-city arrangements with Chinese cities will show already existing connections between the destination and China. Downloadable Chinese information materials concentrating on the main attractions and telling some local stories will also positively influence the website visitors. Few Chinese tourists will have the confidence or see the need to book accommodation or tickets in the destination by Internet from China, with payment being another bottleneck. The function of a website for Chinese tourists is therefore mainly to enhance the status of a destination and to document its welcoming attitude towards Chinese guests. Finally, websites that give information for one location only will be of less interest compared to websites providing information about larger regions more easily identified as a possible destination. Maps on a country or even continent scale should at least help to locate the destination and lists of web-links should help to find information about surrounding areas.

The goal of the increase of the 'boasting factor' of a destination is supported by TV series showing either the destination[11] directly as a tourist destination or using it as a setting for Chinese movies. The material should be filmed by Chinese crews to make sure that features that are interesting from a Chinese point of view are included, such as daily-life scenes or curiously dressed locals, and the presenter is seen as actually being in the destination. The activities of a Chinese sports star playing soccer or basketball or other games in an overseas location also guarantees extended coverage of that place in Chinese media.[12]

For countries or cities with large overseas Chinese communities, strong economic relations or globally recognized sights, the challenge to communicate their 'unique destination proposition' (Morgan, Pritchard, Pride 2004b) is met with less difficulty than for destinations of which Chinese potential culture or nature visitors have no clear image yet or which mainly rely on '3S' (Sea, Sand, Sun) holidaymakers. The status of a 'blank sheet of paper' from a Chinese point of view for many destinations is also, however, an opportunity to build an image without having to fight negative pre-notions. Indeed one of the most common errors found in marketing materials directed at the Chinese market are unnecessary denials to the tune of 'Some people say we are slow-witted but in fact . . .'. Destinations not at the top of the list will be more successful in attracting a large number of Chinese tourists to stay at their place for a short time as part of a greater tour than trying desperately to convince them that this and only this destination should be visited for several days or weeks. Other Chinese tourists are the best argument for more Chinese tourists to visit, to be a well-hidden secret is not a successful marketing strategy for this market. To get things started, possible attractions have to be identified from the Chinese point of view by the destination or outside experts. These will probably differ considerably from those communicated by the destination itself and may include points of reference to connections to China, superlatives of some kind or stories relating local natural phenomena to Chinese legends.

Tourism fairs are a tool to enhance contacts with Chinese travel agencies, which still complain about the lack of sufficient information and brochures in the Chinese language (Chinatravelnews 2003b). In a group-orientated, hierarchical society like China, convincing opinion leaders is more important than talking to the general public. As a positive example, in 2002 the CTC published a 240-page full-colour educational brochure with 10,000 copies to be distributed to tour operators and during the CITM fair. However, 200 hard-cover special editions were distributed at a banquet to high-ranking Chinese tourism officials (CTC 2002). With the Chinese tourism policy still closely connecting inbound and outbound tourism, another useful tool to create political goodwill is the co-promotion of destinations such as the one agreed between the Spanish capital Madrid and the cities of Beijing and Shanghai.

Consequences for selected market segments

Outbound tourism offers by Chinese travel agencies in most cases give three pieces of headline informations: countries visited, main form of activity, and price. With importance in reverse order, price is still the decisive factor: a package tour price as cheap as possible in order to have more money to spend on shopping or as high as possible to show off such as a round-the-world tour for the lucky number price of 88,888 Yuan RMB (8,900 Euros in 2005). Activities concentrate on 'resting', 'culture' or 'round trip'. Compared to western offers, beach or sports holidays are much less prominent. With the diversification of the tourism demand and experience, these forms of holidays and others such as ocean cruises or wellness-orientated spa visits are however in the process of conquering their market niche.

Heritage Tourism

Heritage Tourism is an attractive form of tourism for Chinese outbound tourists, with most offers geographically connected to Europe. Sights connected to stories from films or books, old city centres perceived as 'romantic' and castles and palaces as symbols of power are seen as entertaining and educational at the same time, presenting good photo opportunities. Antiquity and religious symbolism are less important for Chinese visitors, possessing a heritage of their own that is perceived to go back for several thousand years and in most cases with no connection to Christianity. An important enhancement of the attractiveness of heritage sites is reached by connecting them with specific stories[13] and – if possible – by eventizing them with activities Chinese visitors are used to from domestic tourism such as employees in period costumes or the opportunity to have themselves photographed in costumes, etc. Destinations with strong cultural connections to China, for instance Singapore (Henderson 2002; Chang 2004), try to extend the Chinese heritage into their image, offering visits to China's past outside China. An example for this is the *re-sinization* of the Tiger Balm (*Haw Par*) Gardens (Teo, Li 2003; Dann 2005).

Attractions and Event Tourism

Attractions Tourism is rather underutilized in most destinations with regard to offers to Chinese outbound tourists (Arlt 2003). Theme parks and science museums are offering many features that appeal to Chinese tourists: a connection of fun and education, a large number of different activities compressed into a small time–space unit, thematic shopping opportunities, interesting activities for children. The concept of such attractions is, however, in most cases orientated towards the individual visitor or nuclear families and expects visitors to know the limits of 'proper behaviour' in such surroundings. With product adaption and increased awareness of the potential of this market, the importance of attractions within the itineraries of Chinese tourists can be expected to grow. Art museums are often in two minds about the interest of Chinese tourists. With interest centred on icons such as the Mona Lisa in the Louvre only, first-time round trip tourists are more often seen as a disturbance than as welcome visitors. Museums of artists, inventors or politicians well-known in China can be attractive as they provide an understandable story and material information about the age of the person in question in the form of furniture, etc. Such museums are furthermore often not too big in size and in most cases also offer specialized and prestigious shopping opportunities.

Event Tourism also answers many of the demands of Chinese outbound tourists, but encounters several difficulties. The temporary but fixed period of time in which the event is taking place contradicts the short-term decision making process of most outbound tourists from China. Furthermore, hotel rooms, other facilities and entrance fees tend to be more expensive during the staging of major events, complicating the offering of an acceptable price in the competitive Chinese market. However, the main problem is the integration into a given itinerary as especially for long-distance travels an event will not be the sole reason to participate in a tour.

Nature Tourism

China is blessed with a wide variety of landscapes from deserts to tropical islands and from great rivers to high mountains. To spend time, money and effort to travel abroad to 'just' stroll in a park appears to be wasteful for first-time round-trip leisure tourists from China. The expectation of the presentation of nature attractions is furthermore shaped by the national parks in China, which are heavily developed in the buffer zones and in some cases even in core zones of nature reserves, providing shuttle bus or boat transportation, shops and restaurants (Han, Ren 2001). Landscape perception is based on thematized views of certain stones, groups of trees, waterfalls, etc., connected to Chinese legends or historical figures, which are admired and used as photo backdrop one after the other. This kind of spatial consumption of nature reserves contrasts heavily with western ideas of individual exposure to nature including physical exercise. Attracting Chinese visitors to national parks or other nature attractions such as boat cruises on a lake or river involves therefore at least the provision of names and stories for the main features of nature, real or invented for the Chinese visitors. Exploitation to the level normal in China will probably not be possible according to the laws and the self-definition of ecotourism in western countries.

Ethnic Tourism

Visits to 'minority' villages or parks are a favoured pastime for Chinese domestic tourists to watch staged performances of daily life and dances:

> Authenticity is not a prime motivation although many are interested in and prepared to learn about aboriginal cultures. Tourists are aware that certain aboriginal arts, music and dance performance are unique, yet they typically fail to see beyond this façade and seldom comprehend greater, more profound differences.
>
> (Xie 2003: 12)

Equivalent performances are mainly offered in New Zealand's Maori culture villages. Other folklore performances, offered by Hawaiians during Luaus or Bavarian leather-trouser clad dancers, tend to be enjoyed and interpreted in a similar way.

8 Consequences of China's outbound tourism development for tourism studies

Until recently, the vast majority of both tourists and tourism researchers were based in the industrialized 'western' countries of Europe, North America and Oceania. Therefore, the activities and experiences of tourists as well as their analysis were based on occidental, Christian and market-economy values and behaviour patterns. This chapter is a first attempt to discuss the potentially liberating influences of the growing number of Asian – not least Chinese – tourists in international tourism on the way tourism, especially tourist motivation and tourist behaviour, is conceptionalized in tourism theory, hopefully leading towards more holistic insights. The following pages can only strive to give suggestions for further research by looking rather unsystematically at a number of related topics. The main problem beside the novelty of the phenomenon is the fact that currently almost no research results connected to such questions especially from Chinese colleagues both from inside or outside Mainland China are available, with the notable exception of the works of Wang Ning.[1] The reasons for this sorry situation are discussed below.

Roots and consequences of western domination

The beginnings of a scientific treatment of tourism are connected to monetary and statistical approaches and a Central European view. The oldest major texts from Germany, Guyer-Feuler: *Beiträge zu einer Statistik des Fremdenverkehrs*, (1895), Stradner: *Der Fremdenverkehr* (1905) and Schullern zu Schrattenhofen: *Fremdenverkehr und Volkswirtschaft* (1911) concentrated on economic aspects as did the early Italian text of Bodio: *Sul movimento dei forestieri in Italia e sul denaro che vi spendono* (1899) and the first major French text of Picard: *L'Industrie du Voyageur* (1911)[2] (Gross 2004). At the end of the 1920s the economist Robert Glücksmann started in Berlin his *Archiv für Fremdenverkehr* as a periodical publication of his private tourism research institute (Spode 1997). Accordingly in the 1920s at several universities in Austria and Switzerland, 'the first European chairs and research programmes were an outgrowth of departments of economics and, to a lesser extent, the field of economic geography' (Hall 2005: 7). While the war stopped tourism and tourism science development in many countries and Glücksmann's 'Jewish' institute had to close in Nazi Germany, in Switzerland the development continued. In 1941 in Professor Krapf in Berne started the 'Forschungsinstitut für Fremdenverkehr', while in St Gallen, Professor Hunziker became the first director of the 'Seminar für Fremdenverkehr'.[3]

Tourism research in the second half of the twentieth century was – and still is today – dominated by an Anglo-Saxon point of view from researchers working out of North America, Great Britain or Australia and New Zealand.[4] The impact from other, especially non-western, areas is minimal. Indeed Cohen in his review of Wang Ning's *Tourism and Modernity* (2000) stated that 'the book is probably the first major work on Western tourism by a non-Western researcher' (Cohen 2002: 57). Hall and Page acknowledge the existence of a large body of non-English research on tourism geography including Asian work by lamenting the fact that 'unfortunately there is much of the Asian research which is yet to be published in English' (Hall, Page 2002: 27). Swarbrooke and Horner remark in their discussion of tourist typologies that

> there is still bias towards Europe and the USA in the vast majority of typologies. Far less has been published on the types of tourist found in Asia, Africa and the Middle East, for example, which might yield very different results.
>
> (Swarbrooke, Horner 1999: 92)

Disciplines engaged in tourism studies are no longer restricted to economics and geography but have multiplied,[5] even though the quality and depth of tourism research is sharply criticized within the guild. Cooper (2003b) finds four problems, identified by him back in 1989, still existing in 2003: conceptual weakness and fuzziness, a spread of topics and a lack of focus, a predominance of one-off atheoretical case studies, and difficulties with access to quality large-scale data sources. He invites tourism researchers to '*break the meniscus* of the poverty of tourism studies by taking the many conceptual and theoretical approaches to tourism that have yet to be tested' (Cooper 2003b: 3).

Pearce (2004) states that even though the rejection of positivism is a popular and common view, tourism studies have not yet managed to get out of this 'blinding' (Pearce 2004: 60) tradition. He specifically addresses tourism research in the Asia Pacific region, in which he finds many of the problems extant in the global tourism research community as well, compounded however by a further set of specific Asia Pacific tourism research issues. In Pearce's eyes much of the indigenous literature is descriptive, destination-specific and atheoretical:

> This Asian tourism research scenario effectively amounts to a form of academic neo-colonialism where some of the brightest students from

Asia are educated in the traditions of Western social science thinking, and if they return to Asia transmit these ideas in their own setting. Viewed in this way innovation in the theoretical realm is difficult. ... Lacking confidence in their own mastery of the theoretical traditions and post-positivism appraisals, bereft of leadership in their own universities from experienced tourism scholars and needing to cement career places, it is not surprising to see Asia Pacific scholars resort to strong atheoretical and largely North American positivist traditions.

(Pearce 2004: 62–63)

Pearce gives several examples of new and innovative possibilities to stimulate new approaches in Asia Pacific tourism research,[6] without however questioning the approaches in the west for universal theories such as 'authenticity'. This is true for almost all tourism literature available. Even though Cohen has criticized the viewing of the modern western tourist as an essential category as being reductionist, because an ideal typification is seen as universally applicable (Cohen 1995), little thought is given in most cases to the cultural foundations of the underlying *Menschenbild* of tourism studies.

Cultural pitfalls for tourism research

As an example, the study of Junek, Binney and Deery (2004) on Chinese tourists in Australia can be used, not to criticize the authors as such but to demonstrate the many cultural pitfalls of tourism research. Their text starts with the sentence: 'Little is known about the Chinese market in terms of what attractions they expect to see whilst in Australia and what expectations they have of these attractions' (Junek, Binney, Deery 2004: 149). Even though the authors approvingly quote Yu and Weiler (2000) with the statement 'One of the main motivations for Chinese tourists in visiting a destination is the desire to learn about the culture of that destination' on the same page, Junek, Binney and Deery base their research on the – western – idea that travelling is 'to see attractions'. In the same way they see no contradiction in quoting Swarbrooke's (2002) assertion that '[v]isiting attractions is often considered the main reason for visiting a destination' just after musing that 'there may be specific factors unique to the Chinese culture, language and also expectation' (Junek, Binney, Deery 2004: 150). In their research results they list the different offers by various attractions. The fact that audio-guides are not used by Chinese visitors even though Chinese-language versions are provided is explained by the assumption that this is caused by the high price of renting them out. A closer look or indeed a direct contact with the tourists might have revealed that the Chinese tourists expect to be given authoritative interpretation of the sights by their guide – free of charge – and are not necessarily interested in individual technical solutions.

In *Tourism and Postcolonialism* (Hall, Tucker 2004), to take another recent example, it is implicitly assumed that *the tourists* are originating in the former colonial countries, even in the chapter talking about the SAR Hong Kong (Du Cros 2004), which at the time of publication was already overwhelmingly dominated by Mainland Chinese tourists, leading one author to the surprising conclusion that 'as consumers of global culture, tourists from all countries tend to look more and more alike' (Jaakson 2004: 175).

Kirsti Laitinen in her study about Chinese tourists in Finland (Laitinen 2004) discusses classical tourist typologies such as those of Cohen, Plog and others, only to find that 'it is difficult to categorize Chinese travellers based on any of the typologies mentioned above' because they do not consider cultural differences or external influences. Citing Swarbrooke and Horner she resigns that 'we may need as many typologies as there are tourism products, tourism markets, countries and cultures' (Swarbrooke, Horner 1999: 93). But in an earlier chapter of the cited book by Swarbrooke and Horner, the authors list *of course* in their discussion of tourism motivators 'the main factors determining an individual [*sic*!] tourist's motivations' (Swarbrooke, Horner 1999: 55).

Questionnaires, the magical wand of quantitative research, are regularly used to collect data from Chinese tourists, even though in many cases tour groups start to discuss the answers among themselves or adjust their answers according to the status and nationality of the interviewer. As Huang and Hsu criticize: 'Most of the researchers believe tourist motivations are derived from the influence of travellers inner personality, psychographic characteristics and outside social/cultural forces' (Huang, Hsu 2005: 194).

There are, however, also positive examples to cite. Shaw and Williams (2004), while giving an overview of the main tourist motivation literature of the 1980s and 1990s, which clearly shows the predominance of self-esteem and ego-enhancement as the main motivators described, also present the Travel Career Ladder (TCL) developed by Pearce in the 1980s, originally based on Maslow's pyramid. Tourists are supposed to 'climb up' the ladder as they become more experienced travellers. Whereas the lower steps of the ladder show both self-directed and other-directed motives, the uppermost 'fulfilment' step is only concerned with the individual fulfilment of a dream, self-understanding and inner peace and harmony. Pearce himself has changed his concept of the TCL, which was 'conducted in a Western cultural concept' (Pearce 2005: 78) in the meantime to a Travel Career Pattern (TCP) model, employing the results from research among Korean tourists. The TCP is less hierarchical, puts novelty, escape/relax and relationship as core motives and is meant to be applicable across cultures.

Some authors have obviously started to take notice of Kaplan's statement that '[t]he tourist is a specifically Euro-American construct who marks shifting peripheralities through travel in a world of structured economic

asymmetries' (Kaplan 1996: 63 in Oakes 1998: 19) and Clifford's observation: 'Theory is no longer naturally "at home" in the West . . . Or, more
cautiously, this privileged place is now increasingly contested, cut across,
by other locations, claims, trajectories of knowledge articulating racial,
gender and cultural differences' (Clifford 1989: 179).

Uriely in his analysis of conceptualizations of the tourist experience finds
that more attention should be given to non-western tourists because

> most of the generalizing conceptualizations concern the mind of the
> western tourist, while ignoring other voices, whether they are Japanese,
> Singaporean, or Brazilian. Thus, pluralizing depictions of the tourist
> experience, which are sensitive to gender or cultural diversity, seems
> to be appropriate for future research.
>
> (Uriely 2005: 211)

And Pearce summarizes his learning process with the TCL and TCP
models by stating that 'much work on travel motivation has been with US-
based samples. It is important, given the global nature of tourism to test
ideas, concepts and theories developed in Western countries with other
emerging traveller nationalities' (Pearce 2005: 67).

The growing awareness of the specifics of Asian tourism and tourists can
also be judged from the announcements of two conferences concerned with
Asian tourism in 2006 in Leeds and Singapore respectively. The organizers
of the Asia Research Institute in their call for papers explain the background of the attempt of 'Rethinking Tourism in Contemporary Asia':

> Very little attention has been given to the social, cultural and political
> implications of Asia's transformation from mere host destination into
> a region of mobile consumers. . . . To date the majority of academic
> studies on tourism have focused on east/west, north/south encounters
> between westerners, often seen as white, male and travelling alone,
> and their host destinations. . . . With so much tourism theory predi-
> cated on so many universalisms, there are indications that key threads
> of the literature are critically challenged by the praxis of Asian tourists.
> Is a different theoretical vocabulary required to interpret the socio-
> cultural dynamics arising from the consumption of beaches, nature
> reserves, religious sites and ethnically diverse destinations?
>
> (Winter 2005)

China and tourism research

In 2001, Graburn[7] optimistically stated:

> Until recently, the study of tourism in China was relatively less well
> developed. . . . The publication of this book[8] and other recent volumes

(Lew and Yu 1995; Oakes 1998; Yamashita, Eades, and Din 1997) will, I hope, make it impossible for social scientists to ignore tourism in and from China, in either their empirical or their theoretical research.
(Graburn 2001a: 71)

His optimism is unfortunately not reflected in the literature. Not only is the number of published studies of tourism in China beyond positivistic, hospitality-centred papers small, also looking for example at some introductions to tourism used regularly in university courses yields little result. Goeldner and Ritchie (2003) deal with 'Tourism through the Ages' in their second chapter. Starting with the Sumerians, Queen Hatshepsut of Egypt and the Romans, they move to the 'Silk Road', which appears however as having been only used to transport Marco Polo to China and some goods back and forth.[9] Holloway (2002) introduces 'The history of tourism' in a whole chapter. Starting with Babylonian and Egyptian empires, the text moves on to the Greeks and Herodotus as the 'first significant travel writer' (Holloway 2002: 18). Not a word about China. Even Hall in his recent seminal book on tourism mobilities mentions China only once briefly in connection to the Beijing Olympics 2008 (Hall 2005). However, even more astounding is the fact that Reisinger and Turner (2003) fill a book called *Cross-cultural Behaviour in Tourism* without a specific word on Chinese tourists, which are – unlike their Japanese counterparts – grouped under the heading 'Other Asian cultures and values' (Reisinger, Turner 2003: 122).

As stated above, this is not the place to discuss in detail the theory production of western tourism sciences in the light of Chinese tourism; however some remarks can be added. In Chapters 5 and 7 this text discussed the importance of group orientation and Power distance, which leads to many consequences for the expectations and behaviour of Chinese tourists, for instance in the preference of learning over experiencing. Whereas the concept of an experience economy (Pine, Gilmore 1999) is reflected in a wave of experience-based tourism products in recent years, tourism focused on the gaining of cultural capital within the network of the tourist at home is much more concerned with memory-production in the form of proofs for 'been there, done this' than in the individual experiences and potential self-enhancements connected to it.

The differences in nature perception based on the different forms of *reading* landscapes have also been reviewed earlier. Graburn laments rightly that Dann in his 'encyclopaedic' (Graburn 2001a: 71) *Language of Tourism* (1996) does not mention China at all, even though semiotics, an undercurrent in tourism research since MacCannell, provide an important key to understand 'How Asians and Westerners think differently . . . and Why' as, Nisbett subtitles his *Geography of Thought* (Nisbett 2003). MacCannell's famous example of the contentment of American tourists visiting the site of the 'Bonnie and Clyde Shootout Area' in Iowa (MacCannell 1999: 114), which offers nothing but a marker, is well worth comparing with the contentment of Chinese tourists, 'arriving at a renowned tourist site, they find nothing but a stele marking the spot' (Pan 1988: 92). Whereas the contentment in the Iowa case can be supposed to be proportional to the individual memories connected to the story or to seeing the movie, for Chinese tourists, as cited at the beginning of Chapter 7, the historical or legendary importance of a site within the Chinese culture will be decisive for the significance associated with it.[10]

That the 'seventeenth century disease of nostalgia seems to have become a contemporary epidemic' (Urry 2002: 95) informing the tourist gaze, may also need to be looked at more closely in the light of the fact that Chinese tourists travelling outside the 'Chinese world' do not perceive the historical sites as a part of their heritage, especially when connected to the differences in the concept of authenticity discussed earlier. Even Wang (2000) in his 'fundamental inversion' (Cohen 2002: 57) of MacCannell's frustration of objective authenticity-seeking tourists through the concept of existential authenticity stays explicitly within the theoretical boundaries of the late-modern western society without taking into account what could be termed the *spirit*-centred form of authenticity in China.

As an example of how new theoretical approaches can be productive in supporting a deeper understanding of outbound tourism from China in the context of China's tourism policies and development, the embedding of tourism into a wider study of mobility (Urry 2000; Coles, Timothy 2004; Coles, Duval, Hall 2005; Hall 2005) can be cited. 'Tourism is . . . increasingly being interpreted as but one, albeit highly significant, dimension of temporary mobility and circulation' (Hall 2005: 21).[11] The development of China's international tourism relations is not understandable without embedding it into migration and the transnational identity concept of overseas Chinese. Lew and Wong (2004), qualifying tourism first of all as a tool for enhancing social capital, underline that, extending beyond national borders,

> Chineseness can be denied, but it cannot be escaped. . . . Part of what it means to be Chinese is to carry the legacy of a long history of traditional values and obligations that are centred on the family and extended to community and other relationships.
>
> (Lew, Wong 2004: 203)

9 The future of China's outbound tourism

After analysing the development of China's outbound tourism, the last chapter will be devoted to arguing for the observation that it entered a new phase in the middle of the first decade of the twenty-first century and to provide an outlook, answering questions such as those on the possible influence of an economic downturn in China for outbound tourism, the probable development of motivations, forms and destinations of travelling abroad in the coming years, and the influence of China's outbound tourism on the future global tourism.

Outbound tourism entering its third phase

The year 2005 can be defined as the beginning of the third phase of China's modern outbound tourism.

The first phase started in 1983 with the reluctant opening of the gates by the Chinese government to 'family visits', first to Hong Kong and Macao and later to several Southeast Asian countries, ostensibly paid by the receiving side. This policy provided the opportunity for the development of clandestine leisure tourism by offering a way to get the passports, foreign currency and visa necessary for a view of the world outside the People's Republic of China. Quantitatively even more important, at the same time the beginning of China's integration into the world economy resulted in delegations and study groups travelling to the leading economic countries to visit fairs, talk to business partners, attend training programmes and the like. Almost all of these trips had a touristic element, with many simply being pleasure trips in disguise paid by public or government money or in some cases arranged by the foreign business partner.

The second phase started in 1997 with the official recognition of the existence of outbound leisure tourism as opposed to family reunions and business trips in the 'Provisional Regulation on the Management of Outbound Travel by Chinese Citizens at Their Own Expense' and the signing of the first ADS agreements with Australia and New Zealand. Many more ADS MOUs were signed in the following years and the regulations for visits to Hong Kong and Macao relaxed to help the tourism industries of these two newly formed SARs. A stormy development unfolded with the increases in outbound travellers far outstripping the growth of 8–10 per cent per year until 2020 (Dai 2005) planned by the Chinese government. While the government policy still talked about 'moderate, carefully managed growth' and close links to the number of inbound tourists (WTTC 2003), in reality a tripling of the number of outbound travellers occurred between 1999 and 2004. A chaotic and mostly unregulated situation developed with many travel groups organized by non-authorized agencies in the form of Zero-Dollar tours and other undesirable effects such as the outflow of large sums of money into casinos beyond the Chinese borders.[1]

Beside the wish to control the spatial movements of its citizens, the loss of hard currency informed the official reluctance to fully address the outbound market. As a measure to 'rein in the pace of growth in the Chinese outbound market' (Guo 2002: 10), an exit tax was discussed at the beginning of the new century. Dou (2000) proposed the idea of an 'exit tax' on all Chinese outbound tourists in a Qinghua University study. A tax of 100 Yuan RMB (in 2005 approximately 10 Euros) for each outbound traveller to Hong Kong and Macao and of 200 Yuan RMB (in 2005 approximately 20 Euros) for tourists to other outbound destinations was seen as appropriate (Dou, Dou 2001: 47). At 2005 outbound number levels, this would have generated 4 billion Yuan RMB, equalling about half a billion US$. Dai (2005) supported the idea of such an exit tax as a way to bolster the policy of 'limited' growth of outbound travel, proposing a level between 50 Yuan RMB for travel to Hong Kong and Macao and 120 Yuan RMB for travels to the USA. The money earned this way should be invested in promotion activities for inbound tourism, similar to measures taken in South Korea. However, in the same semi-official publication published in early 2005, Zhang (2005) defended the outbound tourism and pointed out that most of the money taken by the Chinese tourists out of Mainland China stays in the SARs Hong Kong and Macao, ergo within the bigger Chinese economy.

Four main arguments can be listed for declaring 2005 as the beginning of the third phase of China's modern outbound tourism. First, against some expectations,[2] the increase of outbound travels in 2005 to 31.03 million stayed within the single-digit bracket, even though this year saw neither outbreaks of SARS, similar health hazards or other internal problems, nor did wars or other major external developments stop Chinese travellers from visiting foreign countries. Quite to the opposite, newly opened ADS countries and the start of operations of Hong Kong Disneyland were thought to provide new pull factors to lure visitors from Mainland China.

Second, the unabashed enthusiasm for Chinese incoming tourists gave way to a number of concerns and irritations. Australia declared that its 'reputation as a desirable holiday destination, and the related healthy growth of Chinese inbound tourism are being threatened by a very small number of tour operators who had incorporated unethical practices in their business activities' (McAllan 2005: 133). Therefore in June 2005, the Australian ADS scheme witnessed a complete overhaul. Only outbound and inbound tour operators who adhere to a strict Code of Business Standards and Ethics are now supported, while any breach of the Code results in losing access to the Australian ADS visa processing scheme. According to McAllan, these activities have resulted in a noticeable downturn in the number of 'overstays', i.e. Chinese visitors disappearing into the local Chinese community. The EU countries reverted in mid-2005 to interviewing a percentage of all ADS visa applicants and to insisting on proof by the tour operators of the return of all members of ADS groups.[3] Singaporean and German retailers described their Chinese customers as 'loud, rude, culturally naïve ... and not-very-polite' (Arnold 2005) or 'loud, boorish and demanding' (Hoffmann 2005).

Third, the waning enthusiasm is also recognizable on the demand side. For the first time Chinese travellers appeared to be choosy. Tour operators told European destinations that African destinations are much less fussy in issuing visa for Chinese visitors, so Egypt and Kenya could be substituted for Germany and Greece, if the EU countries continued their restrictive visa policies (People's Daily 2005d). The number of Chinese tourists to Thailand and Malaysia plunged as a result of the bad image created by the Zero-Dollar tours and official Chinese protests over the treatment of young female tourists by immigration officers and sit-ins by hundreds of Chinese tourists who felt slighted by being provided vouchers with a pig's face stamped on them.[4] Some Chinese tourism experts proclaim it to be 'imperative' for overseas destinations not only to provide Chinese-language signs and hot water for tea but also moon cakes in August and long noodles for birthdays of Chinese travel group members. Taking the example of Britain, destinations are criticized for not being adaptive enough: 'The food, the drinks, as well as the sightseeing places, charming though, are all too British, not to our taste or habit' (Xu, F. 2005: 88).

Last, but certainly not least, the Chinese government seems to have changed its official stance on the question of outbound tourism. In a speech given at the Second International Forum on Chinese Outbound Tourism in Beijing in November 2005, Mr Zhang Jianzhong, director of the Policy and Regulation Department of CNTA outlined the new, more positive approach to outbound tourism that the CNTA has obviously adopted during 2005[5] and stated that for a 'really strong tourism country' the outbound sector also had to be developed. Therefore the policy would change from controlling the total volume of outbound travels towards a more pragmatic policy giving room to the rule of the market with the government neither encouraging nor restraining the development. He even conceded that the sharp increase in demand required the Chinese government to readjust their policies, something seldom heard publicly before from official Chinese sources.

According to Zhang, CNTA would, however, still have to regulate many forms of behaviour, among them the administrative behaviour in the form of amended and more detailed regulations instead of direct control, the enterprise behaviour that sometimes contributes to a negative image of China in the world and does not tackle the problem of overstays enough,[6] and finally the tourists behaviour, which would be asked to promote the good image of a country with 5,000 years of history by being courteous and by respecting local customs and laws – implying the partial absence of such behaviour heretofore. Zhang defined a 'healthy, sustainable and orderly development' as the goal of CNTA's outbound tourism policy, a departure from the insistence of 'controlling' – with the goal of slowing down – the development in earlier years (Arlt 2006c). An exit tax was not mentioned by Zhang and would also not fit with the rule of the market proclaimed as the new paradigm.

However, the new policy stops short of using outbound tourism as a way of redressing the huge trade imbalances in favour of China arising from its export power.[7] With the second biggest currency reserves in the world at more than 700 billion US$, a stable yearly inflow of 60 billion US$ FDI and exports amounting to almost 800 billion US$ in 2005, China could follow the precedents of Taiwan and Japan. The Republic of China Tourism Bureau (ROCTB) stated in its 1992 Annual Report:

> The traditional objective of international tourism policy, to earn foreign exchange reserves has diminished in importance due to the [Republic of China's] own massive accumulation of foreign exchange reserves. Emphasis has now shifted towards outbound tourism that promotes international understanding, improves the image of the [Republic of China], and strengthens substantive international relations.
>
> (ROCTB 1993: 7, quoted in Hall 1994b: 52)

Already 'in 1987 Japan's Ministry of Transport launched the Ten Million Programme to double the number of outbound travellers in five years to expand (!) Japan's tourism deficit' (Morris 1997: 154). The programme was mainly designed to show the willingness of Japan to try to close the gap in its trade balance with the USA. It remains to be seen if China will start to use the same public relations instrument to placate China-bashers in the USA and other countries.

Elasticity, motivations, forms and destinations of China's future outbound tourism

How would an economic crisis in China influence the future development of its outbound tourism? Given the enormous amounts of mis-invested capital, a banking system ridden with bad debts, which looks in some respects like the Japanese banking system before the 1990 crash, and the mounting ecological problems, a slowdown of the Chinese economy has been predicted for years. The *Economist* saw in a survey of China in 2002 the Chinese dragon as being 'Out of puff' (Economist 2002), Yu warned in 2004 that 'while the long-term prospects for China's economic rise is promising, it may be also necessary for the Chinese people to prepare themselves to brace for a harder time ahead' because of 'structural imbalance and a partial overheating' (Yu 2004: 20) of the Chinese economy. Even Michael Porter is quoted in November 2005 cautioning western companies: 'The bloom is starting to come off the rose in terms of China being the mecca for business' (Wehrfritz 2005: 40), despite a GDP growth in China that, according to official figures in 2005, again reached 9.9 per cent (Xinhua 2006a).

If China did enter a severe economic crisis, domestic tourism would most probably be strongly affected by a more parsimonious behaviour of the *laobaixing*, the normal people, not yet being used to perceiving tourism as an integrated part of their life. The quasi-domestic outbound tourism to Hong Kong and Macao might also suffer slightly as the inhabitants of Guangdong might scale back their shopping and gambling trips to the SARs. Real foreign travel and especially transcontinental leisure trips, however, stand a good chance of riding out an economic storm without much damage. The top 5 per cent or so of the population rich or influential enough to engage in holidays overseas should be in a position that will more or less isolate them from the effects of any downturn. A hypothetical severe political crisis however would certainly change the rules of the game thoroughly.[8]

Why will the Chinese travel in the coming years? Besides the knowledge, status and social capital to be gained from the 'middle-class rite of passage' (Naisbitt 1997: 61) of travelling abroad, opportunities for gambling, shopping and affirmation of national pride will continue to be motivators for Chinese tourists to visit other countries.[9]

Eight out of ten outbound tourists are said to visit casinos when travelling outside Mainland China, and junket agents[10] are rumoured to operate in many major Chinese cities (Liu, J.-K. 2005). In the first half of 2005, the Chinese government started a major campaign against gambling especially among Communist Party members and cadres. Since the end of 2004, party members are kicked out of the organization if they enter casinos or other gambling establishments while travelling outside China. Several high cadres were sentenced to death for gambling away up to 1 million US$ during visits to casinos in North Korea, Macao and other places. Under pressure from the Chinese government, more than 100 casinos just beyond the borders of Southwest and Northeast China were closed. With the campaign over and the Chinese government not willing to legalize gambling within China beyond some charity raffles, the attractiveness of gambling will not disappear and the border casinos will return. Indeed not only Macao is building up a whole new casino town on reclaimed land, Singapore has also lifted its ban on casinos to attract more visitors mainly from China. Nevada is lobbying intensively to find a way around the American visa restrictions for visitors to Las Vegas. The representative of Air China in Los Angeles is quoted as saying that 'Las Vegas would benefit hugely, hugely, hugely if some of these restrictions were lifted' (Satish 2005c).

Shopping will remain the favoured activity especially for transcontinental Chinese travellers. Shops and department stores in Europe have reacted in many places by employing special Chinese-speaking salesforces, themed decorations and signs using Chinese characters (Arlt *et al.* 2006). The major part of the 25 billion US$ estimated to have been spent by Chinese travellers in 2004 (EIU 2005) is used for the purchase of branded goods and souvenirs. Figures for the amount per trip per person spent vary depending on the destinations included or excluded. Wang and Liang (2005) found Chinese travelling to Asian destinations and Australia were spending about 760 US$ each, representing 71 per cent of the total expenses outside the package deal. According to statistics by Global Refund, the leading tax-free company, Chinese tourists use 30 per cent of their money to buy fashion and clothing, 16 per cent for watches and jewellery and 13 per cent for souvenirs and gifts with the rest spent in department stores and for other items (Setterberg 2005).

Almost all major brands have already opened shops in China, so the motivation for shopping is no longer a scarcity of goods in China. Because of the higher taxes in China, these items are, however, still cheaper especially in Europe or Singapore. Abroad the risk of buying faked goods is also perceived as being much smaller. These factors, together with the joy of being able to spend money freely and the need to bring back home gifts, will continue to fill the high-street shops and chic boutiques in the major cities of the world with Chinese shoppers (Arlt, Kelemen 2006).

Another motivation for travelling abroad can be traced to the rise of the Chinese authoritarian nationalism in the 1990s, actively changing the culturalistic pride of being a Chinese person into the nationalistic pride of being a citizen of the People's Republic of China, patriotism[11] taking the place of Maoism as justification for the anachronistic continuation of the political system of single-party rule. The earlier insistence on a 'Chinese way' in tourism, a 'tourism with a socialist face' ceased to be voiced in China after 1989 but is still echoed in western sources (Hall 1997; Jackson 2006). However, few today would agree with the verdict of Srirang that China fell 'victim to the massive power of the international tourism industry. China finally ended up in allowing reckless mass tourism for profit-making only' (Srirang 1991: 3).

The assertion of one's cultural and national identity (Edensor 2002) can take different forms. Discovering 'Chineseness' undisturbed by the upheavals of the Cultural Revolution in Singapore's Chinatown,[12] in the restored Tiger Balm Garden (Teo, Li 2003) or in celebrations for Zheng

He on Java (Handayani 2005), were 'attractions originally intended for foreign – usually Western – tourists [that] have become important cultural guides' (Graburn 1997: 201), creating a Chinese outbound form of the 'national imagination' (Petersen 1995), which differs from the 'furosato away from home' (Rea 2000) imagination of their Japanese counterparts. Visiting the remains of western domination of the SARs now back in Chinese hands (Du Cros 2005a) is another form of assertion, as is the satisfaction derived from finding out that the city centres of China's big cities appear to be more modern than those of cities in western countries. Patterns of visits to 'minority' areas within China (Su, Huang 2005) are reproduced in foreign countries as a form of asserting the superiority of the Chinese culture (Nyiri 2005a). It has to be kept in mind that 'the cultural experiences offered by tourism are consumed in terms of prior knowledge, expectations, fantasies and mythologies generated in the tourists' origin culture rather than by the cultural offerings of the destination' (Craik 1997: 118). If theme parks 'assert their localness by celebrating their local nationalistic identity' (Teo, Chang, Ho 2001b: 6), Chinese tourists travelling within the 'Chinese world' expect this identity to be predominantly Han Chinese.

If the growth of nationalism[13] in China keeps its pace of the last decade, this aspect of outbound tourism will further increase in importance. Already 'Red tourism' tours have started to be offered to foreign countries.

What forms of travel are likely to develop? With its development over time, China's outbound tourism becomes more diversified. Individual and small-groups travels are slowly increasing. Books on backpacking have started to appear on the bookshelves of Chinese bookstores in greater numbers, even 'Lonely Planet' guidebooks for Australia, Germany and Great Britain will be published in 2006 in an authorized 'appropriately adapted' Chinese translation (Chinahospitalitynews 2006). A quarter of the travellers interviewed by Wang and Liang (2005) and a third of the Chinese tourism students surveyed by Gu and Schwandner (2005) voiced their preference for Foreign Individual Traveller (FIT) style travel. However, the examples of the outbound tourism of other group-orientated societies such as Japan and Taiwan point to the fact that without bureaucratic constraints such as the ADS regulations in force today for Mainland Chinese, a predominance of group tourism will continue even when the unfamiliarity with the destinations, languages and cultures becomes less of a problem and visas are available for individual leisure travellers.

Specialized and themed tours for the more sophisticated return visitors to a destination are also bound to increase over time. Cruises, golf tours, language courses or a week of speed-limit-free driving on Germany's highways will find increasing numbers of Chinese customers and will enjoy a greater bragging factor.[14] The rushed sightseeing tour covering the highest possible number of countries and sights within the shortest possible time will, however, remain the main form of China's outbound tourism especially to overseas destinations for the foreseeable future. On the one hand, travelling to far away places will still be perceived as a special, maybe once-in-a-lifetime occasion. On the other hand, the main object of such travels, to gain status at home, is not related to getting detailed knowledge or deep personal impressions, so a quick glance – or gaze – is quite enough for each attraction.

Where will the Chinese outbound tourists travel to? Hong Kong and Macao are bound to remain by far the most important destinations for Mainland Chinese. It is nevertheless hard to imagine that the pace of increases in the travels to Hong Kong and Macao of the last years can continue for decades without overstretching the carrying capacity of the two cities. The island of Taiwan may be the next big growth area offering Chineseness, modernity and difference to an even higher degree than Hong Kong and Macao. Obviously many political questions have to be resolved before such a development could happen. Southeast Asian countries, the first destinations outside China to welcome Chinese leisure tourists in the early 1990s, fell from favour in 2004 and 2005. PATA predicts however the biggest average annual growth rate of all Asia Pacific destinations until 2007 for Malaysia (PATA 2005). With better quality control Southeast Asia should regain its position as the main regional destination for relaxation within a 'Chinese world' context. With Europe remaining a culturally interesting and diverse destination, Australia and New Zealand becoming almost familiar areas, and Africa and South America bound to increase visitor numbers from very small bases, North America remains the big uncharted territory for China's leisure tourism. A solution to the visa questions via ADS agreements or other measures will most probably turn the USA and Canada into the 'hottest' destination for Chinese outbound travellers. As for so many aspects of China's outbound tourism, this remains mainly a political question at governmental level and of brand recognition at the level of the Chinese consumers. Price variations triggered among other things by the fluctuations in the exchange rate of the US$ will only exert temporary influence.

An important question for all destinations relying on airborne travel remains furthermore the cost of transportation. On the one hand, the Chinese government will probably have to allow low-cost carriers to operate out of China rather sooner than later. On the other hand, the expected further rises in fuel costs – and a shortage in trained pilots (Koldowski 2005) – might well make it more expensive to travel long distances and thereby support the concentration of China's outbound tourism on neighbouring countries.

The role of China's outbound tourism in future world tourism

Estimates about the future numbers of outbound travellers from China have been revised upwards in the last few years. The often-cited figure of 100 million, predicted by the WTO[15] in 1997 for the year 2020 is based on forecasts that for many destinations had already been exceeded by 2005 (WTO 2000). In 2006, the United Nations World Tourism Organization (UNWTO) reacted by moving the predicted year of reaching the 100 million mark forward by one year to 2019 (TDC 2006). The investment bank CLSA, even though predicting a economic downturn to just 3 per cent GDP growth for China in 2007 (Arnold 2005: 38), sees the 2020 figure reaching 115 million outbound travellers, identifying young career-driven women as the major new segment responsible for the growth (Eturbonews 2005). EIU, however, reduced its 2003 expectation of 58 million outbound travellers in 2008 (EIU 2003) somewhat to the forecast that 'the number of outbound travellers will hit 49m by 2008, 60m by 2010 and 100m by 2015' (EIU 2005). Single destinations mostly share the optimistic view. For example, the Australian Tourism Forecasting Committee predicted in October 2005 an average 16 per cent annual growth rate of Chinese visitors to Australia for the next decade, which would result in a quadrupling of numbers from 250,000 in 2004 to 1.1 million in 2014 (McAllan 2005: 129).

Besides the famous Niels Bors adage that *predictions are hard, especially when they are concerning the future*, all general forecasts suffer from the major statistical problem of China's outbound tourism, the bundling together of visits to Hong Kong and Macao and 'real' international travels. The number of trips to these two SARs jumped from 5 million in 1999 to 20 million in 2004, whereas all other destinations witnessed an increase of visits from China of 4 million in 1999 to 'only' 8 million in 2004. In the first half of 2005 this trend continued. Some 10.5 million out of the 14.5 million outbound trips (72 per cent) from Mainland China already ended in Hong Kong and Macao. For the first time more than 90 per cent of all travellers stayed within Asia[16] (Wang, Liang 2005), many of them day-trippers and petty traders moving within the border regions.

For the role and importance of the Chinese outbound tourism in the world, it is not crucial if the Mainland Chinese continue to throng to the SARs[17] and if border traders continue to move, for more or less legal reasons, to neighbouring countries on a daily basis. Instead the important question is how the number of approximately 6 million international travellers, who moved in 2005 from Mainland China beyond Hong Kong and Macao and the border regions for business, leisure, VFR and study purposes, will develop. Keeping in mind that 6 million represents just half a per cent of the total Chinese population, an increase seems certain. For the size of the increase, an extrapolation of the approximate doubling of the number of 'real' international travellers in the six years between 1999 and 2005 leads, assuming a similar future increase, to a figure of 35 to 40 million non-SAR, non-border Mainland Chinese travellers by the year 2020.

This still seems not too big a number compared to the 1.6 billion international travellers predicted by UNWTO for 2020 or the 1.3 billion population of Mainland China, representing only 2 per cent and 3 per cent respectively. Keeping in mind however, how ubiquitous Japanese outbound tourists appear even though their number has been never higher than 18 million per year,[18] the impact of Chinese outbound tourists for major tourism destinations should not be underestimated. A reason for this is the tendency of most tourists from non-individual-orientated societies to follow the same trails as other members of their group, to visit the most famous places and sights, and to travel in comparatively large groups. The majority of Chinese outbound tourists will not disappear from sight in small summer houses in the countryside or try to blend in with the locals in a metropolis. There might be many more German than Chinese tourists visiting France, still the point will be reached much sooner where the number of Chinese wishing to move up the Eiffel tower will exceed the total carrying capacity of its lifts.

If today 'Rimini is Russian' (Economist 2005i), many destinations in Southeast Asia and Oceania will appear as if they have been effectively taken over by Chinese tourists. Unlike their Japanese predecessors, Chinese tourists are much more openly demanding that their wishes and expectations are fulfilled and their country paid due respect by providing Chinese food, signage, guides and forms of entertainment at rock-bottom prices. Until now, the sinization of destinations has not proceeded with a speed acceptable to the Chinese visitors (Arlt 2005e; Du 2003). In Europe, the ECTW Award has recognized since 2004 efforts by European DMOs, NTOs, tourism companies and retailers, who provide products and online and offline marketing adapted for the Chinese market or organize special training for their staff (Global Refund 2006; Williamson 2006). The ECTW Award winners are organized in a club for information exchanges and the best practice examples are published (Arlt 2006d). No similar initiatives have developed yet in other regions.

As a major and outspoken part of the growing number of non-western international travellers, Chinese tourists will continue to change the appearance and offers of many tourism destinations. The emergence of China's outbound tourism is furthermore supporting the necessary change in the theoretical perception of the history and content of tourism. The common western view of a linear development of tourism from Grand Tour to beach holidays to post-tourism as different forms of the quintessential quest for authenticity and individual self-actualization finds itself challenged by a discussion of its occidental (Chen 2002; Buruma, Mavgalit 2005) roots. With the Chinese successfully claiming their share of the pleasures of international leisure travelling, tourism – and our perception of it – will never be the same again.

Notes

Introduction

1　As a first caveat to the many trap-doors awaiting the unsuspicious *flaneur* entering the world of China-related statistics, Zhang and Heung (2001: 9) give the percentage of outbound travellers for 1998 as 6.7 per cent because they use 125 million instead of 1.25 billion for the population of China. Zhang, Pine and Lam (2005: 66) arrive at 3.5 per cent for 2001, minimizing the error by again underestimating the population by a factor of 10 but at the same time giving the figure for Chinese visitors to Hong Kong (4.4 million) erroneously as the total figure of all outbound travellers (12.1 million) in that year.

1　Framework of China's outbound tourism

1　With the description as blind men, no offence is meant to any visionally challenged person, nor is it in any way meant to deny the fact that there are a large number of excellent women among us.

2　Including the provinces, municipalities and autonomous regions of (from north to south) Liaoning, Hebei, Beijing, Tianjin, Shandong, Jiangsu, Shanghai, Zhejiang, Fujian, Guangdong, Guangxi and Hainan.

3　The border provinces in the Northeast and Southwest are presented as relatively more important for outbound tourism because of the large number of border tourists crossing from China into Russia, Mongolia, Vietnam and Myanmar, many as day trippers, distorting the statistics (compare Chapter 2).

4　*(Gang-Ao-Tai) Gongbao,* compatriots, is the term used for Chinese living in Hong Kong, Macao and Taiwan, who are regarded as integral parts of China with only temporarily specific other forms of administration.

5　This total of 42 million is the fourth-biggest number for any country in the world, but taking into account the existence of 1.3 billion inhabitants, 16 million foreign visitors translate into a ratio of just over one foreign tourist per 100 inhabitants. Even including the compatriots, the equation only rises to three per 100. The corresponding ratios for France or Spain would be more than 1:1, for the USA 1:6 and even for the more xenophobic Japan 1:18.

6　In Germany, for instance, Finck 2004; Indrich 2004; Kelemen 2005; Mrkwicka, Belz 2005; Schaefer 2004; Schmeckenberger 2004; Schuler 2005; Sun 2004; Wen, Zhao 2004; Weyhereter, Yang 2005.

7　As far as the author is aware, by the end of 2005 no book focusing on China's outbound tourism had been published in China, not counting the Research Report of Project Research Team (2003) and the Annual Report of China Outbound Tourism Development of the BISU (Du, Dai 2005).

8 Only since the 2003 edition, the text at least names the countries with ADS and provides figures for travel to selected individual countries.

2 Economic and social development of the People's Republic of China

1 Some scholars in China, doubting the official statistics on which calculations are based, put the figure for the ratio as high as 6 (Economist 2003).
2 The Gini coefficient measures the distribution of income within a society. It ranks on a scale from 0 to 1. A zero ranking would represent a society where all members have equal income, while a ranking of one represents the other extreme, where one person would receive all income in a society. Thus a higher Gini index indicates a wider income gap.
3 Even simple data are politicized. Whereas many figures that are only based on rough estimates are presented to the last decimal, the innocuous number of square kilometres that make up the total area of China is given in official statistical, as well as in popular publications, always as '9.6 million sq km' (NSBC 2004; Li, Li, Huang 2004), thereby avoiding having to discuss disputed border areas with Russia and India, the question of Taiwan, the contested islands off Vietnam, etc.
4 Unlike a simple conversion of local currencies into US dollars, the widely used PPP takes into account the differences in the prices of the same goods and services between countries. Using market exchange rates, the share of China's national economy appears to be only 4 per cent of the global economy.
5 Source: own calculation after NBSC 2003; NBSC 2004.
6 Source: NBSC 2003; China Statistical Bureau 2004; EIU 2005; Economist 2005e.
7 The next following currency holders Taiwan, the Euro area and South Korea, have 'only' less than 250 billion US$ each in their vaults. The total external debt of China was estimated for the end of 2004 as less than a third of the reserves with 233 billion US$ (EIU 2005). The private individual foreign exchange reserves are estimated to amount to another 90 billion US$ (CNTO Canada 2004b).
8 Only companies with sales above 5 million Yuan RMB are included. Source: own calculation after NBSC 2003.
9 Source: Schuler 2005; Data: NBSC 2003.
10 Source: NBSC 2003.
11 Source: own calculations after NBSC 2003.
12 Source: China Statistical Bureau 2004 (using World Bank data).
13 Source: own calculation after NBSC 2003.
14 Source: own calculation after NBSC 2003.
15 Source: own calculation after NBSC 2003.
16 Source: Schuler 2005; Data: NBSC 2003.
17 Source: own calculation after NBSC 2003.
18 Of course the actual percentage is less as some have travelled hundreds of times as border traders or in other professional capacities.
19 For details, see Chapter 4.
20 The figures for the inbound revenue are furthermore to be used with caution as they include all revenue from international air transport and do not account for necessary imports to be paid in foreign currencies (Johst 2001).
21 Compare Chapter 4.
22 Other than monetary, restrictions to own a passport have disappeared in most countries, and outbound restrictions by citizens' own governments have eased

with a few exceptions, such as North Koreans travelling abroad or US citizens travelling to Cuba. Inbound restrictions, the permission to enter another country are, however, still very much in existence with obligatory visas being rather the norm than the exception for citizens of most countries.

3 Government policies and the development of demand in Chinese tourism

1 China does have a legacy of travellers who left behind descriptions of many parts of China of precision and detail, such as Shen Kuo in the eleventh and Xu Xiake in the seventeenth century, but they never left China and are unknown outside their own country.

2 The myth of the 'closed', centrally ruled unified China as the prevalent state of affairs can be traced from 'dynastic cycle' history-writing of Imperial China (Wang 1991) through Marx's 'mummy carefully preserved in a hermetically sealed coffin' (Marx 1853 in Torr 1968) to modern nationalistic propaganda. Echoes of it also find their way into western authors' writing on tourism in China (for example Oudette 1990).

3 More recent acquisitions of Xinjiang (Sinkiang) and Tibet being the obvious exceptions.

4 The ships brought silver from Japan and the Americas to be exchanged for silk, porcelain and tea from China for export to the Iberian Peninsula in the second half of the sixteenth and the first half of the seventeenth century.

5 The inhabitants of those *islands in stormy seas* George III ruled were graciously granted the right to exchange every few years in tributary missions some of their curiosities and knick-knacks for books extolling the wisdom of China, silk, tea – and rhubarb to guard against constipation.

6 To illustrate the range and impact, some notable Chinese should be mentioned here. 'Michael' Shen Fuzong travelled with the Jesuit Father Couplet (1624–1692) to Oxford and was received by Louis XIV and James II. 'Arcadio' Huang helped catalogue Chinese books in the Royal Library in Paris and met Montesquieu. Both did not return to China. 'Louis' Fan Shouyi (1682–1753) was sent with the Jesuit missionary Provana by the Kangxi emperor to the Pope. Provana died on the journey back to China, but Fan returned to brief the emperor on his meetings with the Pope and on the customs and geography of Portugal and Italy. Two well-known Chinese who stayed in France from 1751 to 1766, 'Stephen' Yang Dewang and 'Aloysius' Gao Leisi, became Jesuits and returned to China to work as missionaries (Waley-Cohen 2000). Outside religious circles, in 1854 Rong Hong (Yung Wing), having learned his English in Hong Kong, became the first Chinese Yale graduate and started the movement of students from Southern China to study in the USA (Arlt 1984).

7 This is often further extended to the complete two decades, for instance 'This burgeoning tourism industry ceased to exist during the wars of the 1930s and 1940s' (Guo 2002: 46); 'The first attempt of running tourism business did not survive long because of the prolonged and ruthless Sino-Japanese War and Civil War from the 1930s through the 1940s' (Zhang, Pine, Lam 2005).

8 In comparison, Semmens (2005), for example, describes the successful tourism business in many German destinations until 1942, with even international visitors, assisted by fully operating representative offices in 17 countries, enjoying leisure activities in wartime Nazi Germany.

9 The growth rate predictions – with hindsight – look pathetic: 'In 1986 there were some 27 million domestic travellers, and that number is expected to double in the 1990s' (Richter 1989: 37).

10   The quote continues:

> The first ripple was the growth of domestic tourism within China. The second ripple was outbound travel to Hong Kong. . . . The third ripple effect comprises intra-Asia travel, starting in 1990. The addition of New Zealand and Australia marks the beginning of the fourth ripple, in which the high demand in China for travel beyond Asia, which is still very limited for leisure purposes, will gradually expand to encompass the entire globe.

11   A slightly different periodization is offered by Zhang, Pine, Zhang (2000): 1949–1966: tourism as a part of foreign affairs of the state; 1966–1978: standstill; 1978–1985: tourism as an important economic activity; 1986 onwards: tourism as a significant contribution to the national economy. Zhang, Chong and Ap (1999) divide the last phase further by distinguishing between the phase 1986 to 1991 and a new phase with faster economic reforms starting in 1992.

12   The foreign-language publication about tourism in China (Wang, Mai 1983) states that 'tourism does not have a long history in China. In 1954 CITS was founded' (Wang, Mai 1983: 2). There is no mention of any old traditions of travelling in China or to China and no mention at all in the whole book of domestic tourism or even outbound tourism. 'Tourism' is inbound tourism and nothing else.

13   Literally translated, *tongbao* means co-uteral, coming from the same womb (Dikötter 2005).

14   Services or infrastructure were not included, tourism was meant to cross-finance the other sectors only.

15   Foreign passport holders could also buy scarce and high-quality goods at so-called 'Friendship Stores' similar to Berjozka shops in the Soviet Union or Intershops in East Germany.

16   Even though the desire to present either the modern, advanced aspects of China, the glorious past or the primitiveness of minority cultures is still noticeable in today's tourism in China. The interest of many foreign visitors to see the 'backward' daily life of rural China they perceive as authentic is in contrast greeted with reluctance.

17   He Guangwei was replaced as director and party secretary of CNTA by Shao Qiwei, a former vice-governor of Yunnan Province, in March 2005 (China Aktuell Data Supplement 2005).

18   Excluding in both directions Hong Kong, Macao and Taiwan.

19   Lew unfortunately did renege on this insight, which did not include outbound tourism *from* China in the first place, by dropping the subject in the 2003 edition of *Tourism in China* (Lew *et al.* 2003). An updated version appeared as Lew and Wong (2002). Compare also Lew and Wong (2004).

20   These terms includes *huaqiao*, Chinese citizens living outside China, and *huaren*, foreign nationals of Chinese descent. In most cases, the generic term of *huaqiao huaren* is used in Chinese government texts, to signify that both groups fall within the scope of its overseas Chinese policies. *Huaren* are not considered automatically as *jus sanguinis* Chinese citizens any more since 1955, as they have been since the nationality law of 1909 introduced this principle together with the idea, imported via Japan, of being a 'nation'. However, since the mid-1990s a return to treating all overseas Chinese as Chinese citizens in all but name is observable (Barabantseva 2005). In the official tourism statistics, a small number of visitors (never more than 150,000) were shown in the tourism statistics as 'overseas Chinese' (*huaqiao*) without a definition given in the technical explanations but probably referring to persons holding a Chinese passport but living outside of China (Lew 1995). From 2002, this was discontinued without

any comment, eliminating another differentiation between *huaqiao* and *huaren* (CNTA 2002a).

21   However, no 'creole' communities of Chinese came into existence, no 'New Wuhan' like a New York, because unlike New York, which existed alongside the English city of York, competing with it, the Chinatowns were not competitors, but extensions of the homeland (Anderson 1991).

22   Written with a different character for *qiao* then in *huaqiao*, see Li (1994).

23   The 'new migrant' – an official and media term usually applied to those who have left China since 1978 – is a figure symbolic of the new, globally modern and yet authentically national – even racial – way of being Chinese. He is successful in the global capitalist economy, . . . and yet he (or she) is able to do so precisely because of certain innate Chinese moral qualities, which include loyalty to the Chinese state.

(Nyiri 2005a)

24   For the history of SARS see Mason, Grabowski and Du (2005), and also Overby, Rayburn and Hammond (2004).

25   See also Rea (2000) for Japanese existential travellers.

26   Reports of such visits are full of illustrative stories: The wish to go swimming results in a complete public pool closed to thousands of Chinese in the summer heat for the exclusive use of a group of foreign friends, trains waiting for hours for delayed foreigners, etc. For a Polish visitor's experience in 1956, see for example Konwicki (1976), for a German Maoists' irritations see for example Schon (1972). The simple idea of a visiting group the author was part of in 1978 to use for once the Beijing metro resulted in a whole line being closed to the public so that a special train carrying the group could drive for a few stations back and forth.

27   The 3 million mark was not reached before 1992 and the 5 million hurdle could only be cleared in 1994. A third conference was held in December 1983 (Gao, Zhang 1984), following the first international tourism conference in March 1983 with representatives from Japan, the USA, France, Britain, Switzerland and Hong Kong attending (Gee, Choy 1982; Gee 1983).

28   In line with Haas, who declared that 'the 1990s are the decade of nationalism' (Haas 1997: VII).

29   Or in the translation of Wang (2003b): 'To develop the inbound tourism energetically, to develop the domestic tourism actively and to develop the outbound tourism within limits.'

30   In spite of the fact that the 'WTO Principles as They Relate to China' would qualify ADS as a disguised non-tariff protectionism, which therefore needs to be scrapped (Panitchpakdi, Clifford 2002).

31   In the case of the EU countries in 2004, the opposite partner of China was not one but a whole group of countries.

32   'Overstaying' is the polite term used for this phenomenon (WTO 2003).

33   Typically for China, that something is forbidden does not necessarily mean that it is not happening with official support. In spring 2003, for example, Denmark, Norway and Sweden promoted their countries as tourism destinations with various activities in big Chinese cities without having ADS status at that time (People's Daily 2003d).

34   The minimum number differs for different countries, for Australia for instance nine persons are required as the smallest possible group (Junek, Binney, Deery 2004: 151), whereas five persons are enough for package tours to the EU (EU 2004).

35   Unlike before 1997 no obligations are connected with the invitation of a friend or 'relative' from China, although it was not impossible to get such a letter even when no real contacts existed to the required destination. Some travel

agencies provided the required documents to obtain Private Passports for their customers, if so requested (Roth 1998: 13).

36 For example, in 2004 Chinese Premier Wen Jiabao announced at the China–Africa Cooperation Forum that the Chinese government will grant ADS to eight African countries (Xinhua 2003). In October 2005, the South Pacific Tourism Organization (SPTO) followed the request of China to deny Taiwan membership in the SPTO. 'In return, China will give due consideration to other Pacific Island countries to have them listed as tourist destinations for Chinese tourists' (Sulaiman 2005).

37 A growing number of Chinese citizens do however own internationally recognized credit cards, which make any restriction on the amount of hard currency they are allowed to exchange meaningless.

38 Own presentation.

39 'Travel agents in China have indicated that there is, in practice, no restriction on the destinations that can be offered as long as a visa can be obtained' (Zhang, Heung 2001: 9). The author himself organized in the 1990s more than 100 Chinese inbound travel groups to Europe with most of them being clearly leisure-orientated.

40 For details of ADS experiences of different countries, see Chapter 6.

41 Both Class I and II agencies were owned collectively. 'Collective' covers a wide range of possible ownerships: state-owned companies, trade unions, the Communist Youth League, local, regional or central government organizations and ministries.

42 This led to a temporary fall in the number of travel agencies, as more than 1,500 agencies failed to deposit the required amount of cash (Qian 2003).

43 A list of the original 67 fully licensed ITAs can be found in the appendix of WTO 2003.

44 A list of the 529 fully licensed ITAs in 2004 can be found in CNTA 2004c.

45 With the ADS agreement for Australia, for instance, only 22 ITAs were authorized to establish links with Australian incoming agencies to arrange travel groups, later increased to 31.

46 Source: 1997–2003 (CNTA 1998a–2004a), 2004–2005 (Hu, Graff 2005). These figures, based on the *Yearbook of China Tourism Statistics*, cover the members of the CTA. This official part of the tourism agencies industry is estimated to be responsible for 60 per cent of the total turnover (DZT 2005).

47 Own presentation after Guo 2002: 114.

48 It would be a mistake to subsume all this tours under the category 'MICE', as the differentiation is not only between private or non-private reasons for travelling but also between private or public funding.

49 In the early phase of the reform process, some foreign- or overseas Chinese-owned travel agencies seem to have started business in China, as the State Council called for the closure of such companies, restating that 'no foreign businessmen are allowed to run any tourist operations' (Gao, Zhang 1983: 78).

50 As a constant source of misunderstanding, the transmutation of the former General Agreement on Trade and Tariffs (GATT) into the World Trade Organization led to the appropriation of the acronym WTO, even so this was already used by the World Tourism Organization, a move which can be understood as a sign of the underestimation of the importance of tourism for the world economy.

51 The Shenzhen government was the first to allow WFOEs for hotels immediately after the start of the membership (Zhang, Pine, Lam 2005).

52 Some 51 per cent are owned by TUI AG, 25% by CTS Headquarters Beijing and 24 per cent by Martin Buese MB China Invest GmbH. Mr Buese, a China

tourism veteran who initiated the cooperation and became the first executive director, stepped down from this post in May 2005, but kept his stake in TUI China.

53 Source: own presentation, after Schaefer 2004.
54 This late start is however not restricted to China but a phenomenon found in the whole Asia Pacific region (King, McKercher, Waryszak 2003).
55 According to Hall, five stages of tourism development and especially tourism planning can be distinguished: boosterism, economic approach, physical/ spatial approach, community-orientated approach, sustainable approach. For the obstacles to reach the stage of community-orientation, see Li (2004a).
56 See Chapter 1.
57 Meaning that political correctness in adherence to the Party line is as important as professional knowledge and expertise.
58 In spite of reform and streamlining attempts in 1999, often so-called 'sunset disciplines'(Du 2004e: 488), which are in want of students, jump onto the bandwagon of tourism education, creating 'tourism and calligraphy' and other such rather unusual programmes.

4 Quantitative development of China's outbound tourism

1 In most official publications, the 'old' figures are still used up to the year 1997 and are also the base for statistical information given in this book for regional developments.
2 For 2003, because of the effects of SARS, the figures were lower at 11.2 million jobs and 210 billion Yuan RMB (approximately 21 billion Euros in 2005) direct effect and 48 million jobs and 1,006 billion Yuan RMB (approximately 101 billion Euros in 2005) including indirect effects.
3 Own calculations. Source: CNTA 2004a; CNTO 2005. The overnight figure for 1980 is probably a rough estimate. Figure for compatriots include a small number of 'overseas Chinese' which were counted separately without further explanation until the year 2000.
4 Own calculations. Source: WTO 2003; CNTA 2004a; CNTO 2005. Figures before 1994 are probably rough estimates.
5 Sources: Chen 1998; Arlt 2004; TBP 2004; Du, Dai 2005.
6 Own presentation. Sources: Chen 1998, Arlt 2004, TBP 2004, Du, Dai 2005.
7 Own calculation. Source: Xu, Kruse 2003; China Statistical Bureau 2004; TBP 2005; Guo *et al.* 2005a. For 2003 and 2004 the first figure given is the estimate of TBP 2005, the second figure is taken from *Guo et al* 2005a. For 2003 the third figure is taken from EIU 2005.
8 Own calculation. Source: Guo *et al.* 2005a; alternative estimates: TBP 2004; EIU 2005.
9 Own calculation. Sources as Table 4.5. The domestic expenditure is valued at the exchange rate of 8.28 Yuan RMB for 1 US$, the pegged exchange rate valid from May 1995 until July 2005.
10 Own presentation. Sources: WTO 2003; TBP 2004; Du, Dai 2005.
11 Own calculations. Sources: WTO 2003; CNTA 2003a; TBP 2004; CNTA 2004a; Du, Dai 2005.
12 Own calculations. Sources: WTO 2003; TBP 2004; Du, Dai 2005.
13 Own calculations. Sources: WTO 2003; TBP 2004; Du, Dai 2005.
14 Taking the example of Japan, differences due to different fiscal years cannot explain the discrepancies, as the differences are as high for the years before and following. A use of Japan as a stopover for a different destination or as a the final point of a day-trip excursion is also not very likely. Some differences could be explained by diverting statistical procedures in the inclusion or exclusion of aircraft personnel, sailors, etc., but not to the magnitude of

several hundred thousand persons. Nobody would expect from the Japanese immigration administration or the statistical services anything less than meticulous working and bookkeeping. So the mystery remains.

15 It cannot be denied that the economic, political and social circumstances of Hong Kong and Taiwan differ from those in Mainland China, influencing the possibilities and necessities to travel. However, a comparison of the Hofstede index values for the three entities (see Chapter 5) reveals a significant similarity with regard to the cultural set-up (Hofstede 2001: 500–502). It is the author's guess that these similarities would be even bigger if separate Hofstede values for Southern China had been available.

16 The numbers of visits to China include, of course, visits to parts other than neighbouring Guangdong for leisure reasons, the percentage of these kind of trips is given by various sources as between 20 and 35 per cent (for example, Law, Cheung, Lo 2004; DZT 2005), in any case making China the main leisure destination.

17 Source: 1990 and 1995: Wong, Kwong (2004); 1997–1998: Zhang, Qu, Tang 2004; 1999–2002: Mintel 2004; 2003–2004: Tourism Research HKTB 2005. Most of the visitors of Mainland China are day trippers. The estimates of how many of them are using some tourism industry services as opposed to simple VFR/shopping vary between 25 and 60 per cent (Bailey 2004).

18 Source: Tourism Research HKTB 2005.

19 Exercising is perhaps an unexpected factor. It may be explained by the fact that ordinary people live in a limited space in both Taiwan and Hong Kong and they naturally [*sic*] see Tourist Night Markets as avenue for exercising along with shopping.

(Hsieh, Chang 2006)

20 Source: 1979–1994: Huang, Yung and Huang (1996); 1995: Prideaux (1996), 1998–2003: Mintel (2004).

21 Even in the SARS year 2003, 74 per cent of Hong Kongers planned an overseas trip within the next six months, according to a survey of ITE Hong Kong (2003).

5 Chinese travellers

1 Because of the size and diversity of China, comparisons of Chinese domestic tourism should rather be made with tourism within the EU or even better with Arabic intra-regional travels (Al-Hamarneh 2004). In these forms of tourism we find the same parallel existence of otherness in nature, food, customs, etc., and nearness in written language, basic culture, etc.

2 The only dissenting voice is that of Zhang, Pine and Lam (2005: 9), who argue that: 'Frequent travel away from home is not a Chinese custom, and people who are fond of travel for leisure were considered as good for nothing.' They do, however, give no evidence for their claim. Soffel (2005) discusses the inner conflicts arising from the contradiction between the obligation of filial piety of sons towards their parents, which keeps them at home, vis-à-vis the obligations to travel for career advancement or government duties. This concerned however only some very conservative neo-Confucianists during the Song and early Ming Dynasty.

3 The *Analects* are a collection of Confucius' sayings allegedly compiled by his pupils after his death in 497 BC. Translation by D.C. Lau. The 'classical' translation by Arthur Waley reads: 'Is it not delightful to have friends coming from distant quarters' (Waley 1938).

4 The 'general' and his two 'ministers' in Chinese chess (Xiangqi) never leave the palace area on the Xiangqi board.

5 Gavin Menzies controversively tries to prove that beside the seven voyages of Zheng He, several Chinese fleets actually circumnavigated the globe, accumulating information in the 1420s that found its way into European maps well before Columbus set sail (Menzies 2002).

6 Inspiring Jules Verne to describe the travels of a Chinese in China, using western steamboats as well as traditional wheelbarrows as means of transportation, in 'Les Tribulations d'un Chinoise en Chine' in 1879 (Verne 1879).

7 *Laobaixing*, 100 names, colloquial Chinese expression for the masses.

8 Major destinations were the various mountains connected to Buddhism or Daoism such as Taishan, Emeishan, Jiuhuashan, Putuoshan, Wutaishan, Wudangshan and Qingchengshan.

9 This form of tourism is still today of great importance. A study by Huang and Xiao (2000) found that 78 per cent tourists in Changchun, the capital of Jilin, travelled on 'business' sponsored trip, even though visits to the famous cinema studio and shopping for ginseng and deer antlers were their main activities.

10 The ratio of private cars per head moved only in 2003 above the 1:100 threshold.

11 With roughly a 10 million square kilometre area and 15 million cars, the ample space an average density of less then two cars per square kilometre in China provides is of little consolation for Beijingers enduring traffic jams every day.

12 For 1996 the amount of money saved by private households was estimated at the equivalent of 800 billion US$, for 2005 more than double this amount. The Chinese national household saving rate has been climbing from 10 per cent in 1980 to more than 30 per cent in 1994. From that peak it fell to about 25 per cent since the year 2000, still a very high level in international comparison. The urban household saving rate started to climb continuously from 1988 (5 per cent) to 25 per cent in 2004 (Economist 2005e).

13 After experimental introduction and a State Council decree in June 2000, these 'Golden Weeks' were first announced individually and at short notice; only after 2001 did they become a regular feature (Tuinstra 2003).

14 49 weeks × two-day weekend + 3 weeks 'Golden Week' = 119 days.

15 The differences for the living conditions within the big cities between regular and irregular citizens are however still very big.

16 'For Japan and China, long-distance domestic tourism, in the form of "pilgrimages" or visits to spas for cures, helped tie these countries together in the popular imagination' (Graburn 1997: 201).

17 The reference to the 3,000, 5,000 or even 7,000 years of historical ancestry is not missing in almost any text on China, be it from Chinese or western sources. This kind of change from a dynastic history to linear history was pioneered by the historian Liang Qichao, who in 1902 was the first to see China in a western 'nation-state' linear development, presupposing the Hegelian–Marxist point of view which still today helps the present to shape the past into an uninterrupted lineage of 'China' (Duara 1995).

18 The homogenization of the Han identity, the 'Chineseness of China' (Wang 1991), and the denial of the 'latent multiethnicity' (Schein 2005) within the Han is supported by the rigid and sometimes arbitrary construction of minority identities with the introduction of the official list of minorities in 1958. Two examples: Wu about the Bai: 'Most of the change in the Bai's self-perception has occurred because of their officially named identity; where they previously claimed to be ethnic Chinese, they now claim to be a minority – "non-Chinese"' Wu (1994: 159). Gladney about the Hui:

> Until the 1950s in China, Islam was simply known as the 'Hui religion'.
> ... Until then, any person who was a believer in Islam was a 'Hui religion

discipline'. . . . Nevertheless, they are recognized by the state as one nationality, the Hui, and they themselves now use that self-designation in conversations.

(Gladney 2005: 269)

19   One of the delusions that impaired many Multinational Corporations is the assumption of a Big Emerging Market (BEM) country as a huge and single market. In reality, geographic diversity and economic disparity are prevalent among the BEMs like China, Brazil, India, and Indonesia. While metropolitan areas in these countries such as Shanghai, Sao Paulo, and Bombay have become the hotly contested markets, the vast areas of these countries show quite a different picture. In fact, a BEM usually includes a number of smaller sub-markets that are distinctive from one another in many ways including language, culture, and economic development.

(Cui, Liu 2000: 57)

20   Own presentation, Source: Cui, Liu 2000.
21   That border tourism is less connected to leisure tourism also appears from the tables in Chapter 2, where the provinces engaged in border tourism all show below average levels of ECRS spending.
22   In 2002, Shanghai and Beijing had an urban living expenditure level of above 10,000 Yuan RMB. Other regions, which beside Guangdong, Zhejiang, Tianjin and Fujian had an urban living expenditure level above the national average of 6,030 Yuan RMB (600 Euros in 2005) were Chongqing and Jiangsu (NSBC 2003).
23   Source: Guo *et al.* 2005b.
24   No Chinese ruling emperor ever saw with his own eyes the South China Sea or visited Guangzhou.
25   Own presentation.
26   Source: CTC 2001; Du, Dai 2005.
27   Compare Chapter 2.
28   The difference between total income and disposable income (after income taxes and personal contribution to social security) is very small, disposable income is about 95 per cent of total income.
29   What Wu, Zhu and Xu (2000) call a Recreational Belt Around Metropolis (ReBAM), formed at the end of the twentieth century around most major Chinese cities.
30   Own calculation, data: WTO 2003; TBP 2004.
31   Own calculation, data: CTC 2001 (weighted averages from answers of participants interested to travel to Canada and participants not interested to travel to Canada).
32   Source: Guo 2004. The responses to the ten destination attributes were measured on a 7-point Likert-type scale where 1 was 'strongly disagree', 4 was 'neutral', and 7 was 'strongly agree'.
33   Source: Guo 2004.
34   An extensive discussion of different analytical approaches to cultural dimensions can be found in the literature review of Reisinger and Turner (2003).
35   This is even true now for tourists' cultures themselves, with airport bookshops selling stacks of Alain de Botton's *The Art of Travel* (2002).
36   Nevertheless as customs statistics worldwide record nationality according to passports rather than personal self-definition, the fact that the overseas Chinese society is 'non-coterminous with the boundaries of nation-states' (Urry 2000: 163), fortunately does not create yet another problem with China outbound tourism statistics.

37 The two spheres of everyday life and touristical encounters are of course also interacting, not only in the way described by Hobsbawm and Ranger (1992) and Brown (1996) as *invention of tradition* or production of *genuine fakes*, but more importantly by the introduction of tourism into the cultural mainstream as touristical culture (Picard 1993; 1996; Hitchcock, King, Parnwell 1993b).

38 Therkelsen (1997) has shown that different images for different target markets play a crucial role for tourism expectations and behaviour.

39 Litvin, Crotts and Hefner (2004) claim that Hofstede is the third most cited author in the international business literature and had more than 2,600 citations noted, from 1980 to 2002, in the Social Sciences Citation Index.

40 Own presentation, adapted from Hofstede (2001), Crotts and Litvin (2003), Bowden (2003). Index scores range from a low (= 5) to a high (= 118).

41 Own presentation.

42 The ideographs used have however been extended in recent years by text-message and email emoticons (like ☺ or ☹).

43 Modern Chinese words consist mostly of two or three characters, but in classical Chinese they usually functioned as independent words and most of them have retained their 'lexicalic independence'.

44 Most characters have an internal 'story' rooted in history, from which there is no escape. The character for 'good' (*hao*) is made up of a woman and a male child, the character for immortal (*xian*) is made up of a human being and a mountain. It is impossible to write/think of 'good' outside the idea of patrilinearity, it is impossible to depict the natural home of immortals anywhere else then on hilly ground.

45 For an overview, compare Chandler (2002).

46 The basic name for language is *wen*, writing, like in *zhongwen*, Middle (country) writing, i.e. Chinese or *yingwen,* Hero (country) writing, i.e. English.

47 Nisbett (2003) in *The Geography of Thought* reports a whole series of tests revealing the different mindsets of independence versus interdependence of western and Chinese students.

48 In cases where local food is very prestigious but from a Chinese point of view un-eatable, the consumption in the form of just taking a photo while pretending to eat the delicacy is sometimes the preferred solution, very much to the disdain of French restaurateurs offering their raw oysters with pride.

49 Supporting this interpretation is the experience that for many Chinese outbound tourists the fact that the owner of a Chinese restaurant is a Han Chinese seems to be of higher importance than the quality of the food offered.

50 The world has yet to see the first Chinese tourist buying a T-shirt proclaiming 'My Dad went to xxx and all I got was this lousy T-shirt'.

51 The preference for local food, stated in several studies, seems at odds with the high Uncertainty avoidance wish of the Japanese. An explanation might be found in the fact that after the Meiji reform the Japanese diet was changed completely by the introduction of meat into a predominantly vegetarian cuisine, leading to the display of plastic mock-ups of all dishes in front of restaurants to this day (Blumenstengel 2003). Traditional food like imperial *kaiseki* dishes are eaten rarely, mainly 'to help keep up the Japanese belief that they can identify themselves with their heritage' (Graburn 2001a: 81). Another explanation could be offered with the higher willingness to follow American *politically correct* tourism behaviour.

52 At the end of the year 2003, 80 million users were registered (CNNIC 2003), by mid-2005 the number exceeded 100 million (CNNIC 2005).

53 Figures for 2003 (Li, Li, Huang 2004).

54 Earlier, incoming agencies had to find their own solutions to the lack of guide books. The author initiated the publication of a travel guide to German cities (Gao, Giese 1993) for his customers.
55 It is interesting to note that for instance the guide book *Guide to Business Travelling to Europe* (Xie, He 2002) of a series of books published by Favor (Feiyang) Travel, contains only one chapter with less then ten pages information about business travelling but 200 pages about sightseeing and 'Enjoying the time'. A mini Chinese-English sentence dictionary concentrates on travelling with plane, hotel and restaurant encounters only.
56 The names of the magazines are slightly confusing. 'Xin Lüxing', literally translating into *New Travel*, published by Hubei Press Centre, was called in English *Traveler* in 2004 but changed its name to *Voyage* in 2005. 'Lüxingzhe', *Traveller*, published by Shanghai's People Fine Art Publishing House, is named *World Traveller* in English. 'Lüxingjia', *Travel Specialist*, published by CTS Group since 1996, is called *Traveler Magazine*. 'Lüyou Tiandi', another magazine published by the Shanghai Art Publishing House, would translate as *Tourism World*, but is called *Travelling Scope*.
57 An example is the *Travel and Trade in Europe* magazine, published monthly in Cologne/Germany.
58 This figure does not include 62 purely educational TV channels for long-distance learning programmes.
59 Naisbitt (1997: 60) in his *Megatrends Asia* argues that for the East Asians 'exposure to satellite television has piqued their curiosity about faraway places'.
60 As Bartsch describes the return of one of the 'Baofahu', *people who became rich like in an explosion*: 'The first thing, Lao Li had to tell his friends about when he came back from Hawaii, were the women. "Expensive" he said with a crooked smile, "two hundred dollars, but so good!"' (Bartsch 2005).

6 Destinations of Chinese outbound tourism

1 The three airlines considered as low-cost carriers inside China, Okay Airways, Air Spring and United Eagle, cannot live up to their name as the prices of air tickets are still tightly regulated by the government. For outbound travel, only so-called 'Red Eyes' tickets for night flights can officially be discounted.
2 In 'Zero-Dollar' tours the incoming agency is not charging the tour operator for accomodation, transport and food for the customers but instead tries to finance all costs and his margin from commissions and kick-backs derived from the shopping of the customers in specially selected shops and from facultative programs sold locally to the tourists, both forced upon them with different degrees of coercion.
3 For example: Beijing–Frankfurt 7,800 km distance, 10 h flight-time, Beijing–San Francisco 9,500 km, 11 h, Beijing–Vancouver 8,500 km, 10 h, Beijing–Sydney 8,900 km, 13 h (OAG 2005). The fact that, unlike Europe or America, Oceania is only a few timezones away from China also supports the perception of 'nearness'. For the influence on tourism images in China through the perception of distance see Bao (1996).
4 Literally 'high nose', the common Chinese colloquial term for all non-Asian people.
5 Source: Own research, without offices in Hong Kong. As officially NTO representative offices can only fully operate if the country they represent has been given ADS, the legal status of many offices was or still is shaky, with sometimes contradicting information to be found in China and in the homeland of the NTO. This problem is even more pronounced for the representative offices of cities or regions.

6   These figures do represent the first country of arrival, so an estimated 12 per cent of all visitors to Hong Kong do however continue to travel to a second country, mostly within Southeast Asia.

7   Source: Own calculation. 1995: Roth 1998 (WTO figures); all other Chinese figures: 1999–2000 (WTO 2003); 2001–2003 (TBP 2004); 2004: (Du, Dai 2005).

8   Source: Own presentation.

9   In October 2004 a meeting between the bordering provinces on both sides, Xinjiang Autonomous Region in China and Kashmir and Northern Areas in Pakistan, discussed the use of the Karakorum Highway for Chinese tourists. The representatives of Xinjiang were, however, more interested in inviting Pakistani investors to their region (Ecotourism Society Pakistan 2004).

10 Macao is, according to a recent study, still perceived mainly as a gaming and entertainment destination and not as a leisure holiday and cultural destination by visitors from Mainland China, Taiwan and Hong Kong (Kong, Chen, Zheng 2005).

11 For details see the text about Hong Kong on pp. 135–144.

12 Source: DSEC 2005.

13 *Neidi*, the inner area, is one of the terms used especially in Hong Kong to distinguish between the People's Republic of China, which includes Hong Kong, Macao and theoretically also Taiwan, and the area without these cities and islands (for instance Chong 2005).

14 A fact that was overlooked by early western analysts of Chinese tourism, for example by Richter (1983).

15 Another misconception is the wrong, even though surely brand-enhancing, translation of Hong Kong as 'Fragrant Harbour'. The name-giving small harbour, located close to modern Aberdeen on Hong Kong Island, was used to ship locally produced incense sticks, hence the name 'Incense Harbour'. This activity, based on the cultivation of the Guanxiang (*Aquilaria sinensis* (Lour) Gilg.) tree, ceased after the Ming rulers ordered the evacuation of the coastal areas including Hong Kong island during the 1670s (Arlt 1984).

16 Moving the tourism development closer to the concept of a 'tourism triangle of growth' including Taiwan, Fujian, Guangdong, Macao and Hong Kong, as proposed by Chang (2004).

17 The relatively stable figures for Mainland visits to Hong Kong for 1996 (2.3 million), 1997 (2.3 million) and 1998 (2.6 million) do not support the steep increase as found in the responses to the survey of the HKTB. It might well be that the number of onward travelling Mainland visitors was under-reported before 1997.

18 Own calculation. Source: 1984–2001 (Zhang, Pine, Lam 2005); 2002–2003 (Chong 2005); 2004 (HKTB 2005a).

19 In 2005 services from Pearl River Delta harbours Zhongshan, Dongguan, Fuyong, Shekou and Macao were scheduled for ferries going directly to the Hong Kong airport without the necessity of going through Hong Kong immigration (Airport Authority Hong Kong 2005a). In addition, bus services operated by three bus companies from the airport reach 40 towns and cities in Guangdong with 160 daily services (Airport Authority Hong Kong 2005b).

20 The large number of movies and TV programmes showing Hong Kong as the battlefield of Triads, the Chinese version of the mafia, help to shape the perception of Hong Kong as an unsafe place (Huang, Hsu 2005).

21 As a good example of the arbitrariness of positivistic quantitative research, several items concerning meeting relatives were grouped together by Zhang and Lam with 'Meeting new people' into a factor 'Enhancement of human relationships', even so the results did not support this artificial grouping. The

high score for 'Positive attitude of Hong Kong residents and service staff to mainland tourists' is probably more a reflection of the situation (a Hong Kong researcher asking questions in a face-to-face interview at the railway station) then of the real feelings of the visitors. 'Among those who had visited Hong Kong, quite a number of them expressed their bad feeling about Hong Kong people's self-superiority attitude over them' (Huang, Hsu 2005).

22   Wang Tao is also the author of *Jottings from Carefree Travel*, the first travel book about Europe written by a Chinese scholar, describing France, England and Scotland (Cohen 1988).

23   Translation: McAleavy 1953.

24   Source: Zhang, Pine, Lam 2005.

25   Accordingly the Hong Kong tourism industry is reluctant to concentrate on the guests from Mainland China alone: 'Although Mainland China has been, and will continue to be, a major driver of growth, we remain committed to achieving a balanced portfolio of visitors from all markets worldwide', as HKTB Executive Director Clara Chong is quoted in (HKTB 2005a).

26   Source: HKTB figures (HKTB 2002); (HKTB 2005b); Mainland Chinese figures (Chong 2005).

27   In 2005 direct flights to Thailand were operating from Beijing, Chengdu, Fuzhou, Guangzhou, Guilin, Haikou, Hangzhou, Jinghong, Kunming, Shanghai, Shantou, Shenzhen, Xiamen, Xian and Zhengzhou (OAG 2005).

28   Source: Tourism Authority of Thailand 2005.

29   In the People's Republic of China, officially the term *spring festival* is used, even so it is celebrated according to the Chinese Lunar Calendar in January or February and not according to the equinox, which is based on the sun seemingly moving into the spring point in March.

30   In the aftermath of the tsunami of 26 December 2004, six-day tours from Guangzhou to Thailand were offered for only 240 US$ for spring festival 2005 (Shenzhen Daily 2005).

31   The role of the Spanish Isles for German or of Malta for British tourists could be used as parallel phenomena in European tourism.

32   Teng (2005) conducted a survey of 202 domestic tourists in Hangzhou in 2003. The seven risks, seen as most important for a decision for or against a destination were, in hierarchical order, law and order/safety, hygiene, medical support, accommodation, sightseeing location, transportation safety, weather safety.

33   Source: Tourism Authority of Thailand 2005.

34   Source: Tourism Authority of Thailand 2005.

35   Source: Own presentation.

36   Source: Own calculation after WTO 2003; Du, Dai 2005.

37   The six additional provinces are Chongqing, Hebei, Jiangsu, Shandong, Tianjin and Zhejiang.

38   Sources: Sources: Pan, Laws 2001; Yu, Weiler 2001; Guo 2002; WTO 2003; ATC 2004; Australian Senate 2005; and Du, Dai 2005.

39   Even so, Australia was before 1999 already ranked repeatedly among the top ten preferred destinations for private trips (Guo 2002: 2).

40   These figures seem very high compared to the total number of arrivals. Unfortunately the Australian statistics include students and employees staying less than a year in Australia. For the same reason the average stay for Chinese is given by the ATC as 37 days. The office of the ATC in Shanghai reports an average of Chinese tourists of only ten nights in Australia.

41   As car accidents and drowning were the main reasons for accidental deaths, this reflects the group-travelling behaviour of Asian tourists who are less likely to drive their own car, motorcycle or bicycle and are less involved in water sports.

42 WTO (2003) gives the example of 600 employees of a successful clothing factory being treated to an incentive tour to Australia in 2001.

43   For each city, a total of 500 interviews were conducted with Chinese potential travellers – defined as those adults aged 18–64 years who would either definitely or probably travel outside of Hong Kong, Macau, the Philippines, Singapore, Malaysia, Thailand and Korea in the next 3–4 years.

                      (ATC 2004)

This definition explains the otherwise surprisingly low 'intention' readings for the destinations excluded in the definition.

44 Again, the statistics have to be looked at with care: education fees are making up 14 per cent of the total expenditure of Chinese visitors in 1998, compared to 6 per cent of all visitors (Guo 2002: 4). New South Wales accordingly stated for the period 1999/2000 that a typical Japanese tourist spent 713 US$ on a visit to this Australian state, while the average Chinese tourist spent 2,200 US$, almost four times as much as a Japanese (People's Daily 2001). The inclusion of students obviously distorts the picture, even so their relative weight has been reduced by the faster growth of tourists arrivals in the last years.

45 Yu and Weiler (2001) themselves qualify the reliability of their – and other – surveys:

It is surprising that respondents rate the perceived importance of attending casinos very low, although their level of participation are high. Findings from interviews with tour guides also indicate that Chinese tourists are interested in attending casinos. The reported low rating of importance may not reflect the real opinions of respondents, for the reason that gambling is prohibited in China.

                      (Yu, Weiler 2001: 88)

46 Source: ATC 2004.

47 Junek, Binney and Deery (2004) guess that the reason for this is the high price of renting-out. The do not consider the possibility that they expect to be given interpretation of sights by their guides free of charge as part of their package tour. For a more general discussion of the pitfalls of using western categories on Chinese tourists see Chapter 8.

48 Prideaux's text discusses the Korean visitors of Australia, but this insight is true for the Chinese visitors to Australia as well.

49 Parallel to Australia, the six provinces Chongqing, Hebei, Jiangsu, Shandong, Tianjin and Zhejiang were added to the list in mid-2004.

50 Source: CYTS 2005.

51 Source: TRCNZ 2005.

52 Figures based, if not stated otherwise, on TRCNZ 2005.

53 In comparison the figure for German visitors for Auckland was only 17 per cent of all overnight stays, even though Auckland tops the list for German visitors also. A map showing the places visited in New Zealand in their relative importance would look sprinkled all over for German visitors, whereas for Chinese visitors a big and a medium-sized blob would be accompanied by a few tiny specs only.

54 Source: Becken 2003, using data of the IVS by Tourism New Zealand.

55 Source: TRCNZ 2005.

56 Own calculation. Source: TRCNZ 2005.

57 Source: TRCNZ 2005.

58 Source: Own presentation.

59 Other sources quoted even a figure of envisioned 250,000 Chinese tourists (Travelperu 2005).

60   The next time a Chinese president shows up, somebody should tell him to put the United States and Latin America on his government's list of 'approved' travel destinations. It's bad enough for us to be among the major trading partners of a dictatorship that allows slave-like work conditions, tolerates child labor, represses dissent and controls its people's right to travel. It's even worse when that regime doesn't allow us to benefit from its country's 100 million outbound tourism bonanza.

(Oppenheimer 2004)

61   The statistics of the Office of Travel and Tourism Industries (OTTI) list China as number 13, because they include Hong Kong. If the 300,000 visitors from Taiwan would be included, Chinese visitors would rank on sixth position of source markets, level with South Korea.

62   Both US and Canadian statistics include students, which distort the figures for overnight stays, length of stay, etc.

63   Source: APB Canada 2002; CTC 2005; OTTI 2005.

64   Several one-to-one copies of the White House exist in China, as lodging for some very wealthy nouveau riche (Bartsch 2005).

65   Source: OTTI 2005.

66   American-Born Chinese.

67   The answer to that question is given to the BBC reporter accompanying a group of Chinese tourists in California: 'Given their high expectations, it's not surprising they are disappointed. Even lovely San Francisco doesn't fit the bill. "If that's going to be the end result of China's development, "says one, "then I'm really in despair"' (Dunlop 2004).

68   The 'world's most famous Canadian' is a Chinese TV personality: Mark Rowswell, known as 'Dashan' does Chinese language comedy, presents Saturday night shows and is ubiqutious in commercials (Osnos 2005).

69   Source: OTTI 2005.

70   Source: CTC 2005; Ontario Tourism Marketing Partnership Cooperation 2005.

71   This figure of OTTI (2005) is however bundling Mainland and Hong Kong Chinese together. For Hong Kongers it may be more 'normal' to travel to the USA without having visited other countries in the vicinity before.

72   The figures given by US statistics show that the highest level of spending was reached already in 1998 with 1.14 billion US$, falling below 1 billion to 958 million US$ in 2002 (OTTI 2005).

73   Source: Own presentation, Figures: WTO 2003; Du, Dai 2005.

74   Figures are for 2003. Trade figures are according to European statistics. Chinese customs statistics indicate lower trade figures as they do not include goods exported using Hong Kong as an entrepôt.

75   Membership countries are: Austria, Belgium, Czech Republic, Cyprus, Denmark, Estonia, Finland, France, Germany, Greece, Hungary, Ireland, Italy, Latvia, Lithuania, Luxembourg, Malta, the Netherlands, Poland, Portugal, Slovakia, Slovenia, Spain, Sweden and the United Kingdom.

76   The 15 Schengen countries are: Austria, Belgium, Denmark, Finland, France, Germany, Greece, Iceland, Italy, Luxembourg, the Netherlands, Norway, Portugal, Spain and Sweden. All these countries except Norway and Iceland are EU members. Switzerland is being associated to the agreement.

77   Source: Own presentation.

78   Visa requirements may vary from one Schengen country (embassy) to another. . . . When one embassy (often with the help of its national airline) loosens its requirements, news will quickly spread out and the number of applicants will sharply increase as a result. After the increase continues for half a year or so, problems are bound to arise (the embassies find themselves short-handed or overstay rises) and the embassy will usually change their

policy and the travel agencies will all flock to the next easiest embassy. It is said 2000 was the year of the Austrian Embassy, 2001 was the year of Austria and Finland and this year [2002] Finland (Beijing), France (Shanghai and Guangdong) and Austria (Beijing) are considered the easiest places to get a visa.

(STB 2002: 64)

79 Own calculation. Sources (Chinese figures): WTO 2003; TBP 2004. For EU until 2003: China Daily 2004c; EU 2004; Tiplady 2004. For Malta: Malta 2004. For Switzerland: Switzerland 2005.
80 Where you can have your passport stamped by the tourist office, as Taras Grescoe describes rather unflatteringly: 'Luxembourg, the rest stop of Europe, is tailor-made for bus tours' (Grescoe 2004: 71).
81 Source: DZT 2005.
82 Rumours in the European incoming industry put the figure of illegal immigrants into the region of 5,000 since the introduction of EU-wide ADS visa procedures.
83 Cartoons in an article 'It is not easy to get ADS tourist visa to Europe' depict a blonde consulate officer imagining the harmless Chinese tourists in front of him as a criminal and is shovelling a large number of forms and papers, grinning menacingly, towards a long line of Chinese waiting in front of his counter (Li 2005).
84 Source: Statistisches Bundesamt 2005. Only commercial accomodations with more than eight beds are included.
85 Source: Own calculation, data: Statistisches Bundesamt 2005.
86 Before the German reunification, only very few Chinese lived in the German Democratic Republic (GDR). An agreement from 1986 between China and the GDR to bring 90,000 Chinese to East Germany for working and training programmes could not be realized as the GDR vanished a few years later in 1990 (Gütinger 1998).
87 Source: Kelemen 2005.
88 Source: Kelemen 2005.
89 Source: Hamburg 2005.
90 Source: DZT 2005.
91 Source: Indrich 2004.
92 'A visit to Europe is Cathedral – snoozing in the bus – Cathedral – snoozing in the bus – Cathedral' complained a frustrated Chinese tourist to the author.
93 In 2002, prior to ADS, the STB described the arrangement for leisure tourism groups like this:

> More importantly there is a number of 'black horse agents' in each city (often consisting of a group of five to ten young people who pay a 'management fee' to a travel agent in order to operate as one of the agent's branches). These 'black horse agents' play the role as authorised Europe-bound operators and they collectively control app. 80% of the market to Europe. If caught red-handed by CNTA, they usually get away by paying a fine. . . . If the worst thing happens, namely that their parent agent gets his license suspended – the 'black horse agents' can always get adopted (through a fee) by another agent and continue to play the role of a 'black horse'.
>
> (STB 2002: 63)

94 The seasonality is not necessarily based on climate, as for some source markets groups winter is the equally important (for instance Japan) or even main (for instance Russia) season to visit Northern Europe. For transit passengers, climate realities will be less important.
95 Sources: Finland, Iceland 1998–2000 (WTO 2003); Finland 2003–2004 (MEK 2005); Finland including Hong Kong (MEK 2005); Sweden (STA 2004).

96 H.C. Andersen, an ardent traveller, coined the famous statement: 'Travelling is living' (Andersen 1885 cited in Pinkert, Therkelsen 1997).
97 Figures are given for 2001 and 2002.
98 Source: Finnish Tourist Board (2005: 145)
99 Source: STA 2005.
100 Officially called the Asian–African Conference, it was held in Bandung/ Indonesia, in April 1955. The conference marks the beginning of the Non-Aligned Movement. During the conference, Premier and Foreign Minister Zhou Enlai succeeded in breaking the isolation of the People's Republic within neutral Asian and African countries. In April 2005, a commemorative new Asia–Africa summit was held with over 100 government leaders including Chinese President Hu Jintao attending (Qin 2005).
101 Source: Own presentation.
102 More than 50 per cent are persons travelling from the neighbouring African countries, almost all of them not for leisure purposes.
103 In 2005, approximately 8 Rand equals 1 Euro.
104 Some 90 per cent of all Chinese crossing borders into South Africa arrive by air, the rest by road and sea (SAT 2004a).
105 Source: SAT 2004a, 2004b. Estimate for 2004, Source: DPS 2005a. For 2002, SAT 2004a gives 24,247, SAT 2004b 25,849, CNTA 2004a 25,400. Pick your favorite number.
106 Source: SAT 2004a, figures include Hong Kong visitors.
107 Source: SAT 2004b, figures include Hong Kong visitors.
108 Source: SAT 2004c, figures include Hong Kong visitors.
109 Source: SAT 2004c, figures for third quarter 2003.

7 Product adaptation and marketing of tourism destinations and products for Chinese outbound tourists

1 The ECTW Award was created in 2004 by the China Outbound Tourism Research Project, based in Germany, to collect and distribute knowledge and best practice examples from European tourism destinations and companies working in the field. The award is sponsored by the United Nations World Tourism Organization, Asia Pacific Representative and supported by Global Refund, a supplier of financial services. The award jury is made up of several leading experts on China outbound tourism. Awards are given every year in five categories: Product innovation, Human resource development, Marketing, Use of information technology and Overall performance. Gold medal winners of 2004 and 2005 included the Tourism Boards of Vienna, Germany, Slovakia, Hesse and Switzerland, as well as the companies Accor Hotels, Hotellerie-Suisse and Munich Airport. In 2006, the Gold Medal winners were the Wimbledon Museum, Gassan Diamonds, Visit London, the German Fairy Tale Route and the Tourism Board of Austria. Detailed information about the best practices can be found on www.china-outbound.com/award.htm.
2 In societies, where – unlike China today – tourism has developed into a practice already regularly followed by the parents, a new member of the society is practically *born* as a tourist, having no recollection in adult life of a time *before* travelling for leisure. In such a touristified society every third phase of remembrance is already intertwined with the next phase of preparation, every first phase is informed by earlier third phases without any *status nascendi*.
3 Compare Chapter 6.
4 Source: Adapted from Arlt 2004.
5 Some classical characters were simplified in Mainland China in the 1950s. In Hong Kong, Macao, Taiwan and among overseas Chinese the old writing form is still prevalent.

6　In the academic field these problems are also common. For example, even on the back of the book *Tourism in China* (Lew *et al.* 2003), of which Zhang Guangrui is one of the editors, the first name is used instead of the family name. Hall (1997) quotes the works of Zhang Guangrui extensively, but uses also the first name 'Guangrui' as family name. A text of Han Nianyong and Ren Zhuge in the *Journal of Sustainable Tourism* (2001) is put down in the bibliographical quote as 'H. Nianyong and R. Zhuge'. When accordingly the texts is quoted as 'Nianyong, Zhuge 2001' as do Deng *et al.* (2003), things start to get complicated.

7　Japanese place names are pronounced with the Chinese reading of the characters, changing, for instance, the pronunciation of the 'Eastern Capital', Tokyo, to Dongjing.

8　One of the greatest successes with Chinese tourists ever witnessed by this author was the appearance of a Bavarian lady dressed in a local *Dirndl* costume on stage during the Oktoberfest in Munich – singing two Chinese songs in Chinese.

9　Young Urban Professionals are mostly not called Yuppies in China, rather another term imported from the USA, *BoBo*, Bohemian Bourgeois, is used to describe the post-modern urban middle classes.

10　Warden *et al.* (2003) find for instance that Chinese travellers in other countries do react to service failures in restaurants with the same level of tolerance as at home. If this failure is not followed by any kind of recovery (free drink, apology, etc.) this is taken more seriously than at home. If a recovery is undertaken, the perceived service quality actually improves beyond the normal level. With the recovery a relationship is built and face given to the Chinese guest. Furthermore, as any Chinese outside China perceives him- or herself as a representative of the whole country, rectified service failure also prevent a more general insult.

11　Agencies have to be careful however not to present the merits of the foreign destination in way that could harm Chinese nationalist sentiments. In 2004, Nike had to withdraw a TV spot from Chinese television, which showed an American NBC star beating an animated dragon in a basketball game after audiences had protested against the insult to the Chinese nation as represented by the dragon (China Daily 2004a).

12　Sponsoring the acquisition of such a player for a local club by the responsible DMO would be a clever way of cross-marketing.

13　What Shields calls 'place-myths' (Shields 1991; Crouch, Lübben 2003b).

8　Consequences of China's outbound tourism development for tourism studies

1　One should also mention texts from Ooi Can-Seng, Li Yiping, Philip Feifan Xie and Su Xiaobo even so they mainly only implicitly deal with the questions discussed here.

2　Guyer-Feuler: *Contributions to a Statistic of Tourism*; Stradner: *Tourism*; Schullern zu Schrattenhofen: *Tourism and National Economy*; Bodio: *About the Travels of Foreigners in Italy and their Spending*; Picard: *The Industry of the Traveller*.

3　'Research Institute for Tourism' (Berne) and 'College for Tourism' (St Gallen).

4　To name three examples: Dean MacCannell, John Urry and C. Michael Hall.

5　Hall (2005: 5) names Law, Marketing, Management, Finance and accounting, Hospitality, catering and restaurant administration, Architecture and design, Transport studies, Leisure studies, Ecology, Geography, Urban and Regional planning, Politics and public policy analysis, Sociology and cultural studies, Anthropology, Psychology. Others disciplines, like Semiotics, Ecology, History and Philosophy could easily be added. Hall provides some seminal texts for each discipline he covers, which without fail include only texts published in the Anglo-Saxon world.

6   Kaye Chon, head of the School of Hotel and Tourism Management of the Hong Kong Polytechnic University, in a keynote presentation to the Second PolyU China Tourism Forum and Third China Tourism Academy Annual Conference in December 2005 in Guangzhou echoed the criticism of Pearce but nevertheless envisaged the 'third wave' of tourism and of tourism sciences innovation after Europe and North America now moving to East Asia.

7   Individualistic approaches in western-dominated science can even be seen in such mundane questions such as citation rules that decree the use of initials for authors' first names only. Such a rule causes much confusion in the Asian world, where for instance in Korean texts almost all authors are called either Lee, Park or Kim. Graburn is one of only a few authors who have responded to this by amending the established forms of citation.

8   Tan, Cheung, Yang (2001).

9   Walton (2005): 'The importance of the contribution of history to the understanding of tourism . . . is now beginning to gain recognition within tourism studies, which has been slow to accept that it needs to learn from historical studies' (Walton 2005: 1). Lomine (2005) is one of the few texts treating tourism before the Grand Tour with modern analytical tools, in this case Roman tourism in the Augustan Society (44 BC–AD 69).

10 Encouragingly, in the December 2005 issue of *China Tourism Research*, a research note on semiotics and tourism is included, even though 'the application of semiotics into tourism studies [in China] is rare' (Peng 2005: 482). The text presents some very basic information on the topic but with an interesting Chinese spin, for instance: 'The tourist is a visitor of his destination country or area and thus a guest of that country' (Peng 2005: 477); and 'natural sites also have symbolic meanings, for example, high mountains symbolise heroic figures and primeval tropical forests are of antiquity and full of mystery' (Peng 2005: 479).

11 Hannam, Sheller and Urry, in their editorial for the first issue of the magazine *Mobilities* claim that 'Mobility has become an evocative keyword for the twenty-first century and a powerful discourse that creates its own effects and contexts' (Hannam, Sheller, Urry 2006: 1).

9 The future of China's outbound tourism

1   See Chapter 3 for a detailed discussion of the first two phases.

2   The EIU for instance had prognostizised a number of 34 million in 2005 (EIU 2005).

3   See Chapter 6.

4   Although the casino resort said the drawings were meant only to distinguish their Chinese guests from Muslims, who cannot eat pork – or gamble – the Chinese demonstrated their pique by staging a sit-in in the hotel lobby and belting out their national anthem. It took 40 police officers with dogs to clear them out.

(Arnold 2005)

5   Some sources put the time of the decision of the policy change as August 2005. It is hard to say if the growth of less than 10 per cent in outbound travels, which was known already for the first half of 2005 by that date, helped to support the decision making process.

6   In November 2005, the *China Daily* mentioned the problem of Chinese citizens travelling to Europe under the pretence of tourism with the intent of staying

there. The article blamed 'cunning Snakeheads', human smugglers, as the villains but also admitted bluntly that the 'economic gap serves as a driving force behind illegal immigration. ... Chinese emigrants mainly select European and American countries as their destinations' (Jiang 2005: 2).

7 For 2005, the USA alone incurred a merchandise trade deficit of 200 billion US$ (Economist 2005g: 45).

8 The largest democratic movement of 2005 in China appears to have been an amateur singing TV contest similar to the 'Pop Idol' format in Britain. Some 400 million people, almost a third of the population, watched the final in August 2005, with 8 million of them voting via text message. The show sparked a debate about the alleged 'awakening of democratic consciousness among the younger generation' (Economist 2005h: 56). Otherwise, little progress towards a civil society in China could be witnessed. However, this author, being a Berliner, learned in November 1989 first-hand the historical lesson not to be fooled by the outwardly stable appearance of undemocratic governments.

9 Using foreign countries in a way Farrer described the role of discos in Shanghai for young urbanites as 'not a space to display who they were, but who they wanted to be' (Farrer 2000: 46).

10 Junket agents offer package deals to visit casinos in Las Vegas, Macao or other places like the Northern Marianas, where solvent players are offered discounted or even free transport and accomodation in exchange for a promise to risk an agreed minimum amount of money in the casino.

11 With domestic and inbound tourism as one instrument – and source of financing – for the deification of the glorious past of the motherland (Sofield, Li 1998a) and 'Red tourism' (see Chapter 5) as the political tool.

12 Hui and Wan (2005) point out that it is rather strange to witness Mainland Chinese tourists visiting the 'Chinatown' of Singapore (a city in which anyway three quarters of the population are of Chinese origin), built originally as a fake attraction for western tourists.

13 An example within the literature on China's outbound tourism is the text of Guo (2002), who – surely without any intent of jingoism – faithfully reproduces the Chinese version of historiography by claiming that 'Under the Mongol emperors of Yuan, the Chinese [*sic*] Empire consisted of all China, Korea, Central Asia, India, Persia, and much of Asia Minor, and most of Russia' (Guo 2002: 36).

14 As Feng observes for domestic Chinese leisure behaviour: 'Going bowling, playing tennis and playing golf [are] regarded not merely as a kind of physical exercise, but also as a demonstration of wealth' (Feng 2005: 135).

15 Since 1 December 2005, the official name has been changed to UNWTO United Nations World Tourism Organization to distinguish it from the World Trade Organization.

16 This figure does not include the 350 million travellers to Russia in the first half of 2005, even though most of them are actually border travellers in Siberia and the Russian Far East and not visitors to the European part of Russia. If included, Asia could be seen as the destination of 93 per cent of all Chinese outbound travellers.

17 Like for instance during the spring festival 'Golden Week' in January/February 2006, when within seven days the more than 600,000 Mainland Chinese visitors to Macao clearly outnumbered the total population of Macao of less than 500,000 inhabitants (Xinhua 2006b).

18 The number of Japanese outbound travellers reached 10 million for the first time in 1990. Only in the year 2000 more than 18 million Japanese left their islands to travel abroad, since then the figure stayed below 18 million (OECD 2002; JTM 2006).

Bibliography

All websites in the Bibliography were accessed in November 2005 or later.

Adler, Nancy J. and Doktor, Robert (1986) From the Atlantic to the Pacific Century: Cross-cultural Management Reviewed. In *Journal of Management*, 12: 2: 295–318.

Airport Authority Hong Kong (2005a) *Cross Boundary Ferry Transfer Service Leaflet*. Hong Kong (Airport Authority Hong Kong).

Airport Authority Hong Kong (2005b) *Cross Boundary Coach Service Leaflet*. Hong Kong (Airport Authority Hong Kong).

Aitchison, Cara, MacLeod, Nicola E., Shaw, Stephen J. (2000) *Leisure and Tourism Landscapes. Social and Cultural Geographies*. London and New York (Routledge).

Alcantara, Nelson (2005) Third US Carrier Enters China. In *Travelwirenews*, 16 June 2005, www.travelwirenews.com.

Al-Hamarneh, Ala (2004) Islamischer Tourismus: eine Chance für die Arabische Welt? In Meyer, Günter (ed.) *Die Arabische Welt im Spiegel der Kulturgeographie*. Mainz (CERAW). 339–346.

Allison, Robert E. (ed.) (1989a) *Understanding the Chinese Mind. The Philosophical Roots*. Hong Kong (Oxford University Press).

Allison, Robert E. (1989b) An Overview of the Chinese Mind. In Allison, Robert E. (ed.) *Understanding the Chinese Mind. The Philosophical Roots*. Hong Kong (Oxford University Press). 1–25.

AlSayyad, Nezar (ed.) (2001) *Consuming Tradition, Manufacturing Heritage. Global Norms and Urban Forms in the Age of Tourism*. London (Routledge).

Andersen, Hans Christian (1855) *Mit Livs Eventyr*. Copenhagen (Nordisk Forlag).

Andersen, Vivien, Prentice, Richard C., and Watanabe, Kazumasa (2000) Journeys for Experiences: Japanese Independent Travellers in Scotland. In *Journal of Travel and Tourism Marketing* 9: 1/2: 129–151.

Anderson, Benedict (1991) *Imagined Communities. Reflections on the Origin and Spread of Nationalism*. Revised edition. London (Verso).

Anderson, Benedict (1998) *The Spectre of Comparisons: Nationalism, Southeast Asia and the World*. London (Verso).

Ap, John and Mok, Connie (1996) Motivations and Barriers to Vacation Travel in Hong Kong. Paper presented at the Asia Pacific Tourism Association Conference, Townsville 1996.

APB Canada (ed.) (2002) Canada Must Move on China's Growing Outbound Tourism Market. In *Asia Pacific Bulletin*, 6 September.

APEC (2004) Chile is an Authorizised Destination for Chinese Tourists, www. apec2004.cl.

Apostolopoulos, Yiorgos, Leivida, Stella and Yiannakis, Andrew (eds) (1996) *The Sociology of Tourism. Theoretical and Empirical Investigations*. New York, London (Routledge).

Appleton, Kate and Yu, Doris (2005) Hey, Big Spenders. In TFWA AC Nielsen Press release, 19 May 2005.

Aramberri, Julio (2001) The Host Should Get Lost. Paradigms in the Tourism Theory. In *Annals of Tourism Research*, 28: 3: 738–761.

Arlt, Wolfgang Georg (1984) Die ökonomischen und politischen Beziehungen zwischen Hong Kong und dem übrigen China bis 1898. FU Berlin, Institute for East Asian Studies, Master's thesis.

Arlt, Wolfgang Georg (2001) The Tumen River Economic Development Programme in the Tumen River Area 1990–2000. PhD thesis, FU Berlin.

Arlt, Wolfgang Georg (2002a) Die Eingeborenen sind wir. Ostasiaten als Inbound-Touristen. In *Voyage Jahrbuch für Reise- & Tourismusforschung*, 5: 144–153.

Arlt, Wolfgang Georg (2002b) Internet as a Cross Culture Marketing Tool for Inbound Tourism. A Comparison of Non-German Language Websites of German, Austrian and Swiss-German Tourism Promotion Institutions. In Leisure Futures (ed.) *Proceedings of Leisure Future Conference, Innsbruck 2002*.

Arlt, Wolfgang Georg (2002c) Internet as a Cross Culture Incoming Tourism Communication Tool. In Wales Tourism Alliance (ed.) *Proceedings Wales Tourism Alliance Conference, Cardiff 2002.*

Arlt, Wolfgang Georg (2003) Marketing European Attractions to Chinese Visitors. Paper presented at TiLE Conference Berlin.

Arlt, Wolfgang Georg (2004) Chinesischer Outbound Tourismus in Deutschland. Entwicklung – Perspektiven – Herausforderungen. In Maschke, Joachim (ed.) *Jahrbuch für Fremdenverkehr 2004*. München (DWIF). 7–34.

Arlt, Wolfgang Georg (2005a) Tourists' Motivation in Comparison: Chinese vs German. Why Do Germans Travel to China and Chinese Travel to Germany. In Zhang, Guangrui, Wei Xiaoan and Liu Deqian (eds) *Green Book of China's Tourism 2003–2005 China's Tourism Development: Analysis and Forecast (2003–2005)*. Beijing (Social Sciences Academic Press). (*Green Book of China's Tourism No. 4*) 427–436 (in Chinese).

Arlt, Wolfgang Georg (2005b) 'A VIRTUAL HUANYING, SELAMAT DATANG AND HERZLICH WILLKOMMEN!' The Internet as a Cross-cultural Promotion Tool for Tourism. In Haven-Tang, Claire and Jones, Eleri (eds) *Tourism SMEs, Service Quality and Destination Competitiveness: International Perspectives*. Wallingford, CT (CAB International). 325–336.

Arlt, Wolfgang Georg (2005c) Sustainable Tourism Development in Japan, China and Europe. Paper presented at APU Asia Pacific University Beppu/Japan 26 January.

Arlt, Wolfgang Georg (2005d) Sustainable Tourism in Japan. In Burns, Peter (ed.) *Proceedings of the 4th International Symposium on Aspects of Tourism, Eastbourne 2005.*

Arlt, Wolfgang Georg (2005e) Product Development for Chinese Outbound Tourists. Paper presented at the Second International Forum on Chinese Outbound Tourism, Beijing 20–21 November 2005. In *Proceedings of the Second International Forum on Chinese Outbound Tourism, Beijing, 2005*. 91–98.

252   *Bibliography*

Arlt, Wolfgang Georg (2006a) (forthcoming) Not Very Willkommen – The Internet as a Marketing Tool for Attracting German-speaking Tourists to Non-European Destinations. In Pease, Wayne, Rowe, Michelle and Cooper, Malcolm (eds) *Information and Communication Technologies in Support of the Tourism Industry*.

Arlt, Wolfgang Georg (2006b) (forthcoming) Sustainability in China and Europe. Paper for the German-Chinese Zhangjiajie National Park Conference 2004. In Job, Hubert (ed.) *Proceedings of the German-Chinese Zhangjiajie Conference October 2004*.

Arlt, Wolfgang Georg (2006c) 2nd International Forum on Chinese Outbound Tourism (IFCOT), Beijing, 20–21 November 2005. In *China Outbound Project Newsletter* January 2006, www.china-outbound.com.

Arlt, Wolfgang Georg (2006d) (forthcoming) Communicating Best Practices Examples from Europe – The ECTW European Chinese Tourists Welcoming Award. In *China Tourism Research*.

Arlt, Wolfgang Georg and Domnick, Heinz-Joachim (1994) *China – Cradle of Knowledge. 7000 Years of Inventions and Discoveries*. Berlin (Verlag Ute Schiller).

Arlt, Wolfgang Georg and Kelemen, Marcel (2006) *Chinese Travellers Shopping Behaviour*. Berlin (COTRP).

Arlt, Wolfgang Georg, Kelemen, Marcel, Kröher, Henry *et al.* (2006) (eds): *How to Enter China's Tourism Market*. Berlin (COTRP).

Arnold, Wayne (2005) Chinese Tourists Getting a Bad Image. In *New York Times*, 23 October 2005.

Asquith, Pamela J. and Kalland, Arne (eds): *Japanese Images of Nature*. Richmond (Curzon). 221–235.

Artman, Heikki (ed.) (2005) *Helsinki Tourism Statistics 1/2005*. Helsinki: City of Helsinki Tourism and Conventions Bureau.

ATC (ed.) (2004) Chinese Tourists Activities, www.tourism.australia.com.

AUMA Ausstellungs- und Messeausschuss der Deutschen Wirtschaft e.V. (ed.) (2004) *Exhibition Market China 2004/2005*. Stuttgart (local global Medien).

Australian Bureau of Statistics (ed.) (2003) China's Tourists, www.abs.gov.au.

Australian Senate (ed.) (2005) Australia's Relationship with China. Proof Hansard Transcript of Senate Hearing, Foreign Affairs, Defence and Trade References Committee, Canberra, 21 June.

Bailey, Murray (2004) Outbound Markets in Asia. Paper for the ITB Berlin 2004.

Bao, Jigang (1996) *Study on the Tourism Development – Theory, Methodology, Practice*. Beijing (Sciences Press) (in Chinese).

Barabantseva, Elena (2005) Trans-nationalising Chineseness: Overseas Chinese Policies of the PRC's Central Government. In *Asien*, 95: 7–28.

Bardsley, Alex G. (2003) The Politics of 'Seeing Chinese' and the Evolution of a Chinese Idiom of Business. In Jomo, Kwame Sundaram and Folk, Brian C. (eds): *Ethnic Business. Chinese Capitalism in Southeast Asia*. Abingdon (Routledge). 26–51.

Bartsch, Bernhard (2005) Im Land der Goldkragen. In *Berliner Zeitung*, 19 May, www.berlin.online.de.

BAT (2003) BAT Freizeit-Forschungsinstitut (ed.) *Deutsche Tourismusanalyse 2003*. Hamburg (BAT).

Baudrillard, Jean (1988) *Selected Writings*. Cambridge (Polity Press).

Becken, Susanne (2003) Chinese Tourism to New Zealand. Unpublished discussion paper.

Becken, Susanne (2005) Towards Sustainable Tourism Transport. An Analysis of Coach Tourism in New Zealand. In *Tourism Geographies*, 7: 1: 23–42.

Beer, Jennifer (1993) Packaging Experiences: Japanese Tourism in Southeast Asia. PhD dissertation in Anthropology. Berkeley: University of California.

Beijing Gongye University (ed.) (2005) *66 Places to See in a Lifetime*. Beijing (Beijing Gongye University Publishing House) (in Chinese).

Belaunde, Enrique and Rios, Juan Carlos (2004) Perfil del Turista Chino. Estrategias para promover la offerta turistica del Peru en China. Shanghai (Consulate General).

Bender, Barbara and Winer, Margot (2001) *Contested Landscapes. Movement, Exile and Place*. New York, Oxford (Berg).

Benton, Gregor and Pieke, Frank (eds) (1998) *The Chinese in Europe*. London (Macmillan).

BHG (ed.) (2005) *Andere Länder – andere Sitten. Interkulturelle Kommunikation für Hoteliers, Gastronomen und Touristiker*. Munich (BHG).

Blok, Anders (2002) China Social Anthropology Study. Scandinavia Tourism Research report, November. Tokyo (Scandinavian Tourist Board).

Blume, Georg (2003) Einige werden zuerst reich. In *DIE ZEIT* No. 39, 18 September.

Blumenstengel, Peter (ed.) (2003) *Marktinformation Japan 2003*. Tokyo (DZT).

Bocock, Robert (1993) *Consumption*. London (Routledge).

Bond, Michael H. and Hwang, K.K. (eds) (1986) *The Psychology of the Chinese People*. New York (Oxford University Press).

Bond, Michael H., Leung, Kwok and Wan, K.C. (1982) How Does Cultural Collectivism Operate? In *Journal of Cross Cultural Psychology*, 15: 3: 337–352.

de Botton, Alain (2002) *The Art of Travel*. London (Hamish Hamilton).

Bourdieu, Pierre (1979) *La distinction. Critique sociale du jugement*. Paris (Les editions de minuit).

Bowden, Jiaolan (2003) A Cross-cultural Anlysis of International Tourist Flows within China's Three International Gateway Cities – Beijing, Shanghai and Guangzhou. Paper presented at the Tourism Modelling and Competitiveness Congress, Paphos/Cyprus.

Brameld, Theodor B. (1977) *Tourism as Cultural Learning*. Washington, DC (University Press of America).

Breitung, Werner (2001) *Hong Kong und der Integrationsprozess. Räumliche Strukturen und planerische Konzepte*. Basel (Wepf).

Brown, David (1996) Genuine Fakes. In Selwyn, Tom (ed.) *The Tourism Image: Myths and Myth Making in Tourism*. Chichester (John Wiley & Sons). 33–48.

Buckley, Christopher (1999) How a Revolution Becomes a Dinner-party: Stratification, Mobility and the New Rich in Urban China. In Pinches, Michael (ed.) *Culture and Privilege in Capitalist Asia*. London (Routledge). 208–229.

Burns, Peter (1999) *An Introduction to Tourism and Anthropology*. London (Routledge).

Burns, Peter (ed.) (2005) *Proceedings of the 4th International Symposium on Aspects of Tourism, Eastbourne 2005*.

Burns, Peter and Holden, Andrew (1995) *Tourism – A New Perspective*. Hemel Hempstead (Prentice Hall Europe).

Buruma, Ian and Margalit, Avishai (2005) *Occidentalism. The West in the Eyes of Its Enemies*. London (Penguin).

Cai, Liping A., Boger, Carl and O'Leary, Joseph T. (1999) The Chinese Travellers to Singapore, Malaysia and Thailand: A Unique Chinese Outbound Market. In *Asia Pacific Journal of Tourism Research*, 3: 2: 2–13.

Cai, Liping A., Hu, Bo and Feng, Ruomei (2001) Domestic Tourism Demand in China's Urban Centres: Empirical Analysis and Marketing Implications. In *Journal of Vacation Marketing*, 8: 1: 64–74.

Cai, Liping A., You, Xinran and O'Leary, Joseph T. (2001) Profiling the US-Bound Chinese Travelers by Purpose of Trip. In *Journal of Hospitality & Leisure Marketing*, 7: 4: 3–17.

Cairncross, Frances (1997) *The Death of Distance. How the Communications Revolution Will Change our Lives*. London (Orion Publishing Group).

Cameron, Nigel (1978) *Hong Kong – The Cultured Pearl*. Oxford (Oxford University Press).

CanadaTourism (2005) CTC Opens New Office in Beijing and Welcomes Move to Grant Canada Approved Destination Status, www.canadatourism.com.

Cang, Xin (2002) *Existence in Translation*. Hong Kong (timezone8).

CBS (ed.) (2005) *Migration and Tourism Statistics 2005*. Tel Aviv (CBS).

CCPH (ed.) (2001) *Atlas of Popular Tourist Routes in China. Attached with Tourist Guide to Singapore, Malaysia and Thailand*. Tianjin (CCPH) (in Chinese).

CEN (2005a) Tourism Expo is Feast for Global Visitor's Eyes. In *China Economic Net*, 29 September, en.ce.cn.

CEN (2005b) Shanghai Welcomes Visa-free Russian Tour. In *China Economic Net*, 22 September, en.ce.cn.

CEN (2005c) Mainlanders to Travel the World. In *China Economic Net*, 27 September, en.ce.cn.

Chai, Joseph C.H., Kueh Y.Y. and Tisdell, Clement A. (eds) (1997) *China and the Asia-Pacific Economy*. New York (Nova Science Publishers).

Chan, Henry (2000) From Quong Tarts to Victor Changs: Being Chinese in Australia in the Twentieth Century. CSCSD public seminar at the Australian National University, 24 May, Sydney.

Chandler, Daniel (2002) *Semiotics*. London (Routledge).

Chang, Ton Chuang (2004) Tourism in a 'Borderless' World: The Singaporean Experience. In *Asia Pacific Issues – Analysis from the East-West Center*, 73.

Chen, Aimin, Liu, Gordon G. and Zhang, Kevin H. (eds) (2004) *Urban Transformation in China*. Aldershot (Ashgate Publishing).

Chen, Changfeng (1998) Rising Chinese Overseas Travel Market and Potential for the United States. In *Proceedings of 3rd Annual Graduate Education and Graduate Students Research Conference in Hospitality & Tourism, Houston 1998*, www. hotel-online.comTrendsAdvancesinhospitalityresearch.

Chen, Jian (1994) *China's Road to the Korean War. The Making of the Sino-American Confrontation*. New York (Columbia University Press).

Chen, Xiaomei (2002) *Occidentalism: A Theory of Counter-discourse in Post-Mao China*. Second revised and expanded edition. Lanham, MD (Rowman & Littlefield Publishers).

Cheung, Yin-Wong, Chinn, Menzie D. and Fujii, Eiji (2006) The Chinese Economies in Global Context: The Integration Process and its Determinants. In *Journal of the Japanese and International Economies*, 20: 1: 128–153.

Chi, Xiong Biao (2003) Shenzhen: Endeavor for the Biggest Outbound Tourism Distributing Center in China. Abstract in Wang, Xinjun (ed.) *Proceedings of 1st*

International Forum on Chinese Outbound Tourism and Marketing Shenzhen 17–18 Nov. 2003 Shenzhen. 133–136.

China Aktuell Data Supplement (2005) Institut fuer Asienkunde (ed.) *China Aktuell Data Supplement 2/2005.* Hamburg (China Aktuell Data Supplement).

China Business Weekly (2004) China Ahead of India. In *China Business Weekly,* 1 March.

China Daily (2000) Forging Ahead to Make China a World's Tourism Power. In *China Daily,* 29 March, www.chinadaily.com.cn.

China Daily (2004a) China Bans Nike TV Ad as National Insult. In *China Daily,* 7 December, www.chinadaily.com.cn.

China Daily (2004b) Chinese Travellers Ready for Tour in Europe. In *China Daily,* 14 October, www.chinadaily.com.cn.

China Daily (2004c) EU Attractive Destination in Years to Come. In *China Daily,* 17 February, www.chinadaily.com.cn.

China Daily (2004d) 700 Complaints. In *China Daily,* 23 March, www.chinadaily.com.cn.

China Daily (2004e) Green Light Given, Next Stop: Europe. In *China Daily,* 27 February 2004, www.chinadaily.com.cn.

China Daily (2005) Income Gap in China Widens in First Quarter. In *China Daily,* 19 June, www.chinadaily.com.cn.

Chinahospitalitynews (2006) Lonely Planet Starts New China Travel Publishing Partnership. In *Chinahospitalitynews,* 10 January 2006, www.chinahospitalitynews.com.

Chinanews (2005) 600 billion Yuan 'Gambling' Outflow. In *Chinanews*, 25 January, www.chinanews.cn.

China Statistical Bureau (ed.) (2004) *International Statistical Yearbook 2004.* Beijing (China Statistics Press).

Chinatravelnews (2003a) Little Interest Seen in New Tourism Venues. In *Chinatravelnews*, 20 October, www.chinatravelnews.com.

Chinatravelnews (2003b) Lack of Information Hinders Outbound Tourism Development. In *Chinatravelnews*, 29 September, www.chinatravelnews.com.

Chinaview (2004) US Group Tour Requires Big Deposit. In *Chinaview*, 9 December, www.chinaview.cn.

Chinaview (2005) Brazil Opens to Chinese Tourists. In *Chinaview*, 11 July, www.chinaview.cn.

Choi, Tat Y. and Chu, Raymond (2000) Levels of Satisfaction among Asian and Western Travellers. In *International Journal of Quality & Reliability Management,* 17: 2: 116–131.

Chong, King (2005) CEPA and New Opportunities for Hong Kong's Tourism Development. In Zhang, Guangrui, Wei, Xiaoan and Liu, Deqian (eds) *Green Book of China's Tourism 2003–2005 China's Tourism Development: Analysis and Forecast (2003–2005).* Beijing (Social Sciences Academic Press). (*Green Book of China's Tourism No. 4*) 382–401 (in Chinese).

Chow, Chung-yan and Lai, Chloe (2005) All Geared Up to Bus in the Crowds. In *South China Morning Post,* 10 September.

Chow, W.S. (1988) Open Policy and Tourism between Guangdong and Hong Kong. In *Annals of Tourism Research*, 15: 2: 205–218.

Choy, Dexter J.L. and Gee, Chuck Y. (1983) Tourism in the PRC – Five Years after China Open its Gates. In *Tourism Management*, 4: 2: 85–93.

Choy, Dexter J.L., Guan, Li Dong and Zhang, Wen (1986) Tourism in PR China: Market Trends and Changing Policies. In *Tourism Management*, 7: 3: 197–201.

Chung, Ki-Han and Shin, Jae-Ik (2004) The Relationship Between Destination Cues of Asian Countries and Korean Tourist Images. In *Asia Pacific Journal of Marketing and Logistics*, 16: 2: 82–100.

Civil Aviation Administration Finland (ed.) (2004) *Air Traffic Statistics 2003*. Vantaa (Civil Aviation Administration Finland).

Clifford, James (1988) *The Predicament of Culture: Twentieth Century Ethnography, Literature and Art*. Cambridge, MA (Harvard University Press).

Clifford, James (1989) Notes on Theory and Travel. In *Inscriptions*, 5: 177–188.

Clifford, James (1992) Travelling Cultures. In Grossberg, Lawrence, Nelson, Cary and Treichler, Paula (eds) *Cultural Studies*. London (Routledge). 96–116.

CNNIC (ed.) (2003) *China's Internet Development and Usage Report*. 13th issue. Beijing (CNNIC).

CNNIC (ed.) (2005) *16th Statistical Survey Report on the Internet Development in China*. July 2005. Beijing (CNNIC).

CNTA (ed.) (1998a) *The Yearbook of China Tourism Statistics*. Beijing (China Tourism Press).

CNTA (ed.) (1998b) *The Yearbook of China Tourism Statistics (Supplement) 1998*. Beijing (China Tourism Publishing House).

CNTA (ed.) (1999a) *The Yearbook of China Tourism Statistics*. Beijing (China Tourism Press).

CNTA (ed.) (1999b) *The Yearbook of China Tourism Statistics (Supplement) 1999*. Beijing (China Tourism Publishing House).

CNTA (ed.) (2000a) *The Yearbook of China Tourism Statistics*. Beijing (China Tourism Press).

CNTA (ed.) (2000b) *The Yearbook of China Tourism Statistics (Supplement) 2000*. Beijing (China Tourism Publishing House).

CNTA (ed.) (2001a) *The Yearbook of China Tourism Statistics*. Beijing (China Tourism Press).

CNTA (ed.) (2001b) *The Yearbook of China Tourism Statistics (Supplement) 2001*. Beijing (China Tourism Publishing House).

CNTA (ed.) (2002a) *The Yearbook of China Tourism Statistics 2002*. Beijing (China Tourism Publishing House).

CNTA (ed.) (2002b) *The Yearbook of China Tourism Statistics (Supplement) 2002*. Beijing (China Tourism Publishing House).

CNTA (ed.) (2003a) *The Yearbook of China Tourism Statistics 2003*. Beijing (China Tourism Publishing House).

CNTA (ed.) (2003b) *The Yearbook of China Tourism Statistics (Supplement) 2003*. Beijing (China Tourism Publishing House).

CNTA (ed.) (2004a) *The Yearbook of China Tourism Statistics 2004*. Beijing (China Travel Publishing House).

CNTA (ed.) (2004b) *The Yearbook of China Tourism Statistics (Supplement) 2004*. Beijing (China Tourism Publishing House).

CNTA (ed.) (2004c) *Travel Services in China 2004–2005*. Beijing (China Industry & Commerce Associated Press) (in Chinese).

CNTO (ed.) (2005) *China Tourism Statistics 2004*, www.cnto.org.

CNTO Canada (ed.) (2004a) The Newest Development of China Tourism Industry, www.tourismchina-ca.com.

CNTO Canada (ed.) (2004b) China Emerging as an Important Tourist Generating Country, www.tourismchina-ca.com.

Cohen, Eric (1979) A Phenonemology of Tourist Experiences. In *Sociology*, 13: 179–201.

Cohen, Eric (1995) Contemporary Tourism – Trends and Challenges: Sustainable Authenticity or Contrived Post-modernity? In Butler, Richard W. and Pearce, Philip L.: *Change in Tourism: People, Places, Processes*. London (Routledge).

Cohen, Eric (2002) Wang Ning: Tourism and Modernity (Review). In *Journal of Retailing and Consumer Services*, 9: 53–58.

Cohen, Paul A. (1988) *Between Tradition and Modernity: Wang T'ao and Reform in Late Ch'ing China*. Cumberland (Harvard University Press).

Cohen, Robin and Kennedy, Paul (2000) *Global Sociology*. London (Macmillan Press).

Coles, Tim and Timothy, Dallen J. (eds) (2004) *Tourism, Diasporas and Space*. Abingdon (Routledge).

Coles, Tim, Duval, David Timothy and Hall, C. Michael (2005) Tourism, Mobility, and Global Communities: New Approaches to Theorising Tourism and Tourist Spaces. In Theobald, William F. (ed): *Global Tourism*. Third edition. London (Elsevier). 463–481.

Collins, Mike (ed.) (1996) *Leisure in Different Worlds. Leisure in Industrial and Post-industrial Societies*. Eastbourne (Leisure Studies Association Publications).

Connolly, Ellen and D'Costa, Desiree (2005) $25 to walk on Bondi Beach. In *The Sunday Telegraph*, 26 June, www.news.com.au.

Cooper, Chris (ed.) (2003a) *Classic Reviews in Tourism*. Clevedon (Channel View Publications).

Cooper, Chris (2003b) Progress in Tourism. In Cooper, Chris (ed.) *Classic Reviews in Tourism*. Clevedon (Channel View Publications). 1–8.

Craik, Jennifer (1997) The Culture of Tourism. In Rojek, Chris and Urry, John (eds) *Touring Cultures: Transformations of Travel and Theory*. London (Routledge). 113–136.

Crang, Mike (1998) *Cultural Geography*. London (Routledge).

Crang, Mike, Crang, Phil and May, Jon (eds) (1999) *Virtual Geographies. Bodies, Space and Relations*. London (Routledge).

CRI (2005) *Red Tour Around China*. en.chinabroadcast.cn.

Crockett, Shane R. and Wood, Leiza J. (2004) Western Australia: Building a State Brand. In Morgan, Nigel, Pritchard, Annette and Pride, Roger (eds) *Destination Branding. Creating the Unique Destination Proposition*. Oxford (Elsevier). 185–206.

Crotts, John C. (2004) The Effect of Cultural Distance on Overseas Travel Behaviours. In *Journal of Travel Research*, 43: 1: 83–88.

Crotts, John C. and Litvin, Stephen W. (2003) Cross-cultural Research: Are Researchers Better Served by Knowing Respondents' Country of Birth, Residence, or Citizenship? In *Journal of Travel Research*, 42: 2: 186–190.

Crouch, David and Lübben, Nina (eds) (2003a) *Visual Culture and Tourism*. Oxford, New York (Berg).

Crouch, David and Lübben, Nina (2003b) Introduction. In Crouch, David and Lübben, Nina (eds) *Visual Culture and Tourism*. Oxford, New York (Berg). 1–20.

CTC (ed.) (2001) *Research on the Chinese Outbound Travel Market Report*. Ottawa (CTC).

CTC (ed.) (2002) *China Report First Quarter 2002*. Ottawa (CTC).

CTC (ed.) (2005) *Market Intelligence Profile of China. Market Research Report 2005–1*. Ottawa (CTC).

Cui, Geng and Liu, Qiming (2000) Regional Market Segments of China: Opportunities and Barriers in a Big Emerging Market. In *Journal of Consumer Marketing*, 17: 1: 55–72.

CYTS (ed.) (2005) A Brief Introduction to the Latest Developments and Trends of the Chinese Outbound Market. Power-Point presentation in New Zealand 2005.

Dai, Bingran and Zhang, Jikang (1997) The Features and Trends of Chinese Outbound Tourism Market in 1990s. In *Journal of Financial and Trade Research*, 6: 16–18 (in Chinese).

Dai, Xuefeng (2005) Preliminary Study on China's Outbound Tourism Policy. In Zhang, Guangrui, Wei Xiaoan and Liu Deqian (eds) *Green Book of China's Tourism 2003–2005 China's Tourism Development: Analysis and Forecast (2003–2005)*. Beijing (Social Sciences Academic Press). (*Green Book of China's Tourism No. 4*) 231–244 (in Chinese).

Dann, Graham M.S. (1993) Limitations in the Use of Nationality and Country of Residence Variables. In Pearce, Philip L. and Butler, Richard W. (eds) *Tourism Research. Critiques and Challenges*. London (Routledge). 88–112.

Dann, Graham M.S. (1996) *The Language of Tourism. A Sociolinguistic Perspective*. Walllingford (CAB International).

Dann, Graham M.S. (2005) Nostalgia in the Noughties. In Theobald, William F. (ed.) *Global Tourism, Third Edition*. London (Elsevier). 32–51.

Davidson, Rob, Hertrich, Sylvie and Schwandner, Gerd (2004) How Can Europe Capture Chinese MICE? In *Proceedings of the APTA Conference Nagasaki, 4–7 July 2004*.

Davies, Roger J. and Ikeno, Osamu (2002) *The Japanese Mind. Understanding Contemporary Japanese Culture*. North Clarendon (Tuttle).

Davis, Deborah S. (ed.) *The Consumer Revolution in Urban China*. Berkeley (University of California Press).

Deng, Jinyang *et al.* (2003) Assessment on and Perception of Visitors' Environmental Impacts of Nature Tourism: A Case Study of Zhangjiajie National Forest Park, China. In *Journal of Sustainable Tourism*, 11: 6: 529–549.

Dikötter, Frank (2005) Race in China. In Nyiri, Pal and Breidenbach, Joana (eds) *China Inside Out Contemporary Chinese Nationalism and Transnationalism*. Budapest (CEU Press). 177–204.

Doorne, Stephen, Ateljevic, Irena and Bai, Zhihong (2003) Representing Identities Through Tourism: Encounters of Ethnic Minorities in Dali,Yunnan Province. In *International Journal of Tourism Research* 4: 1–11.

Dou, Qun (2000) A Study of Chinese Mainland Outbound Tourist Market, Institute of Architecture and Urban Study, School of Architecture, Tsinghua University, Peking.

Dou, Qun and Dou, Jie (2001) A Study of Mainland China Outbound Tourism Markets. In *Tourism Review*, 56: 12: 44–46.

DPS (ed.) (2005a) *DPS Express*, May 2005, Beijing (DPS).

DPS (2005b) 7 Pct Citizens in Beijing, Shanghai, Guangzhou Plan to Travel Abroad Within Year. In: *DPS Express*, August 2005, Beijing (DPS).

DPS (ed.) (2005c) *DPS Express*, August 2005, Beijing (DPS).

Droz, Pierre (2004) Erwartungen und Potentiale für den Schweizer Tourismus, aufgezeigt am neuen Markt China. Diploma thesis, Swiss Hotel School Sierre, 18 June.

DSEC (ed.) (2005) *Macao Statistics*. Macao (DSEC).

Du, Jiang (2003) An Analysis of the Features and Evaluation of Consumption of Chinese Outbound Tourists. Abstract in Wang, Xinjun (ed.) *Proceedings of 1st International Forum on Chinese Outbound Tourism and Marketing Shenzhen 17–18 Nov. 2003 Shenzhen*. 124–125.

Du, Jiang (ed.) (2004a) *Selected Works of Tourism Research*. Beijing (Tourism Education Publishing House).

Du, Jiang (2004b) Economic Development in China and Changes in Concepts of Leisure of the Chinese. Paper presented at World Leisure Congress Amsterdam 1992. In Du, Jiang (ed.) *Selected Works of Tourism Research*. Beijing (Tourism Education Publishing House). 28–36.

Du, Jiang (2004c) On the Development of Tourist Service Infrastructure in China. Paper originally presented at Hanyang Tourism Global Forum 2001. In Du, Jiang (ed.) *Selected Works of Tourism Research*. Beijing (Tourism Education Publishing House). 332–356.

Du, Jiang (2004d) On the Development of the Travel Agency Industry in China. In Du, Jiang (ed.) *Selected Works of Tourism Research*. Beijing (Tourism Education Publishing House). 55–76.

Du, Jiang (2004e) Reform and Development of Higher Tourism Education in China. In Du, Jiang (ed.) *Selected Works of Tourism Research*. Beijing (Tourism Education Publishing House). 484–498.

Du, Jiang and Dai, Wu (eds) (2005) *Annual Report of China Outbound Tourism Development 2004*. Beijing (BISU).

Du, Jiang *et al.* (2002) An Analysis on the Changing Trends for Chinese Outbound Tourism. In *Tourism Tribune*, 17: 3: 44–48 (in Chinese).

Duara, Prasenjit (1995) *Rescuing History from the Nation. Questioning Narratives of Modern China*. Chicago (University of Chicago Press).

Duara, Prasenjit (1996) De-constructing the Chinese Nation. In Unger, Jonathan (ed.) *Chinese Nationalism*. Armonk, NY (Sharpe). 31–55.

Duara, Prasenjit (2005) The Legacy of Empires and Nations in East Asia. In Nyiri, Pal and Breidenbach, Joana (eds) *China Inside Out. Contemporary Chinese Nationalism and Transnationalism*. Budapest (CEU Press). 35–54.

Du Cros, Hilary (2004) Postcolonial Conflict Inherent in the Involvement of Cultural Tourism in Creating New National Myths in Hong Kong. In Hall, C. Michael and Tucker, Hazel (eds) *Tourism and Postcolonialism. Contested Discources, Identities and Representations*. Abingdon (Routledge). 153–168.

Du Cros, Hilary (2005a) The concept of Western Exoticism and its Utility in the Development of Cultural Tourism Products for Asian Tourists in Asian Destinations. Paper presented to the Conference on Destination Branding and Marketing Regional Tourism Development, Macao 8–10 December 2005.

Du Cros, Hilary *et al.* (2005b) Cultural Heritage Assets in China as Sustainable Tourism Products: Case Studies of the Hutongs and the Huanghua Section of the Great Wall. In *Journal of Sustainable Tourism*, 13: 2: 171–194.

Dunlop, Fuchsia (2004) California Dreaming through Chinese Eyes. Typescript of broadcast on Radio 4, 11 December, www.bbc.co.uk.

DZT (ed.) (2003) *Travel Guide Germany 2003/2004*. Frankfurt (DZT) (in Chinese).

DZT (ed.) (2004) *Travel Guide Germany 2004/2005*. Frankfurt (DZT) (in Chinese).

DZT (ed.) (2005) *Marktinfo Südchina und Hongkong*. Hong Kong (DZT).

Ebner, Christian (2004) Frankfurt profitiert von reichen Chinesen. In *Frankfurter Rundschau*, 20 July.

Economist (2002) Out of Puff. A Survey of China. In *The Economist*, 15 June 2002. 1–16.

Economist (2003) Rich Man, Poor Man. In *Economist*, 27 September.

Economist (2004) The Dragon and the Eagle. In *Economist*, 2 October.

Economist (2005a) The Dragon Comes Calling. In *Economist*, 3 September.

Economist (2005b) Managing Unrest. In *Economist*, 23 April.

Economist (2005c) Chinese Demographics. In *Economist*, 26 February.

Economist (2005d) Human Development Index. In *Economist*, 17 September.

Economist (2005e) The Frugal Giant. In *Economist*, 24 September.

Economist (2005f) Official Reserves. In *Economist*, 27 August.

Economist (2005g) The New Face of Globalisation. In *The Economist*, 19 November 2005. 45.

Economist (2005h) Democracy Idol. In *The Economist*, 10 September 2005. 56.

Economist (2005i) Russia's Travellers: No Place Like Home. In *The Economist*, 6 August 2005. 27.

Ecotourism Society Pakistan (ed.) (2004) Pakistan and China Decide to Constitute Committee to Explore New Avenues of Trade and Tourism. In *ESP News*, 6 October, www.ecoclub.com.

Edensor, Tim (2002) National Identity, Popular Culture and Everyday Life. New York (Berg).

EIU (ed.) (2003) *Executive Briefing: China*. London (EIU).

EIU (ed.) (2005) *China Tourism: The Doors Open Wider*. London (EIU).

El Kahal, Sonia (2001) *Business in Asia Pacific: Text and Cases*. New York (Oxford).

Erez, Miriam and Earley, P. Christopher (1993) *Culture, Self Identity and Work*. New York (Oxford University Press).

van Ess, Hans (2003) Ist China konfuzianisch? In *China Analysis*, 23.

ETOA (2005) ETOA/European Commission Seminar on the China ADS Agreement. Minutes of seminar held on 20 June in Brussels, Belgium.

Eturbonews (2004) An Accidental Beggar. In *Eturbonews*, 18 October, www.travel wirenews.com.

Eturbonews (2005) Firm Expects Outbound Chinese Tourists to top 115 m. In *eturbonews*, 27 September 2005, www.travelwirenews.com.

EU (ed.) (2004) European Union Signs Landmark Tourism Accord with China Today in Beijing, europa.eu.int.

EUCCC (ed.) (2001) *European Union Chamber of Commerce in China Tourism Industry Working Group Position Paper 2001–2002*, www.travelwirenews.com.

EUCCC (2003) *European Union Tourism Working Group: China Tourism Paper 2003*. Brussels (EUCCC).

European Commission (ed.) (2004) *European Union Signs Landmark Tourism Accord with China Today in Beijing*, www.europa.eu.int.

Ewen, Michael (ed.) (2000) Peak Performance in Tourism and Hospitality Research, Australian Tourism and Hospitality Research Conference, 2–5 February, La Trobe University, Melbourne.

Fan, C. Simon (1997) Overseas Chinese and Foreign Investment in China: An Application of the Transaction Cost Approach. In Chai, Joseph C.H., Kueh Y.Y. and Tisdell, Clement A. (eds) *China and the Asia-Pacific Economy*. New York (Nova Science Publishers). 145–160.

Farrer, James (2000) Dancing Through the Market Transition: Disco and Dance Hall Sociability in Shanghai. In Davis, Deborah S. (ed.) *The Consumer Revolution in Urban China*. Berkeley (University of California Press). 226–249.

Feng, Chongyi (2005) From Barrooms to Teahouses: Commercial Nightlife in Hainan Since 1988. In Wang, Jing (ed.) *Locating China. Space, Place and Popular Culture*. Abingdon (Routledge). 133–149.

Feng, Kathy and Page, Stephen J. (2000) An Exploratory Study of the Tourism, Migration-Immigration Nexus: Travel Experiences of Chinese Residents in New Zealand. In *Current Issues in Tourism*, 3: 3: 46–281.

Feng, Ruomei, Morrison, Alastair M. and Ismail, Joseph A. (2003) East Versus West: A Comparison of Online Destination Marketing in China and the USA. In *Journal of Vacation Marketing*, 10: 1: 43–56.

Finck, Roswitha (2004) Chinas Outgoing Tourismus am Beispiel der Destination Deutschland. Diploma thesis, Bremen University of Applied Sciences.

Finnish Tourist Board (ed.) (2005) *Rajahaastattelututkimus – Border Interview Survey. Part 17 – Foreign Visitors in Finland 2004*. Helsinki (Finnish Tourist Board).

Foucault, Michel (1973) *The Birth of the Clinic*. London (Routledge).

Frey-Ridgway, Susan (1997) The Cultural Dimension of International Business. In *Collection Building*, 16: 1: 12–23.

Fu, Yunxin (ed.) (2004) *An Introduction to Tourism*. Beijing (Jinan University Press) (in Chinese).

Gang, Yue (2005) From Shambhala to Shangri-La: A Travelling Sign in the Era of Global Tourism. In Tao, Dongfeng and Jin, Yuanpu (eds) *Cultural Studies in China*. Singapore (Marshall Cavendish). 165–183.

Gao, Dichen and Zhang, Guangrui (1983) China's Tourism: Policy and Practice. In *Tourism Management*, 4: 2: 75–84.

Gao, Dichen and Zhang, Guangrui (1984) China's Third Nationwide Seminar on Tourism Economics in December 1983. In *Tourism Management*, 5: 2: 153–155.

Gao, Shunli (2005) 'Red Tourism' in China: Evolution, Features and Counter-measures. In Zhang, Guangrui, Wei Xiaoan and Liu Deqian (eds) *Green Book of China's Tourism 2003–2005 China's Tourism Development: Analysis and Fore-cast (2003–2005)*. Beijing (Social Sciences Academic Press). (*Green Book of China's Tourism No. 4*) 164–172 (in Chinese).

Gao, Yingjun and Giese, Karsten (1993) *Städteführer Deutschland in chinesischer Sprache*. Berlin (Verlag Ute Schiller) (in Chinese).

Gee, Chuck Y. (1983) Beijing Meeting Puts China on the Tourist Map. Conference report. In *Tourism Management*, 4: 2: 140.

Gee, Chuck Y. and Choy J.L. Dexter (1982) The First China International Travel Conference. In *Annals of Tourism Research*, 9: 267–269.

Gee, Chuck Y., Makens, James C. and Choy J.L. Dexter (1997) *The Travel Industry: Third Edition*. New York (John Wiley & Sons).

Geertz, Clifford (1984) From the Native's Point of View: On the Nature of Anthro-pological Understanding. In Shweder, Richard A. and Levine, Robert A. (eds) *Culture Theory: Essays on Mind, Self and Emotion*. Cambridge (Cambridge University Press). 123–136.

Gerstlacher, Anna, Krieg, Renate and Sternfeld, Eva (eds) (1991) *Tourism in the People's Republic of China. A Case Study.* Bangkok (Ecumenical Coalition on Third World Tourism).

Giddens, Anthony (1990) *The Consequences of Modernity.* Cambridge (Polity Press).

Giese, Karsten (2005) Antijapanischer Nationalismus – Bedingter Reflex und gefährliches Kalkül. In *China Aktuell*, 34: 3: 3–10.

Gilbert, David and Tsao, Jenny (2000) Exploring Chinese Cultural Influences and Hospitality Marketing Relationships. In *International Journal of Contemporary Hospitality Management*, 12: 1: 45–53.

Gladney, Dru C. (2005) Alterity Motives. In Nyiri, Pal and Breidenbach, Joana (eds) *China Inside Out Contemporary Chinese Nationalism and Transnationalism.* Budapest (CEU Press). 237–292.

Global Refund (ed.) (2003) *Shopping Guide Germany.* Düsseldorf (Global Refund Germany) (in Chinese).

Global Refund (ed.) (2004) *Shopping Guide Germany.* Düsseldorf (Global Refund Germany) (in Chinese).

Global Refund (ed.) (2006) *Shopping Guide Germany.* Düsseldorf (Global Refund Germany) (in Chinese).

Go, Frank M. (1997) Asian and Australasian Dimensions of Global Tourism Development. In Go, Frank M. and Jenkins, Carson L. (eds) *Tourism and Economic Development in Asia and Australasia.* London (Pinter). 3–34.

Go, Frank M. and Jenkins, Carson L. (eds) (1997) *Tourism and Economic Development in Asia and Australasia.* London (Pinter).

Goeldner, Charles R. and Ritchie, J.R. Brent (2003) *Tourism. Principles, Practices, Philosophies.* Ninth edition. Hoboken (John Wiley & Sons).

Goffman, Eric (1959) *The Presentation of Self in Everyday Life.* New York (Doubleday).

Gormsen, Eberhard (1995) Travel Behaviour and the Impacts of Domestic Tourism in China. In Lew, Alan A. and Yu, Lawrence (eds) *Tourism in China: Geographical, Political and Economic Perspectives.* Boulder, CO (Westview). 131–140.

Graburn, Nelson H.H. (1983) *To Pray, Play and Pay: The Cultural Structure of Japanese Domestic Tourism.* Aix-en-Provence (Centre des Hautes Etudes Touristiques).

Graburn, Nelson H.H. (1997) Tourism and Cultural Development in East Asia and Oceania. In Yamashita, Shinji, Din, Kadir H. and Eades, James S. (eds) *Tourism and Cultural Development in Asia and Oceania.* Bangi (Penerbit Universiti Kebangsaan Malaysia). 194–227.

Graburn, Nelson H.H. (1998) Work and Play in the Japanese Countryside. In Linhart, Sepp and Frühstück, Sabine (eds) *The Culture of Japan as Seen through its Leisure.* New York (State University of New York Press). 195–212.

Graburn, Nelson H.H. (2001a) Tourism and Anthropology in East Asia Today: Some Comparisons. In Tan, Chee-Beng, Cheung, Sidney C.H. and Yang, Hui (eds) *Tourism, Anthropology and China.* Bangkok (White Lotus). (Studies in Asian Tourism No. 1) 71–92.

Graburn, Nelson H.H. (2001b) Learning to Consume: What is Heritage and When is it Traditional? In AlSayyad, Nezar (ed.) *Consuming Tradition, Manufacturing Heritage Global Norms and Urban Forms in the Age of Tourism.* London (Routledge). 68–89.

Graff, Roy (2005) China's Affair with Travel Fairs. Supplement article of *China Outbound Tourism Newsletter*, August 2005. London.

Greenlees, Donald (2005) The Subtle Power of Chinese Tourists. In *International Herald Tribune*, 6 October.

Grescoe, Taras (2004) *The End of Elsewhere. Travels Among the Tourists*. London (Serpents Tail).

Gross, Sven (2004) Die Entstehung einer Tourismuswissenschaft im deutschsprachigen Raum – Status Quo und Anforderungen an eine eigenständige Tourismuswissenschaft. In *Tourismus Journal* 8: 2: 243–264.

Grossberg, Lawrence, Nelson, Cary and Treichler, Paula (eds) (1992) *Cultural Studies*. London (Routledge).

Gu, Zheng (2003) The Chinese lodging Industry: Problems and Solutions. In *International Journal of Contemporary Hospitality Management* 15: 7: 368–392.

Guo, L. (1994) Some Issues about China's Outbound Tourism. In *Business Economics and Administration*, 3: 77–80 (in Chinese).

Guo, Wenbin (2002) Strategies for Entering the Chinese Outbound Travel Market. School of Applied Economics, Faculty of Business and Law, Victoria University PhD thesis.

Guo, Wenbin and Turner, Lindsay W. (2001) Entry Strategies into China for Foreign Travel Companies. In *Journal of Vacation Marketing*, 8: 1: 49–63.

Guo, Wenbin, Turner, Lindsay W. and King, Brian E.M. (2002) The Emerging Golden Age of Chinese Tourism and its Historical Antecedents: A Thematic Investigation. In *Tourism, Culture & Communication*, 3: 131–146.

Guo, Yingzhi (2004) A Study on Market Positioning of China's Outbound Travel Destinations. In *Tourism Tribune*, 19: 4: 27–32 (in Chinese).

Guo, Yingzhi *et al.* (2005a) A Study on the Development Features of Mainland Chinese Outbound Tourism Market. Paper presented at second PolyU China Tourism Forum, Guangzhou, 16–17 December.

Guo, Yingzhi *et al.* (2005b) An Empirical Study on Economic Impacts of Mainland Chinese Outbound Tourism by Pleasure Travelers. In *Proceedings of the Fourth Annual Asia Pacific Forum for Graduate Student Research in Tourism*, University of Hawaii at Manoa, August.

Gütinger, Erich (1998) A Sketch of the Chinese Community in Germany Past and Present. In Benton, Gregor and Pieke, Frank (eds) *The Chinese in Europe*. London (Macmillan). 197–208.

Haas, Ernst B. (1997) *Nationalism, Liberalism, and Progress: The Rise and Decline of Nationalism*. Ithaca, NY (Cornell University Press).

Hall, C. Michael (1994a) *Tourism and Politics. Policy, Power and Place*. New York (John Wiley & Sons).

Hall, C. Michael (1994b) *Tourism in the Pacific Rim*. Sydney (Longman).

Hall, C. Michael (1996) Gender and Economic Interest in Tourism Prostitution: The Nature, Development and Implications of Sex Tourism in South-East Asia. In Apostolopoulos, Yiorgos, Leivida, Stella and Yiannakis, Andrew (eds) *The Sociology of Tourism Theoretical and Empirical Investigantions*. New York, London (Routledge). 265–280.

Hall, C. Michael (1997) *Tourism in the Pacific Rim. Developments, Impacts and Markets*. Second edition. Melbourne (Longman).

Hall, C. Michael (2000) *Tourism Planning: Policies Processes and Relationships*. Harlow (Prentice Hall).

Hall, C. Michael (2005) *Tourism: Rethinking the Social Sciences of Mobility.* Harlow (Pearson Prentice Hall).

Hall, C. Michael and Page, Stephen (eds) (2000) *Tourism in South and South East Asia. Issues and Cases.* Oxford (Butterworth-Heinemann).

Hall, C. Michael and Page, Stephen J. (2002) The Geography of Tourism and Recreation. Second edition. London (Routledge).

Hall, C. Michael and Tucker, Hazel (2004) *Tourism and Postcolonialism. Contested Discources, Identities and Representations*, Abingdon (Routledge).

Hall, Edward T. and Hall, Mildred Reed (1990) *Understanding Cultural Differences.* Yarmouth (Intercultural Press).

Hamblyn, Richard (2001) *The Invention of the Clouds. How an Amateur Meteorologist Forged the Language of the Skies.* London (Picador).

Hamburg Tourismus GmbH (ed.) (2005) *Beherbergungsstatistik der Magic Cities*, www.hamburg-tourismus.de.

Han, Kehua (1982) Chinese-Style Tourism. In *Beijing Review*, 25: 29: 20–22.

Han, Kehua (1994) *China: Tourism Industry.* Beijing (Modern China Press).

Han, Nianyong and Ren, Zhuge (2001) Ecotourism in China's Nature Reserves: Opportunities and Challenges. In *Journal of Sustainable Tourism*, 9: 3: 228–242.

Handayani, Conny (2005) The Influence of China and Asian Countries as Target Markets in the Development of Tourism Industry in Central Java Province. In Song, Haiyan and Bao, Jigang (eds) *Proceedings of the Second PolyU China Tourism Forum and Third China Tourism Academy Annual Conference.* 92.

Hannam, Kevin, Sheller, Mimi and Urry, John (eds) (2006) Editorial: Mobilities, Immobilities and Moorings. In *Mobilities* 1: 1: 1–22.

Hansen, Chad (1989) Language in the Heart-Mind. In Allison, Robert E. (ed.) *Understanding the Chinese Mind. The Philosophical Roots.* Hong Kong (Oxford University Press). 75–124.

Hansen, Valerie (2000) *The Open Empire. A History of China to 1600.* New York (Norton).

Hao, Li (2005) Coping with Widening Income Gap. In: *Beijing Portal*, 15 March, www.beijingportal.com.

Harty, Maura (2005) Press Conference of Assistant Secretary of State for Consular Affairs. Beijing 2 March, www.usembassy-china.org.cn.

Hashimoto, Atsuko (2000a) Environmental Perception and Sense of Responsibility of the Tourism Industry in Mainland China, Taiwan and Japan. In *Journal of Sustainable Tourism*, 8: 2: 131–146.

Hashimoto, Atsuko (2000b) Young Japanese Female Tourists: An In-depth Understanding of a Market Segment. In *Current Issues in Tourism*, 3: 1: 35–50.

Haven-Tang, Claire and Jones, Eleri (eds) (2005) *Tourism SMEs, Service Quality and Destination Competitiveness: International Perspectives.* Wallingford, CT (CAB International).

He, Guangwei (ed.) (1992) *China Tourism System Reform.* Dalian (Dalian Press) (in Chinese).

He, Guangwei (2004) *Travel Services in China 2004–2005.* Beijing (China Industry & Commerce Associated Press) (in Chinese).

Heilmann, Sebastian (2002) Grundelemente deutscher Chinapolitik. In *China Analysis*, 14.

Henderson, Joan C. (2002) Heritage Attractions and Tourism Development in Asia: A Comparative Study of Hong Kong and Singapore. In *International Journal of Tourism Research*, 4: 337–344.

Hendry, Joy (2000) *The Orient Strikes Back. A Global View of Cultural Display.* Oxford, New York (Berg).

Herbig, Paul A. (1998) *Handbook of Cross-cultural Marketing.* Binghamton (International Business Press).

Heung, Vincent C.S. (2000) Satisfaction Levels of Mainland Chinese Travelers with Hong Kong Hotels Services. In *International Journal of Contemporary Hospitality Management*, 12: 5: 308–315.

Heung, Vincent C.S. and Chu, Raymond (2000) Important Factors Affecting Hong Kong Consumers' Choice of a Travel Agency for All-inclusive Package Tours. In *Journal of Travel Research*, 39: 1: 52–59.

Hevia, James L. (1995) *Cherishing Men from Afar. Qing Guest Ritual and the Macartney Embassy of 1793.* Durham, NC (Duke University Press).

Hing, Nerilee (1997) A Review of Hospitality Research in the Asia Pacific Region 1989–1996: A Thematic Perspective. In *International Journal of Contemporary Hospitality Management*, 9: 7: 241–253.

Hirsch, E. and O'Hanlon, M. (eds) (1995) *The Anthropology of the Landscape.* Oxford (Oxford University Press).

Hitchcock, Michael, King, Victor T. and Parnwell, Michael J.G. (eds) (1993a) *Tourism in South-East Asia.* London (Routledge).

Hitchcock, Michael, King, Victor T. and Parnwell, Michael J.G. (1993b) Tourism in South-East Asia: Introduction. In Hitchcock, Michael, King, Victor T. and Parnwell, Michael J.G. (eds) *Tourism in South-East Asia.* London (Routledge). 1–31.

HKTB (ed.) (2002) *2002 Hong Kong Tourism Statistics in Brief.* Hong Kong (HKTB).

HKTB (ed.) (2003) *HKTB Around the World – Regional Reports 2002–2003.* Hong Kong (HKTB).

HKTB (ed.) (2005a) December Arrivals Set New High of 2.08 m to Cap a Record-breaking Year. Press release 27 January.

HKTB (2005b) *2004 Hong Kong Tourism Statistics in Brief.* Hong Kong (HKTB).

HKTDC (ed.) (2003) *CEPA and Opportunities for Hong Kong*, www.tdctrade.com.

Hobsbawm, Eric and Ranger, Terence (eds) (1992) *The Invention of Tradition.* Cambridge (Canto). (First edition: 1983.)

Hoffmann, Catherine (2005) Die Chinesen kommen. In *Frankfurter Allgemeine Zeitung*, 13 November 2005.

Hofstede, Geert (1980) *Culture's Consequences.* Thousand Oaks (Sage).

Hofstede, Geert (2001) *Culture's Consequences. Comparing Values, Behaviors, Institutions, and Organizations across Nations.* Second edition. Thousand Oaks (Sage).

Hofstede, Geert and Hofstede, Gert Jan (2005) *Cultures and Organizations – Software of the Mind.* Second edition. New York (McGraw Hill).

Holloway, J. Christopher (2002) *The Business of Tourism.* Sixth edition. Harlow (Prentice Hall).

Holmes, David (ed.) (2001) *Virtual Globalization. Virtual Spaces Tourist Spaces.* London, New York (Routledge).

Hotelleriesuisse, Switzerland Tourism (eds) (2004) *Swiss Hospitality for Chinese Guests.* Berne (Hotelleriesuisse).

Hsieh, An-Tien and Chang, Janet (2006) Shopping and Tourist Night Markets in Taiwan. In *Tourism Management*, 27: 1: 138–145.

Hsieh, Sheauhsing, O'Leary, J.T. and Morrison, Alistair M. (1992) Segmenting the International Travel Market by Activity. In *Tourism Management*, 13: 2: 209–223.

HTA (ed.) (2003) Identifying and Analyzing the Chinese Outbound Market for Hawaii. Honolulu (HTA).

Hu, Huaming and Graff, Roy (2005) *The China Outbound Travel Handbook 2005*. London (CContact).

Hu, Xiaoling and Watkins, David (1999) The Evolution of Trade Relationships between China and the EU since the 1980s. In *European Business Review*, 99: 3: 154–161.

Huang, Anmin and Xiao, Honggen (2000) Leisure-based Tourist Behavior: A Case Study of Changchun. In *International Journal of Contemporary Hospitality Management* 12: 3: 210–214.

Huang, Can (2000) The Dynamics of Overseas Chinese Invested Enterprises in South China. In Huang, Can, Zhuang, Guotu and Kyoko, Tanaka (eds) *New Studies on Chinese Overseas and China*. Leiden (International Institute for Asian Studies). 191–209.

Huang, Can, Zhuang, Guotu and Kyoko, Tanaka (eds) (2000) *New Studies on Chinese Overseas and China*. Leiden (International Institute for Asian Studies).

Huang, Chih-Wen and Tai, Ai-Ping (2003) A Cross-cultural Comparison of Customer Value Perceptions for Products: Consumer Aspects in East Asia. In *Cross Cultural Management*, 10: 4: 43–60.

Huang, Chun-Te, Yung, Chi-Yeh and Huang, Jen-Hung (1996) Trends in Outbound Tourism from Taiwan. In *Tourism Mangagement*, 17: 3: 223–228.

Huang, Rui (ed.) (2003) *Beijing 798. Reflections on Art, Architecture and Society in China*. Beijing (timezone8 + Thinking Hands).

Huang, Songshan and Hsu, Cathy H.C. (2005) Mainland Chinese Residents' Perceptions and Motivations of Visiting Hong Kong: Evidence from Focus Group Interviews. In *Asia Pacific Journal of Tourism Research*, 10: 2: 191–206.

Huang, Ying and Fang, Qiang (2003) Travel in the Heart of Europe. Design of a Tourism Program for a German Tourism Company. Term paper FHW Berlin, Euroasia MBA Program.

Hui, Tak-Kee and Wan, David (2005) Factors Affecting Chinese Tourists Visiting Singapore. In Song, Haiyan and Bao, Jigang (eds) *Proceedings of the Second PolyU China Tourism Forum and Third China Tourism Academy Annual Conference*. 95.

Hummel, Manfred (2005) Heftiges Werben um chinesische Touristen. In *Süddeutsche Zeitung*, 25 May.

Hun, Tongchao (2005) *A Tour Guide to North Europe*. Guangzhou (Guangdong Travel Press) (in Chinese).

Huntington, Samuel P. (1996) *The Clash of the Civilizations and the Remaking of World Order*. New York (Simon & Schuster).

Huyton, Jeremy R. and Ingold, Anthony (1997) Some Considerations of Impacts of Attitude to Foreigners by Hotel Workers in the Peoples Republic of China on Hospitality Service. In *Progress in Tourism and Hospitality Research*, 3: 107–117.

Huyton, Jeremy R. and Sutton, John (1996) Employee Perceptions of the Hotel Sector in the People's Republic of China. In *International Journal of Contemporary Hospitality Management*, 8: 1: 22–28.

Indrich, Elias (2004) Chinesische Touristen in Deutschland. Master's thesis (DWIF).

Inglis, Fred (2000) *The Delicious History of the Holiday*. London and New York (Routledge).

IPK (ed.) (2003) *World Travel Trends 2003–2004*. San Guiliano (IPK).

Ipsen, Detlev (2004) High Speed Urbanismus. In *archplus*, 168: 28–30.

ITE Hong Kong (ed.) (2003) SARS Affected Travel Habits and Choice of Destinations of Hong Kong Travellers. Hong Kong (ITE).

Iverson, Thomas J. (1997) Decision Timing: A Comparison of Korean and Japanese Travelers. In *International Journal of Hospitality Management*, 16: 3: 209–219.

Ivy, Marilyn (1995) Discourses of the Vanishing. Modernity – Phantasm – Japan. Chicago (University of Chicago Press).

Jaakson, Reiner (2004) Globalisation and Neocolonialist Tourism. In Hall, C. Michael and Tucker, Hazel: *Tourism and Postcolonialism. Contested Discources, Identities and Representations*, Abingdon (Routledge). 169–183.

Jackson, Julie (2006) Developing Regional Tourism in China: The Potential for Activating Business Clusters in a Socialist Market Economy. In *Tourism Management*, 27: 4: 695–706.

Jafari, Jafar and Ritchie, J.R. Brent (1981) Towards a Framework for Tourism Education Problems and Prospects. In *Annals of Tourism Research*, 8: 13–34.

Jameson, Fredric (1993) Cultural Studies. In *Social Studies*, 34.

Jang, SooCheong Shawn and Wu, Chi-Mei Emily (2006) Seniors' Travel Motivation and the Influential Factors: An Examination of Taiwanese Seniors. In *Tourism Management*, 27: 2: 306–316.

Jang, SooCheong, Yu, Larry and Pearson, Thomas E. (2003) Chinese Travellers to the United States: A Comparison of Business Travel and Visiting Friends and Relatives. In *Tourism Geographies*, 5: 1: 87–108.

Jaworski, Adam and Pritchard, Annette (eds) (2005) *Discourse, Communication and Tourism*. Clevedon (Channel View).

Jeffrey, Douglas and Xie, Yanjun (1995) The UK Market for Tourism in China. In *Annals of Tourism Research*, 22: 4: 857–876.

Jenkins, Carlson L. and Liu, Zhen-Hua (1997) China: Economic Liberalization and Tourism Development – The Case of the People's Republic of China. In Go, Frank M. and Jenkins, Carson L. (eds) *Tourism and Economic Development in Asia and Australasia*. London (Pinter). 103–122.

Jensen, Lionel M. (1997) *Manufacturing Confucianism*. Durham (Duke University Press).

Jiang, Zhuqing (2005) Snakeheads More Cunning, Say Officials. In *China Daily*, 22 November 2005. 2.

Jim, C.Y. and Xu, Steve S.W. (2002) Stifled Stakeholders and Subdued Participation: Interpreting Local Responses Toward Shimentai Nature Reserve in South China. In *Environmental Management* 30: 3: 327–341.

Jin, Liangpo (2000) *Germany*. Beijing (Tourism Education Publishing House) (in Chinese).

Job, Hubert (ed.) (2006) (forthcoming) *Proceedings of the German-Chinese Zhangjiajie Conference October 2004*. München.

Johnston, Robert (1989) The Customer as Employee. In *International Journal of Operations & Production Management*, 9: 5: 15–23.

Johst, Roland (2001) Tourismus in der Volksrepublik China. Trier University, Diploma thesis.

Jomo, Kwame Sundaram and Folk, Brian C. (eds) (2003) *Ethnic Business. Chinese Capitalism in Southeast Asia*. Abingdon (Routledge).

JTM (Japan Tourism Marketing Co.) (ed.) (2006) *Japanese Outbound Tourism Statistics*, www.tourism.jp/english/statistics/outbound.

Junek, Olga, Binney, Wayne and Deery, Marg (2004) Meeting the Needs of the Chinese Tourist – The Operators' Perspective. In *ASEAN Journal of Hospitality and Tourism*, 3: 2: 149–161.

Kajiwara, Kageaki (1997) Inward-Bound, Outward-Bound: Japanese Tourism reconsidered. In Yamashita, Shinji, Din, Kadir H. and Eades, James S. (eds) *Tourism and Cultural Development in Asia and Oceania*. Bangi (Penerbit Universiti Kebangsaan Malaysia). 164–177.

Kao, John (1993) The Worldwide Web of Chinese Business. In *Harvard Business Review*, 71: 2: 24–35.

Kaplan, Caren (1996) *Questions of Travel. Postmodern Discourses of Displacement*. Durham, NC (Duke University Press).

Karwacki, Judy, Deng, Shengliang and Chapdelaine, Colin (1997) The Tourism Markets of the Four Dragons – A Canadian Perspective. In *Tourism Management*, 18: 6: 373–383.

Kaynak, Erdener and Kucukemiroglu, Orsay (1993) Foreign Vacation Selection Process in an Oriental Culture. In *Asia Pacific Journal of Marketing and Logistics*, 5: 1: 21–41.

Kelemen, Marcel (2003) *The Characteristics of Outbound Tourism. Market: Hong Kong & South China*. Hong Kong (DZT).

Kelemen, Marcel (2005) The Shopping Behaviour of Chinese Tourists on the Example of the China Incoming Tourism to Germany. University of Applied Sciences Stralsund, Bachelor thesis 2005.

Kiefer, Thomas (2004) Trade Fairs in the Consciousness of Chinese Companies. In AUMA Ausstellungs- und Messeausschuss der Deutschen Wirtschaft e.V. (ed.) *Exhibition Market China 2004–2005*. Stuttgart (local global Medien). 32–36.

Kim, Donghoon, Pan, Yigang and Park, Heung Soo (1998) High- Versus Low-Context Culture: A Comparison of Chinese, Korean, and American Cultures. In *Psychology & Marketing*, 15: 6: 507–521.

Kim, Samuel Seongseop and Morrison, Alastair M. (2005) Change of Images of South Korea among Foreign Tourists after the 2002 FIFA World Cup. In *Tourism Management*, 26: 233–247.

Kim, Samuel Seongseop and Prideaux, Bruce (2003) A Cross-cultural Study of Airline Passengers. In *Annals of Tourism Research*, 30: 2: 489–492.

King, Brian, McKercher, Bob and Waryszak, Robert (2003) A Comparative Study of Hospitality and Tourism Graduates in Australia and Hong Kong. In *International Journal of Tourism Research*, 5: 409–420.

Klawitter, Nils (2004) Kickdown ins Paradies. In *Der Spiegel*, 36: 86–87.

KNTO (ed.) (2005) Chinese Visitors in Korea, www.knta.or.kr.

Koldowski, John (2003) The China Outbound Market. In Wang, Xinjun (ed.) *Proceedings of 1st International Forum on Chinese Outbound Tourism and Marketing Shenzhen 17–18 Nov. 2003 Shenzhen*. 167–178.

Koldowski, John (2005) China's Importance in Asia Pacific Travel and Tourism. Paper presented at the Second International Forum on Chinese Outbound Tourism, Beijing 20/21 November 2005. In *Proceedings*. 105–112.

Kong, Yunxien, Chen, Chunqian and Zheng Xiangmin (2005) Tourists' Image of Macau: Assessment and Analysis. In *China Tourism Research*, 1: 1: 53–67.

Konwicki, Tadeusz (1976) *Kalendarz i Klepsydra*. Warszawa (Wyd. Czytelnik).

Koshar, Rudy (2000) *German Travel Cultures*. Oxford, New York (Berg).

Kotkin, Joel (1993) *Tribes: How Race, Religion, and Identity Determine Success in the New Global Economy*. New York (Random House).

Krongkaew, Medhi (2004) The Development of the Greater Mekong Subregion (GMS): Real Promise or False Hope? In *Journal of Asian Economics*, 15: 977–998.

Lai, Lai-Hsin and Graefe, Alan R. (2000) Identifying Market Potential and Destination Choice Factors of Taiwanese Overseas Travelers. In *Journal of Hospitality & Leisure Marketing*, 6: 4: 45–65.

Laitinen, Kirsti (2004) The Perception of Tour Guides of Finland as a Travel Destination for Chinese tourists. MA European Tourism Management, Bournemouth University 2004.

Lam, Terry and Hsu, Cathy H.C. (2004) Theory of Planned Behaviour: Potential Travellers from China. In *Journal of Hospitality & Tourism Research*, 28: 4: 463–482.

Lam, Terry and Xiao, Honggen (2000) Challenges and Constraints of Hospitality and Tourism Education in China. In *International Journal of Contemporary Hospitality Management* 12: 5: 291–295.

Lam, Terry and Xiao, Honggen (2000) Challenges and Constraints of Hospitality and Tourism Education in China. In *International Journal of Contemporary Hospitality Management*, 12: 5: 291–295.

Latelinenews (2001) China's Holiday Tourism Booming, www.latelinenews.com.cn.

Lau, D.C. (1989) *Confucius: The Analects*. London (Penguin Classics).

Lau, Justine (2004) China's Cash Won't Always Be There. In *Financial Times*, 19 October.

Law, Rob, Cheung, Catherine and Lo, Ada (2004) The Relevance of Profiling Travel Activities for Improving Destination Marketing Strategies. In *International Journal of Contemporary Hospitality Management*, 16: 6: 355–362.

Laws, Eric and Pan, Grace Wen (eds) (2002) *Tourism Marketing: Quality and Service Management Perspectives*. London (Continuum).

Leisure Futures (ed.) (2002) *Proceedings of Leisure Futures Conference Innsbruck/Austria 2002*. Innsbruck.

Leung, Maggi W.H. (2002) From Four-course Peking Duck to Take-away Singapore Rice. An Inquiry into the Dynamics of the Ethnic Chinese Catering Business in Germany. In *International Journal of Entrepreneurial Behaviour & Research*, 8: 12: 134–147.

Leung, Paggie (2005) Tourism Boom Sees Increase in Complaints. In *South China Morning Post*, 10 September.

Levathes, Louise (1994) *When China Ruled the Seas. The Treasure Fleet of the Dragon Throne 1405–1433*. New York (Oxford University Press).

Lew, Alan A. (1995) Overseas Chinese and Compatriots in China's Tourism Development. In Lew, Alan A. and Yu, Lawrence (eds) *Tourism in China: Geographical, Political and Economic Perspectives*. Boulder, CO (Westview). 155–175.

Lew, Alan A. (2000) China: A Growth Engine for Asian Tourism. In Hall, C. Michael and Page, Stephen (eds) *Tourism in South and South East Asia Issues and Cases*. Oxford (Butterworth-Heinemann). 268–285.

Lew, Alan A. and Wong, Alan (2002) Tourism and the Chinese Diaspora. In Williams, A.M. and Hall, C. Michael (eds) *Tourism and Migration*. London (Routledge). 205–220.

Lew, Alan A. and Wong, Alan (2003) News from the Motherland: A Content Analysis of Existential Tourism Magazines in Southern China. In *Tourism, Culture & Communication*, 4: 83–94.

Lew, Alan A. and Wong, Alan (2004) Sojourners, Guanxi and Clan Associations. Social Capital and Overseas Chinese Tourism to China. In Coles, Tim and Timothy, Dallen J. (eds) *Tourism, Diasporas and Space*. Abingdon (Routledge). 202–214.

Lew, Alan A. and Yu, Lawrence (eds) (1995) *Tourism in China: Geographical, Political and Economic Perspectives*. Boulder, CO (Westview).

Lew, Alan A. *et al.* (eds) (2003) *Tourism in China*. New York (Haworth).

Li, Conghua (1998) *China: The Consumer Revolution*. New York (Deloitte).

Li, Deshui (2004) Speech at the NBSP Press Conference on 2003 GDP figures, 20 January, www.stats.gov.cn.

Li, Hanzhong (2005) It is Not Easy to get ADS Tourist Visa to Europe. In *Voyage* (Xin Luxing), 9: 42–43 (in Chinese).

Li, Lan, Bai, Billy and McCleary, Ken (1996) The Giant Awakes: Chinese Outbound Travel. In *Australian Journal of Hospitality Management*, 3: 2: 59–68.

Li, Ji and Wright, Philip C. (2000) Guanxi and the Realities of Career Development: A Chinese Perspective. In *Career Development International*, 5: 7: 369–378.

Li, Jianguo, Li, Ning and Huang, Bao (eds) (2004) *China Facts and Figures 2004*. Beijing (New Star Publishing House).

Li, Shi and Yue, Ximing (2004) A Yawning Urban-rural Income Gap. In *Caijing*, 20 February, www.caijing.com.cn (in Chinese).

Li, Victor Hao (1994) From Qiao to Qiao. In Tu, Wei-ming (eds) *The Living Tree. The Changing Meaning of Being Chinese Today*. Stanford, CA (Stanford University Press). 213–220.

Li, Xiaoyan (2002) *The Essential Guides: Berlin*. Originally published by Automobile Association Britain. Beijing (Zhonghua Book Company) (in Chinese).

Li, Yiping (2004) Exploring Community Tourism in China: The Case of Nanshan Cultural Tourism Zone. In *Journal of Sustainable Tourism*, 12: 3: 175–193.

Lim, Christine and Pan, Grace W. (2005) Inbound Tourism Developments and Patterns in China. In *Mathematics and Computer in Simulation*, 68: 499–507.

Lin, Guijin and Schramm, Ronald M. (2003) China's Foreign Exchange Policies since 1979: A Review of Developments and an Assessment. In *China Economic Review*, 14: 246–280.

Lin, Li (2005) Passenger Transport during the Spring Festival: An Astronomical and Special Mobility in China. In Burns, Peter (ed.) *Proceedings of the 4th International Symposium on Aspects of Tourism, Eastbourne 2005*.

Lin, Ling (2005) BITTM 2005 – A Personal View. In *China Outbound Tourism Research Project Newsletter*, 2 June.

Lindberg, Kreg, Tisdell, Clem and Xue, Dayuan (2003) Ecotourism in China's Nature Reserves. In Lew, Alan A. *et al.* (eds) *Tourism in China*. New York (Haworth). 103–125.

Ling, Jin and Wang, Xinjun (2003) Market Analysis for the Chinese Official and Business Outbound Tour – A Case Study of Shanghai. In Wang, Xinjun (ed.) *Proceedings of 1st International Forum on Chinese Outbound Tourism and Marketing Shenzhen 17–18 Nov. 2003 Shenzhen*. 141–146.

Linhart, Sepp and Frühstück, Sabine (eds) (1998) The Culture of Japan as Seen Through its Leisure. New York (State University of New York Press).

Literature Research Center of the CPCCC (ed.) (2000) *Deng Xiaoping on Tourism*. Beijing (Literature Research Center of the Communist Party of China Central Committee) (in Chinese).

Littrell, Romie F. (2002) Desirable Leadership Behaviours of Multi-cultural Managers in China. In *Journal of Management Development*, 21: 1: 5–74.

Litvin, Stephen W., Crotts, John C. and Hefner, Frank L. (2004) Cross Cultural Tourist Behavior: Revisiting Hofstede's Uncertainty Avoidance Dimension. In *International Journal of Tourism Research*, 6: 1: 29–37.

Liu, Deqian (2005) China's Domestic Tourism (2003–2005): Present Situation and Future Perspectives. In *Green Book of China's Tourism 2003–2005 China's Tourism Development: Analysis and Forecast (2003–2005)*. Beijing (Social Sciences Academic Press). (*Green Book of China's Tourism No. 4*) 87–111 (in Chinese).

Liu, Jen-Kai (2005) Grosskampagne gegen illegales Glücksspiel auf dem Festland. In *China aktuell*, 34: 4: 31–36.

Lockwood, Andrew and Medlik, S. (Rik) (eds) (2001) *Tourism and Hospitality in the 21st Century*. Burlington, CT (Butterworth- Heinemann).

Löfgren, Orvar (1999) *On Holiday. A History of Vacationing*. Berkeley (University of California Press).

Lomine, Loykie (2005) Tourism in the Augustan Society (44 BC–AD 69). In Walton, John K. (ed.) *Histories of Tourism. Representation, Identity and Conflict*. Clevedon (Channel View Publications). 71–87.

Lommatzsch, Horst (2004) Zukunftsmarkt China. Marketing- und Vertriebsaktivitäten der DZT in China. Power-Point presentation DSFT seminar on China Outbound Tourism, 8 November, Berlin.

Long, Ting (2003) On Present Trends and Solutions of China's Outbound Tourism. In *Journal of Jiangxi Science and Technology Teachers College*, 6: 47–49 (in Chinese).

Ma, Jennifer Xiaoqiu, Buhalis, Dimitrios and Song, Haiyan (2003) ICTs and Internet adoption in China's Tourism Industry. In *International Journal of Information Management*, 23: 451–467.

McAleavy, Henry (1953) Wang T'ao: The Life and Times of a Displaced Person. Lecture delivered at the China Society of London, 22 May. London.

McAllan, Greig (2005) Growth and Sustainable Development for Australia. Paper presented at the Second International Forum on Chinese Outbound Tourism, Beijing 20/21 November 2005. In *Proceedings of the Second International Forum on Chinese Outbound Tourism, Beijing 2005*. 111–116.

MacCannell, Dean (1992) *Empty Meeting Grounds. The Tourists Papers*. London (Routledge).

MacCannell, Dean (1999) *The Tourist. A New Theory of the Leisure Class*. Berkeley (University of California Press) (First edition 1976).

McGiffert, Carola (ed.) (2005) *Chinese Images of the United States*. Washington (CSIS).

McGirk, Jan (2004) Chinese Flocking to Gaming Tables of Burma's Sin City. In *The Independent*, 18 May.

Magick, Samantha (2005) China Syndrome. Is China the Answer for Pacific Tourism? In *Pacific Magazine*, 4.

Malta Tourism Board (ed.) (2004) *Tourism Statistics*. Valletta (Malta Tourism Authority).

Marx, Karl (1853) Revolution in China and in Europe. In *New York Daily Tribune*, 14 June. In Torr, Dona *Marx on China, 1853–1860: Articles from the New York Daily Tribune*. London (Lawrence & Wishart).

Maschke, Joachim (ed.) (2004) *Jahrbuch für Fremdenverkehr 2004*. München (DWIF).

Mason, Peter, Grabowski, Peter and Du, Wei (2005) Severe Acute Respiratory Syndrome, Tourism and the Media. In *International Journal of Tourism Research* 7: 11–21.

Master, Hoba and Prideaux, Bruce (2000) Culture and Vacations Satisfaction: A Study of Taiwanese Tourists in South East Queensland. In *Tourism Management*, 21: 5: 445–449.

Meentemeier, Doris (1988) Die Tourismusindustrie in der Volksrepublik China. FU Berlin, Bachelor thesis.

Meethan, Kevin (2001) *Tourism in Global Society. Place, Culture, Consumption*. Basingstoke (Palgrave).

Mehmetoglu, Mehmet (2004) Tourist or Traveller? A Typological Approach. In *Tourism Review*, 59: 3: 33–39.

MEK (ed.) (2005) *Travel Facts 2005*. Helsinki (MEK).

Menzies, Gavin (2002) *1421 – The Year China Discovered the World*. London (Bantam Press).

Merrill Lynch (ed.) (2005) *World Wealth Report 2004*. New York (Merrill Lynch).

Meyer, Günter (ed.) (2004) *Die Arabische Welt im Spiegel der Kulturgeographie*. Mainz (CERAW).

Min, Hui Neo (2005) Raring to Go, Willing to Spend. Global tourism prepares for Chinese visitors. In *EuroBiz Magazine*, February, www.sinomedia.net.

Mintel International Group Ltd (ed.) (2004) Outbound Travel – Asia. In *Travel and Tourism Analyst*, 4: 1–55.

MOFCOM (ed.) (2005) An Introduction to China's Tourism Industry, www.cnc forumenglish.mofcom.gov.cn.

Mok, Connie C. and Armstrong, Robert W. (1995) Leisure Travel Destination Choice Criteria of Hong Kong Residents. In *Journal of Travel & Tourism Marketing*, 4: 1: 99–104.

Mok, Connie C. and DeFranco, Agnes L. (1999) Chinese Cultural Values: Their Implications for Travel and Tourism Marketing. In *Journal of Travel & Tourism Marketing*, 8: 2: 99–114.

Mok, Henry M.K. (1985) Tourist Expenditures in Guangzhou, PR China. In *Tourism Management*, 6: 4: 272–279.

Moon, Okpyo (1997) Marketing Nature in Rural Japan. In Asquith, Pamela J. and Kalland, Arne Kalland (eds) *Japanese Images of Nature*. Richmond (Curzon). 221–235.

Morgan, Nigel, Pritchard, Annette and Pride, Roger (eds) (2004a) *Destination Branding. Creating the Unique Destination Proposition*. Oxford (Elsevier).

Morgan, Nigel, Pritchard, Annette and Pride, Roger: (2004b) Introduction. In Morgan, Nigel, Pritchard, Annette and Pride, Roger (eds) *Destination Branding. Creating the Unique Destination Proposition*. Oxford (Elsevier). 3–16.

Morris, Stephen (1997) Japan: The Characteristics of the Inbound and Outbound Markets. In Go, Frank M. and Jenkins, Carson L. (eds) *Tourism and Economic Development in Asia and Australasia*. London (Pinter). 150–175.

Mrkwicka, Anja and Belz, Marita (2005) Market Analysis Kempinski. International University of Applied Sciences Bad Honnef/Bonn, thesis.

Muecke, Nina and Sommer, Angelika (ed.) (2003) *Rituale in der zeitgenössischen Kunst*. Berlin (Akademie der Künste Berlin).

Munzeer, Delerine (2005) China Looks South. In *TTG Asia*, 10 June, www.ttgasia. com.

Muqbil, Imtiaz (2005) 10 Trends that Will Shape the Future of Asia Pacific Travel. Press release ITB Berlin 2005, Berlin.

Muzi (2004) China makes Peru Official Tourism Destination, www.muzi.com, 30 December.

Naisbitt, John (1997) *Megatrends Asia. The Eight Asian Megatrends that are Changing the World*. London (Nicolas Brealey).

NBSC (ed.) (2003) *China Statistical Yearbook 2003*. Beijing (China Statistics Press).

NBSC (ed.) (2004) *China Statistical Yearbook 2004*. Beijing (China Statistics Press).

News Guangdong (2004) Outbound Tourism Market Becomes Spotlight of Industry. In *News Guangdong,* 23 July, www.newsgd.com.

Nisbett, Richard E. (2003) *The Geography of Thought. How Asians and Westerners Think Differently . . . and Why*. New York (Free Press).

Norway (2004) New Drive for More Business. In Norway – The Official Site in China, 26 July, www.norway.cn.

Norway (2005) Norway and China Sign Tourist Agreement. In Norway – The Official Site in China, 3 June, www.norway.cn.

Nyiri, Pal (2005a) *Scenic Spot Europe*, www.EspacesTemps.net.

Nyiri, Pal (2005b) The 'New Migrant': State and Market Constructions of Modernity and Patriotism. In Nyiri, Pal and Breidenbach, Joana (eds) *China Inside Out Contemporary Chinese Nationalism and Transnationalism*. Budapest (CEU Press). 141–176.

Nyiri, Pal and Breidenbach, Joana (eds) (2005) *China Inside Out. Contemporary Chinese Nationalism and Transnationalism*. Budapest (CEU Press).

OAG (ed.) (2005) *OAG Executive Flight Guide China June 2005*. Dunstable (OAG).

Oakes, Tim (1995) Tourism in Guizhou: The Legacy of Internal Colonialism. In Lew, Alan A. and Yu, Lawrence (eds) *Tourism in China: Geographical, Political and Economic Perspectives*. Boulder, CO (Westview). 203–222.

Oakes, Tim (1998) *Tourism and Modernity in China*. London (Routledge).

Oakes, Tim (2005) Land of Living Fossils: Scaling Cultural Prestige in China's Periphery. In Wang, Jing (ed.) *Locating China. Space, Place and Popular Culture*. Abingdon (Routledge). 31–51.

OECD (Organization for Economic Co-operation and Development) (ed.) (2002) *National Tourism Policy Review of Japan*. Geneva (OECD).

Office of the United Nations Resident Coordinator (ed.) (2003) *Millennium Development Goals. China's Progress*. Beijing (Office of the United Nations Resident Coordinator).

Ontario Tourism Marketing Partnership Cooperation (ed.) (2005) *China Market Profile June 2005*. Ontario (Ontario Tourism Marketing Partnership Cooperation).

Ooi, Can-Seng (2002) *Cultural Tourism and Tourism Cultures. The Business Mediating Experiences in Copenhagen and Singapore*. Copenhagen (Copenhagen Business School).

Open China (2004) *Open China: Chinese Tourism Growth*, www.openchina.com.au.

Oppenheimer, Andres (2004) Why Isn't US Angry over China's Tourism Policies? In *Herald*, 12 September.

Osnos, Evan (2005) Talking in Tongues. In *South China Morning Post* 12 December 2005.

OTTI (ed.) (2005) *Statistics Inbound Travel to the US*, www.tinet.ita.doc.gov.

Oudiette, Virginie (1990) International Tourism in China. In *Annals of Tourism Research*, 17: 1: 123–132.

Overby, John, Rayburn, Mike and Hammond, Kevin (2004) The China Syndrome: The Impact of the SARS Epidemic in Southeast Asia. In *Asia Pacific Journal of Marketing and Logistics* 16: 1: 69–94.

Page, Stephen J. (2003) *Tourism Management. Managing for Change*. Oxford (Butterworth-Heinemann).

Palmer, Catherine (1998) Tourism and the Symbols of Identity. In *Tourism Management*, 20: 313–321.

Pan, Grace Wen (2002) Chinese Tourism: Emerging Markets. In Laws, Eric and Pan, Grace Wen (eds) *Tourism Marketing: Quality and Service Management Perspectives*. London (Continuum). 113–125.

Pan, Grace Wen and Laws, Eric (2001) Tourism Marketing Opportunities for Australia in China. In *Journal of Vacation Marketing*, 8: 1: 39–48.

Pan, Lynn (1988) *The New Chinese Revolution*. London (Sphere Books).

Pang, Chee Keen, Roberts, Diane and Sutton, John (1998) Doing Business in China – The Art of War? In *International Journal of Contemporary Hospitality Management* 10: 7: 272–282.

Panitchpakdi, Supachai and Clifford, Mark L. (2002) *China and the WTO: Changing China, Changing World Trade*. Singapore (John Wiley & Sons).

PATA (ed.) (2005) *China (PRC) Outbound Tourism. A PATA Snapshot*. Bangkok (PATA).

Pearce, Philip L. (2004) Theoretical Innovation in Asia Pacific Tourism Research. In *Asia Pacific Journal of Tourism Research*, 9: 1: 57–70.

Pearce, Philip L. (2005) *Tourist Behaviour: Themes and Conceptual Schemes*. Clevedon (Channel View Publications).

Pearce, Philip L. and Butler, Richard W. (eds) (1993) *Tourism Research. Critiques and Challenges*. London (Routledge).

Pease, Wayne, Rowe, Michelle and Cooper, Malcolm (eds) (2006) (forthcoming) *Information and Communication Technologies in Support of the Tourism Industry*.

Peng, Dan (2005) A Study of Semiotics and Its Meaning in Tourism. In *China Tourism Research,* 1: 4: 474–493.

People's Daily (2000) Forging Ahead to Make China a World's Tourism Power. In *People's Daily online*, 29 March, english.people.com.cn.

People's Daily (2001) Chinese Tourists Spend More than Japanese in Australia. In *People's Daily online*, 29 July, english.people.com.cn.

People's Daily (2003a) Chinese Outbound Tourists Travel to Australia. In *People's Daily online*, 27 October, english.people.com.cn.

People's Daily (2003b) Finland Aims at 260,000 Visitors from China by 2010. In *People's Daily online*, 17 January, english.people.com.cn.

People's Daily (2003c) Misbehaviour of Chinese Outbound Tourists Offends Local Citizens in Vietnam. In *People's Daily online*, 22 September, english.people.com.cn.

People's Daily (2003d) Information Given to Outbound Tourists. In *People's Daily online*, 17 March, english.people.com.cn.

People's Daily (2004) Outbound Tourism Market Becomes Spotlight of Industry. In *People's Daily online*, 20 July, english.people.com.cn.

People's Daily (2005a) Five Hundred Chinese Tourists to Spend Spring Festival in Egypt. In *People's Daily online*, 6 January, english.people.com.cn.

People's Daily (2005b) Nevada Hopes to Attract More Chinese Tourists. In *People's Daily online*, 26 June, english.people.com.cn.

People's Daily (2005c) President Hu Urges Closer Ties with Denmark. In *People's Daily online*, 28 February, english.people.com.cn.

People's Daily (2005d) Namibia Added to China's Tourism Destinations, www.english.people.com.cn.

Petersen, Ying Yang (1995) The Chinese Landscape as a Tourist Attraction: Image and Reality. In Lew, Alan A. and Yu, Lawrence (eds) *Tourism in China: Geographical, Political and Economic Perspectives*. Boulder, CO (Westview). 141–154.

Picard, Michel (1993) Cultural Tourism in Bali: National Integration and Regional Differentiation. In Hitchcock, Michael, King, Victor T. and Parnwell, Michael J.G. (eds) *Tourism in South-East Asia*. London (Routledge).

Picard, Michel (1996) *Bali. Cultural Tourism and Touristic Culture*. Singapore (Archipelago Press).

Pieke, Frank N. (2002) *Recent Trends in Chinese Migration to Europe: Fujianese Migration in Perspective* (IOM Migration Research Series No. 6). Geneva (International Organization for Migration).

Pinches, Michael (ed.) (1999) *Culture and Privilege in Capitalist Asia*. London (Routledge).

Pine, Joseph B. II and James H. Gilmore (1999) *The Experience Economy. Work is Theatre and Every Business a Stage*. Boston (Harvard Business School Press).

Pine, Ray, Zhang, Hanqin Qiu and Qi, Pingshu (2000) The Challenges and Opportunities of Franchising in China's Hotel Industry. In *International Journal of Contemporary Hospitality Management* 12: 5: 300–307.

Pinkert, Ernst-Ullrich and Therkelsen, Anette (eds) (1997) *Intercultural Encounters in Tourism* (Language and Cultural Contact No. 21). Aalborg (Aalborg University Press).

Pitta, Dennis A., Fung, Hung-Gay and Isberg, Steven (1999) Ethical Issues across Cultures: Managing the Differing Perspectives of China and the USA. In *Journal of Consumer Marketing*, 16: 3: 240–256.

Pizam, Abraham and Jeong, Gang-Heon (1996) Cross Cultural Tourist Behavior. In *Tourism Management*, 17: 4: 277–286.

Pizam, Abraham and Sussman, S. (1995) Does Nationality Affect Tourist Behaviour? In *Annals of Tourism Research*, 22: 4: 901–917.

Poon, Wai-Ching and Yong, Gu-Fie David (2005) Comparing Satisfaction Levels of Asian and Western Travellers using Malaysian Hotels. In *Journal of Hospitality & Tourism Management*, 12: 1: 64–80.

Prideaux, Bruce (1996) Recent Developments in the Taiwanese Tourist Industry – Implications for Australia. In *International Journal of Contemporary Hospitality Management*, 8: 1: 10–15.

Prideaux, Bruce R. (1998) Korean Outbound Tourism: Australia's Response. In *Journal of Travel & Tourism Marketing*, 7: 1: 93–102.

Prideaux, Bruce *et al.* (2004) Exotic or Erotic. Contrasting Images for Defining Destinations. In *Asia Pacific Journal of Tourism Research*, 9: 1: 5–17.

Project Research Team (2003) *Research on Consumption Behaviour Model of Chinese Outbound Tourists*. Beijing (Tourism Education Press) (in Chinese).

Project Team (2003) An Analysis on the Tourism Expenditure Evaluation and Relationship between Host and Guest for Mainland Chinese Outbound Market. In *Transaction of Beijing Second Foreign Language University*, 1: 1–18 (in Chinese).

Qian, Wei (2003) Travel Agencies in China at the Turn of the Millennium. In Lew, Alan A. *et al.* (eds) *Tourism in China*. New York (Haworth). 143–164.

Qin, Jize (2005) Leaders Relive Bandung Spirit in Walk. In *China Daily*, 25 April.

Qu, Hailin and Lam, Sophia (1997) A Travel Demand Model for Mainland Chinese Tourists to Hong Kong. In *Tourism Management*, 18: 8: 593–597.

Qu, Hailin and Li, Isabella (1997) The Characteristics and Satisfaction of Mainland Chinese Visitors to Hong Kong. In *Journal of Travel Research*, spring: 37–41.

Qu, Hailin and Zhang, Hanqin Qiu (1997) The Projection of International Tourist Arrivals in East Asia and the Pacific. In Go, Frank M. and Jenkins, Carson L. (eds) *Tourism and Economic Development in Asia and Australasia*. London (Pinter). 35–47.

Qu, Hanqin Zhang and Lam, Terry (2004) Human Resources Issues in the Development of Tourism in China: Evidence from Heilongjiang Province. In *International Journal of Contemporary Hospitality Management*, 16: 1: 45–51.

Qu, Riliang, Ennew, Christine and Sinclair, M. Thea (2005) The Impact of Regulation and Ownership Structure on Market Orientation in the Tourism Industry in China. In *Tourism Management*, 26: 6: 939–950.

Rea, Michael H. (2000) A *furusato* Away from Home. In *Annals of Tourism Research*, 27: 3: 638–660.

Reisinger, Yvette and Turner, Lindsay (1997) A Cultural Analysis of Japanese Tourists: Challenges for Tourism Marketers. In *European Journal of Marketing*, 33: 1112: 1203–1227.

Reisinger, Yvette and Turner, Lindsay W. (2003) *Cross-cultural Behaviour in Tourism. Concepts and Analysis*. Oxford (Butterworth and Heinemann).

Reisinger, Yvette and Waryszak, Robert Z. (1994) Tourists' Perceptions of Service in Shops. In *International Journal of Retail and Distribution Management*, 22: 5: 20–28.

Ren, Feng (1988) Travelling and Fleeing the Country. In *Jiushi Niandai* (The Nineties). 67–69 (in Chinese).

Richards, Greg (2001) Marketing China Overseas: The Role of Theme Parks and Tourist Attractions. In *Journal of Vacation Marketing*, 8: 1: 28–38.

Richter, Linda (1983) Political Implications of Chinese Tourism Policy. In *Annals of Tourism Research*, 10: 395–413.

Richter, Linda (1989) *The Politics of Tourism in Asia*. Honolulu (University of Hawaii Press).

Robinson, Mike (1998) Cultural Conflicts in Tourism: Inevitability and Inequality. In Robinson, Mike and Boniface, Priscilla (eds) *Tourism and Culture Conflicts*. Wallingford, CT (CAB International).

Robinson, Mike (2001) Tourism Encounters: Inter- and Intra-Cultural Conflicts and the World's Largest Industry. In AlSayyad, Nezar (ed.) *Consuming Tradition, Manufacturing Heritage Global Norms and Urban Forms in the Age of Tourism*. London (Routledge). 34–67.

Robinson, Mike and Boniface, Priscilla (eds) (1998) *Tourism and Culture Conflicts*. Wallingford, CT (CAB International).

ROCTB (1993) *Annual Report 1992*. Taibei (ROCTB).

Rojek, Chris and Urry, John (eds) (1997) *Touring Cultures: Transformations of Travel and Theory*. London (Routledge).

Roth, Silvia (1998) *The Chinese Outbound Travel Market. Austria's Position in the Outbound Travel to Europe*. Wien (Austrian National Tourist Office).

Ryan, Chris (ed.) (2002) *The Tourist Experience*. London, New York (Continuum).

Ryan, Chris and Mo, Xiaoyan (2001) Chinese Visitors to New Zealand – Demographics and Perceptions. In *Journal of Vacation Marketing*, 8: 1: 13–27.

Sachs, Jeffrey D. and Woo, Wing Thye (2003) China's Economic Growth After WTO Membership. In *Journal of Chinese Economic and Business Studies*, 1: 1: 1–31.

SAT (ed.) (2004a) *2003 Annual Tourism Report*. Johannesburg (SAT).

SAT (ed.) (2004b) *Marketing South Africa in China*. Johannesburg (SAT).

SAT (ed.) (2004c) *South African Tourism Quarterly Report China. Third Quarter 2004*. Johannesburg (SAT).

Satish, G. (2005a) China and Thailand Have Broad Prospects for Tourism Cooperation. In *Travelwirenews*, 25 October, www.travelwirenews.com.

Satish, G. (2005b) Chinese Tourism Officials Pay Taiwan a Visit. In *Travelwirenews*, 3 November, www.travelwirenews.com.

Satish, G. (2005c) Chinese Airlines Eye Nonstop Flights between China and Las Vegas. In *Travelwirenews*, 31 October 2005, www.travelwirenews.com.

Schaefer, Kay (2004) Erfolgsfaktoren für Unternehmen auf dem Outbound-Tourismusmarkt China. Lüneburg University, Diploma thesis.

Schein, Louisa (2005) Minorities, Homelands and Methods. In Nyiri, Pal and Breidenbach, Joana (eds) *China Inside Out Contemporary Chinese Nationalism and Transnationalism*. Budapest (CEU Press). 99–140.

Schirmer, Dominique (2004) *Soziologie und Lebensstilforschung in der Volksrepublik China. Perspektiven einer Mikrotheorie gesellschaftlichen Wandels*. Bielefeld (transcript).

Schmeckenberger, Pia (2004) Determinanten des Marketing für den chinesischen Incoming-Tourismus in Deutschland. University of Applied Sciences Ludwigshafen, Diploma thesis.

Schon, Jenny (1972) *China: Im Vertrauen auf die eigene Kraft. Reisebericht einer Genossin*. Berlin (Oberbaumverlag).

Schuler, Alexander (2005) Modernisierung und Tourismus: Transformationsprozesse in der chinesischen Gesellschaft und deren Bedeutung für den Reiseverkehr. Potsdam University, Bachelor thesis.

Schwandner, Gerd, Gu, Huimin (2005) Beer, Romance, and Chinese Airlines. Mindsets and Travel Expectations of Chinese Tourism Students. In Suh, Seung-Jin and Hwang, Yeong-Hyeon: *New Tourism for Asia-Pacific, Conference Proceedings*. Seoul (Asia Pacific Tourism Association). 110–118.

Selwyn, Tom (1995) Landscapes of Liberation and Imprisonment: Towards an Anthropology of the Israeli Landscape. In Hirsch, E. and O'Hanlon, M. (eds) *The Anthropology of the Landscape*. Oxford (Oxford University Press). 114–134.

Selwyn, Tom (ed.) (1996) *The Tourism Image: Myths and Myth Making in Tourism*. Chichester (John Wiley & Sons).

Semmens, Kristin (2005) *Seeing Hitler's Germany. Tourism in the Third Reich*. Houndmills (Palgrave Macmillan).

Setterberg, Per (2005). Purchase Behaviour Analysis of Chinese Outbound Tourists. Paper presented at the Second International Forum on Chinese Outbound Tourism, Beijing, 20/21 November 2005. In *Proceedings of the Second International Forum on Chinese Outbound Tourism, Beijing 2005*. 117–120.

Shambaugh, David (2004) China and Europe: The Emerging Axis. In *Current History*, September 2004. 243–248

Shanghai Daily (2004) Western Australia Wants More Chinese Tourists. In *Shanghai Daily*, 3 July.

Sharpley, Richard (1999) *Tourism, Tourists and Society*. Second edition. Huntingdon (ELM Publications).

Sharpley, Richard and Telfer, David J. (eds) (2002) *Tourism and Development. Concepts and Issues*. Clevedon (Channel View Publications).

Shaw, Gareth and Williams, Allan M. (2004) *Tourism and Tourism Spaces*. London (Sage).

Shen, Fuwei (1996) *Cultural Flow Between China and Outside World Throughout History*. Beijing (Foreign Languages Press).

Shenzhen Daily (2005) Price of Thailand Tour Down to US$ 240 in South China. In *Shenzhen Daily*, 10 January.

Shields, Rob (1991) *Places on the Margin: Alternative Geographies of Modernity*. London (Routledge).

Shu, Tassei (1995) The Establishment and Development of Tourism in China. In Umesao, Tadao, Befu, Harumi and Ishimori, Shuzo (eds) *Japanese Civilization in the Modern World. IX Tourism* (Senri Ethnological Studies No. 38). Osaka (National Museum of Ethnology). 155–167.

Shweder, Richard A. and Levine, Robert A. (eds) (1984) *Culture Theory: Essays on Mind, Self and Emotion*. Cambridge (Cambridge University Press).

Singh, Nitish, Zhao, Hongxin and Hu, Xiaorui (2005) Analyzing the Cultural Content of Web Sites. A Cross-national Comparision of China, India, Japan, and US. In *International Marketing Review*, 22: 2: 129–146.

Smith, Anthony D. (1991) *National Identity*. Las Vegas (University of Nevada Press).

Smith, Kevin J. (1997) Cross-cultural Schemata and Change in Modern China: 'First Fresh Air Comes In and also Flies Come In'. In *Language and Education*, 11: 4: 260–270.

Smith, Valene L. (2003) East to West: The New Wave of Tourism. Paper presented at the Conference on Global Frameworks and Local Realities, University of Brighton, September, Eastbourne.

So, Siu-Ian Amy and Morrison, Alistair M. (2004) The Repeat Travel Market for Taiwan: A Multi-stage Segmentation Approach. In *Asia Pacific Journal of Tourism Research*, 9: 1: 71–87.

Soffel, Christian (2005) Reisehemmungen und Reiseangst im Konfuzianismus der Song- und Ming-Dynastie. Paper presented to the 16th Annual Conference of the German Association for China Studies, Berlin, 2–4 December.

Sofield, Trevor H.B. and Li, Fung Mei Sarah (1998a) Tourism Development and Cultural Policicies in China. In *Annals of Tourism Research*, 25: 2: 362–392.

Sofield, Trevor H.B. and Li, Fung Mei Sarah (1998b) Historical Methodology and Sustainability: An 800-year-old Festival From China. In *Journal of Sustainable Tourism*, 6: 4: 267–292.

Solomon, M.R. (1996) *Consumer Behavior*. Third edition. Upper Saddle River (Prentice-Hall).

Song, Haiyan (2005) Editorial. In *China Tourism Research*, 1: 1: 2–3.

Song, Haiyan and Bao, Jigang (eds) (2005) *Proceedings of the Second PolyU China Tourism Forum and Third China Tourism Academy Annual Conference*. Hong Kong (Hong Kong Polytechnic University and Sun Yat-sen University Guangzhou).

Song, Haiyan and Witt, Stephen F. (2006) Forecasting International Tourist Flows to Macau. In *Tourism Management*, 27: 2: 214–224.

Spence, Jonathan (1997) *God's Chinese Son. The Taiping Heavenly Kingdom of Hong Xiuquan.* London (Flamingo).

Spode, Hasso (1997) Voyage – Wohin die Reise geht. In *VOYAGE Jahrbuch für Reise- und Tourismusforschung*, 1: 1–11.

Srirang, Koson (1991) Preface. In Gerstlacher, Anna, Krieg, Renate and Sternfeld, Eva (eds) *Tourism in the People's Republic of China: A Case Study.* Bangkok (Ecumenical Coalition on Third World Tourism). 3–4.

STA (ed.) (2004) *Kina/China.* Stockholm (STA).

STA (ed.) (2005) *Kina/China.* Stockholm (STA).

Statistikbanken (ed.) (2005) *Denmark Tourism Statistics – China*, www.statistik-banken.dk.

Statistisches Bundesamt (ed.) (2005) *Ankünfte und Übernachtungen aus der VR China und Hong Kong in Deutschland*, www.destatis.de.

STB (ed.) (2002) *Chinese Outbound Travel Market Report 2002.* Tokyo (STB).

STB (ed.) (2004) *Chinese Outbound Travel Market 2004 Update.* Tokyo (STB).

Stone, Mike and Wall, Geoffrey (2003) Ecotourism and Community Development: Case Studies from Hainan, China. In *Environmental Management* 33: 1: 12–24.

Strittmatter, Kai (2004) *Gebrauchsanweisung für China.* München (Piper).

Strizzi, Nicolino (2001) An Overview of China's Inbound and Outbound Tourism Markets. Canadian Tourism Commission research report 2001–2005. Ottawa.

STV (ed.) (2004) *Schweizer Tourismus in Zahlen 2004.* Bern (STV).

Su, Xiaobo and Huang, Chunyuan (2005) The impacts of Heritage Tourism on Public Space in Historic Towns: A Case Study of Lijiang Ancient Town. In *China Tourism Research* 1: 4: 401–442.

Suh, Seung-Jin and Hwang, Yeong-Hyeon (2005) *New Tourism for Asia-Pacific, Conference Proceedings.* Seoul (Asia Pacific Tourism Association).

Sulaiman, Y. (2004) Demand High for Chinese Prostitutes. In *Eturbonews*, 10 July.

Sulaiman, Y. (2005) Taiwan Denied Tourism Association Membership. In *Travelwirenews*, 27 October, www.travelwirenews.com.

Sun, Jing (2004) Möglichkeiten der Kundengewinnung aus dem touristischen Quellmarkt China für europäische Destinationen, insbesondere Deutschland/Mecklenburg-Vorpommern. University of Applied Sciences Stralsund, Bachelor thesis.

Sun, Wangning (2002) *Leaving China: Media, Migration and Translational Imagination.* Lanham, MD (Rowman and Littlefield Publishers).

Sun, Yat-sen (1922) *The International Development of China.* New York (n.p.).

Sun, Yuqing (2003) Current Situation and Strategies of China's Outbound Tourism after China's Entering WTO. In *Jiangxi Social Sciences*, 4: 238–241 (in Chinese).

Sun, Yuqing and Dong, Sihua (2003) Solutions to Development of China's Outbound Tourism. In *Journal of Jiangxi University of Finance and Economics*, 4: 58–59 (in Chinese).

Sussebach, Henning (2004) Im Reich der Stille. In *Die Zeit*, 31: 45.

Swain, Margaret Bryne (1995) A Comparison of State and Private Artisan Production for Tourism in Yunnan. In Lew, Alan A. and Yu, Lawrence (eds) *Tourism in China: Geographical, Political and Economic Perspectives.* Boulder, CO (Westview). 223–233.

Swarbrooke, John (2002) *The Development and Management of Visitor Attractions.* Oxford (Butterworth-Heinemann).

Swarbrooke, John and Horner, Susan (1999) *Consumer Behaviour in Tourism.* Oxford (Butterworth-Heinemann).

Swissinfo (ed.) (2003) Sonnenaufgang für Schweizer Tourismus. In *Swissinfo*, 20 November, www.swissinfo.com.

Switzerland Tourism Board (ed.) (2005) *Tourism Statistics.* Zurich (Switzerland Tourism Board).

Tafarodi, Romin W. *et al.* (2004) The Inner Self in Three Countries. In *Journal of Cross Cultural Psychology*, 35: 1: 97–117.

Tairo, Apolinari (2005) Tanzania Promotes Tourism in Chinese Language Website. In *Travelwirenews*, 25 July, www.travelwirenews.com.

Tan, Chee-Beng, Cheung, Sidney C.H. and Yang, Hui (eds) (2001) *Tourism, Anthropology and China.* Bangkok (White Lotus). (Studies in Asian Tourism No. 1).

Tan, Manni (1989) Die Entwicklung des Tourismus in China. In *China im Aufbau*, August: 38–41.

Tao, Dongfeng (2005) The Commercialised Revolutionary Culture in Contemporary China. In Tao, Dongfeng and Jin, Yuanpu (eds) *Cultural Studies in China.* Singapore (Marshall Cavendish). 69–84.

Tao, Dongfeng and Jin, Yuanpu (eds) (2005) *Cultural Studies in China.* Singapore (Marshall Cavendish).

TBP (ed.) (2004) The China Outbound Travel Market. In *Travel Markets*, special issue, 21: October/November: 1–36.

TDC (ed.) (2006) China to Become World's Biggest Inbound Tourism Nation by 2019, www.tdc-trade.com.

Teng, Weifeng (2005) Risks Perceived by Mainland Chinese Tourist Towards Southeast Asia Destinations: A Fuzzy Logic Model. In *Asia Pacific Journal of Tourism Research*, 10: 1: 97–116.

Teo, Peggy and Li, Lim Hiong (2003) Global and Local Interactions in Tourism. In *Annals of Tourism Research*, 30: 287–306.

Teo, Peggy, Chang, T.C. and Ho, K.C. (eds) (2001a) *Interconnected Worlds. Tourism in Southeast Asia.* Oxford (Elsevier).

Teo, Peggy, Chang, T.C. and Ho, K.C. (2001b) Globalisation and Interconnectedness in Southeast Asian Tourism. In Teo, Peggy, Chang, T.C. and Ho, K.C. (eds) *Interconnected Worlds. Tourism in Southeast Asia.* Oxford (Elsevier). 1–13.

Theobald, William F. (ed.) (2005) *Global Tourism.* Third edition. London (Elsevier).

Therkelsen, Anett (1997) Tourism Marketing and the Role of National Images. In Pinkert, Ernst-Ullrich and Therkelsen, Anette (eds) *Intercultural Encounters in Tourism* (Language and Cultural Contact No. 21). Aalborg (Aalborg University Press). 11–30.

Thiem, Marion (2001) Tourismus und kulturelle Identität. *Aus Politik und Zeitgeschichte* 47: 37–41.

Timothy, Dallen J. (2001) *Tourism and Political Boundaries.* London (Routledge).

Tiplady, Rachel (2004) The Year of the Chinese Tourist. In *Business Week Online*, 13 January, www.businessweek.com.

Tisdell, Clement A. and Wen, Julie (1991) Foreign Tourism as an Element in PR China's Economic Development Strategy. In *Tourism Management*, 12: 1: 55–67.

Torr, Dona (1968) *Marx on China, 1853–1860: Articles from the New York Daily Tribune.* London (Lawrence & Wishart).

TourismAustralia (ed.) (2004) *Hong Kong Outbound Travel Snapshot*. Melbourne (TourismAustralia).

Tourism Authority of Thailand (ed.) (2005) *International Tourist Arrivals to Thailand 1987–2004 – China*. Bangkok (Tourism Authority of Thailand).

Tourism Research HKTB (ed.) (2005) *Visitors Arrival Statistics – January 2005*. Hong Kong (Tourism Research HKTB).

TourismVancouver (2005) Vancouver Positioned as Key Gateway to China. In *TourismVancouver*, 22 January, www.tourismvancouver.com.

Travel Daily News (2004) Outbound Travel: China Takes the Lead. In *Travel Daily News*, 22 April, traveldailynews.com.

Travelperu (2005) Peru Becomes Chinese Tourists' Destination. In *Travelperu*, 10 January, www.travelperu.info.

Travelwirenews (2005a) 500 Chinese Celebrate Chinese New Year at the Footsteps of the Pyramids. In *Travelwirenews*, 13 February, www.travelwirenews.com.

TRCNZ (2005) *Tourism Research Centre New Zealand*, www.trcnz.govt.nz.

Tsang, Nelson and Qu, Hailin (2000) Service Quality in China's Hotel Industry: A Perspective from Tourists and Hotel Managers. In *International Journal of Contemporary Hospitality Management*, 12: 5: 316–326.

Tu, Wei-ming (ed.) (1994a) *The Living Tree. The Changing Meaning of Being Chinese Today*. Stanford, CA (Stanford University Press).

Tu, Wei-ming (1994b) Cultural China: The Periphery as the Center. In Tu, Wei-ming (ed.) *The Living Tree. The Changing Meaning of Being Chinese Today*. Stanford, CA (Stanford University Press). 1–34.

Tuinstra, Fons (2003) *Het andere Oosten*. Amsterdam (Contact).

Turner, Bryan (1989) Research Note: From Orientalism to Global Sociology. In *Sociology*, 23: 4: 629–638.

Umesao, Tadao, Befu, Harumi and Ishimori, Shuzo (eds) (1995) *Japanese Civilization in the Modern World. IX Tourism* (Senri Ethnological Studies No. 38) Osaka (National Museum of Ethnology).

Unger, Jonathan (ed.) (1996) *Chinese Nationalism*. Armonk, NY (Sharpe).

Uriely, Natan (2005) The Tourist Experience. Conceptual Developments. In *Annals of Tourism Research* 32: 1: 199–216.

Urry, John (2000) *Sociology beyond Society. Mobilities for the Twenty-first Century*. London, New York (Routledge).

Urry, John (2002) *The Tourist Gaze*. Second edition. London (Sage).

Urry, John (2003) *Global Complexity*. Cambridge (Polity Press).

Urry, John (2005) The 'Consuming' of Places. In Jaworski, Adam and Pritchard, Annette (eds) *Discourse, Communication and Tourism*. Clevedon (Channel View). 19–27.

Uysal, Muzatter, Lu, Wei and Reid, Leslie M. (1986) Development of International Tourism in PR China. In *Tourism Management*, 7: 2: 113–119.

Verhelst, Veronique (2003) Study of the Outbound Tourism Industry of the People's Republic of China. The Probability of a Bilateral ADS Agreement between the PRC and the Shengen Area. Faculty of Arts, Department of Oriental and Eastern European Studies, Katholieke Universiteit Leuven. Dissertation submitted for obtaining the Master's degree in Chinese Studies.

Verne, Jules (1977) Les Tribulations d'un Chinoise en Chine 1879 Paris (Magnier).

Vienna Tourist Board (ed.) (2005) *Top Destinations in Europe*, www.b2b.wien.info.

Visit Britain (ed.) (2003) *Market Profile China 2003*. London (Visit Britain).

Visit London (ed.) (2005) *China Mission Report 2005*. London (Visit London).

Voyage (2002) Richter, Dieter (ed.) *Voyage, Jahrbuch für Reise- & Tourismus-forschung 2002. Reisen & Essen*. Köln (Dumont).

Wales Tourism Alliance (ed.) (2002) *Proceedings Wales Tourism Alliance Conference, Cardiff 2002*.

Waley, Arthur (1938) *Confucius: The Analects*. New York (Macmillan).

Waley-Cohen, Joanna (2000) *The Sextants of Beijing. Global Currents in Chinese History*. New York (W.W. Norton).

Wall, Geoffrey and Xie, Philip Feifan (2005) Authenticating Ethnic Tourism: Li Dancers' Perspectives. In *Asia Pacific Journal of Tourism Research*, 10: 1: 1–22.

Walton, John K. (ed.) (2005) *Histories of Tourism. Representation, Identity and Conflict*. Clevedon (Channel View Publications).

Wang, Gungwu (1991) *The Chineseness of China. Selected Essays*. Hong Kong (Oxford University Press).

Wang, Huanqian and Mai, Zhanxiong (eds) (1983) *Tourism*. Beijing (Foreign Language Press).

Wang, Jing (ed.) (2005) *Locating China. Space, Place and Popular Culture*. Abingdon (Routledge).

Wang, Jisi (2005) From Paper Tiger to Real Leviathan. China's Images of the United States since 1949. In McGiffert, Carola (ed.): *Chinese Images of the United States*. Washington (CSIS). 9–22.

Wang, Ning (1999) Rethinking Authenticity in Tourism Experience. In *Annals of Tourism Research*, 26: 2: 349–370.

Wang, Ning (2000) *Tourism and Modernity. A Sociological Analysis*. Amsterdam (Pergamon).

Wang, Shuliang (1998) *History of China's Tourism*. Beijing (Tourism Educational Publishing House) (in Chinese).

Wang, Suosheng and Qu, Hailin (2004) A Comparison Study of Chinese Domestic Tourism: China vs the United States. In *International Journal of Contemporary Hospitality Management*, 16: 2: 108–115.

Wang, Wei (2003c) *Temporary Space. An Experiment by Wang Wei*. Beijing (25000 Cultural Transmissions Centre).

Wang, Xinjun (ed.) (2003a) *Proceedings of 1st International Forum on Chinese Outbound Tourism and Marketing. Shenzhen 17–18 Nov. 2003, Shenzhen*.

Wang, Xinjun (2003b) A Study on the Development of China's Outbound Tourism Market. In Wang, Xinjun (ed.) *Proceedings of 1st International Forum on Chinese Outbound Tourism and Marketing. Shenzhen 17–18 Nov. 2003, Shenzhen*. 86–101.

Wang, Xinjun and Liang, Zhi (2005) The Current Situation and the Characteristics of China's Outbound Tourism Market. In *Proceedings of the Second International Forum on Chinese Outbound Tourism (IFCOT)*. Beijing, 20/21 November 2005, Beijing. 66–72.

Warden, Clyde A. *et al.* (2002) Service Failures Away from Home: Benefits in Intercultural Service Encounters. In *International Journal of Service Industry Management*, 14: 4: 436–457.

Wassmann, Jürg (ed.) (1998) *Pacific Answers to Western Hegemony. Cultural Practices of Identity Construction*. New York (Berg).

Watts, Jonathan (2005) Satellite Data Reveals Beijing as Air Pollution Capital of World. In *Guardian*, 31 October.

Wehrfritz, Georg (2005) Second Thoughts. China's Too-hot Economy is Prompting Firms to Look Elsewhere to Invest. In *Newsweek* CXLVI, 28 November 2005, Asian edition. 22.

Wei, Feng and Wei, Chen (2005) A Study on the Trends in China Outbound Tourism Market. In *Proceedings of APTA Conference South Korea 2005, Seoul*.

Weiermair, Klaus (2000) Tourists' Perceptions Towards and Satisfaction with Service Quality in the Cross-cultural Service Encounter: Implications for Hospitality and Tourism Management. In *Managing Service Quality*, 10: 6: 397–409.

Wen, Chen (2004) Chinesische Touristen haben Urlaub in Europa im Visier. In *Beijing Rundschau* 13.

Wen, Jie (1998) Evaluation of Tourism and Tourism Resources in China. Existing Methods and their Limitations. In *International Journal of Social Economics*, 25: 2–4: 467–485.

Wen, Julie Jie and Tisdell, Clement A. (2001) *Tourism and China's Development. Policies, Regional Economic Growth and Ecotourism*. Singapore (World Scientific Publishing).

Wen, Ying and Zhao, Jing (2004) Quellmarkt China – ein Zukunftsmarkt in Deutschland. Technical University Dresden, thesis.

Wen, Zhang (1997) China's Domestic Tourism: Impetus, Development and Trends. In *Tourism Management* 18: 8: 565–571.

Weyhereter, Björn and Yang, Kezhu (2005) Die Chinesen kommen. Chinas Outbound-Tourismus. University of Applied Sciences Heilbronn, Diploma thesis.

Wien Tourismus (ed.) (2004) *Wien Tourismus Marketing 2005*. Wien (Wien Tourismus).

Wilks, Jeffrey, Pendergast, Donna and Wood, Maryann (2003) Accidental Deaths of Overseas Visitors in Australia 1997–2000. In *Journal of Hospitality and Tourism Management*, 10: 1: 79–90.

Williams, A.M. and Hall, C. Michael (eds) (2002) *Tourism and Migration*. London (Routledge).

Williams, Stephen (1998) *Tourism Geography*. London (Routledge).

Williamson, Hugh (2003) Chinese Tourists Touch Down in Germany. In *Financial Times*, 6 April.

Williamson, Hugh (2006) Chinese Bring New Challenges for Europe's Tourist Industry. In *Financial Times*, 11 March 2006.

Winter, Tim (2005) *Call for Papers for Workshop on 'Of Asian Origin: Rethinking Tourism in Contemporay Asia'*. Singapore (Asia Research Institute).

Wong, Chak-keung Simon and Kwong, Wai-Yan Yan (2004) Outbound Tourists' Selection Criteria for Choosing All-inclusive Package Tours. In *Tourism Management*, 25: 581–592.

Wong, Kwan Yiu (1982) *Shenzhen Special Economic Zone: China's Experiment in Modernization*. Hong Kong (Hong Kong Geographical Association).

Wong, Simon and Lau, Elaine (2001) Understanding the Behaviour of Hong Kong Chinese Tourists on Group Tour Packages. In *Journal of Travel Research*, 40: 1: 57–67.

WTO (ed.) (2000) *Tourism 2020 Vision*. Volume 3. East Asia and the Pacific. Madrid (WTO).

WTO (ed.) (2003) *Study Into Chinese Outbound Tourism*. Madrid (WTO).

WTTC (ed.) (2003) *The Impact of Travel and Tourism on Jobs and the Economy. China and China Hong Kong SAR*. London (WTTC).

Wu, Bihu, Zhu, Hong and Xu, Xiaohuan (2000) Trends in China's Domestic Tourism Development at the Turn of the Century. In *International Journal of Contemporary Hospitality Management*, 12: 5: 296–299.

Wu, David Yen-ho (1994) The Construction of Chinese and Non-Chinese Identities. In Tu, Wei-ming (ed.) *The Living Tree. The Changing Meaning of Being Chinese Today*. Stanford, CA (Stanford University Press). 148–167.

Wu, Kuang-Ming (1989) Chinese Aesthetics. In Allison, Robert E. (ed.) *Understanding the Chinese Mind. The Philosophical Roots*. Hong Kong (Oxford University Press). 236–264.

Wu, Ximing and Perloff, Jeffrey M. (2004) China's Income Distribution over Time: Reasons for Rising Inequality. Paper of the Department of Economics, University of Guelph, Ontario.

Xiao, Honggen (2000) China's Tourism Education into the 21st Century. In *Annals of Tourism Research*, 27: 4: 1052–1055.

Xiao, Honggen (2003) Leisure in China. In Lew, Alan A. *et al.* (eds) *Tourism in China*. New York (Haworth). 263–276.

Xiao, Honggen and Huyton, Jeremy R. (1996) Tourism and Leisure: An Integrative Case in China. In *International Journal of Contemporary Hospitality Management*, 8: 6: 18–24.

Xiao, Jiancheng and Ren, Jiangmin (2003) An Analysis of the Current Chinese Outbound Tourism Market and Development Policies. In *Journal of Yunnan Geographical Environment Research*, 15: 1: 2–11 (in Chinese).

Xie, Philip Feifan (2003) The Bamboo-beating Dance in Hainan, China: Authenticity and commodification. In *Journal of Sustainable Tourism*, 11: 1: 5–16.

Xie, Philip Feifan and Wall, Geoffrey (2002) Visitor's Perceptions of Authenticity at Cultural Attractions in Hainan, China. In *International Journal of Tourism Research*, 4: 5: 353–366.

Xie, Shuguang and He, Kan (2002) *A Guide to Business Travelling in Europe: From Europe to Mediterranean*. Beijing (Social Sciences Documentation Publishing House) (in Chinese).

Xinhua (2003) 8 More African States Become Approved Destinations for Chinese Tourists. In *Xinhua News*, 16 December 2003, www.china.org.cn.

Xinhua (2004) Eiffel Tower Illuminated in Red to Honor China. In *Xinhua News*, 25 January, www.china.org.cn, Beijing.

Xinhua (2005) Egypt Eyes Broad Tourism Cooperation with China. In *Xinhua News* 26 January, www.china.org.cn, Beijing.

Xinhua (2006a) *China's GDP Grows 9.9% in 2005*, www.news.xinhuanet.com.

Xinhua (2006b) *Macao Sees Stronger Visitor Flow in Spring Festival*, www.china. org.cn. 7 February 2006.

Xu, Chongyun and Shao, Zheng (1984) Socio-cultural Impacts of Tourism. In *Tourism Economy*, May: 7–12 (in Chinese).

Xu, Fan (2005) Marketing on China Outbound Tourism. In *Proceedings of the Second International Forum on Chinese Outbound Tourism (IFCOT)*. Beijing, 20/21 November 2005, Beijing. 83–90.

Xu, Gang (1999) *Tourism and Local Economic Development in China. Case Studies of Guilin, Suzhou and Beidaihe*. Richmond (Curzon Press).

Xu, Gang and Kruse, Claudia (2003) Economic Impact of Tourism in China. In Lew, Alan A. *et al.* (eds) *Tourism in China*. New York (Haworth). 83–102.

Xu, Q. and Chen, H. (2003) A Probe into the Tourism Marketing Policies of the Chinese Aged. In *Transaction of Guilin Tourism Higher Technological Academy*, 4: 36–39 (in Chinese).

Xu, Shengli (ed.) (2002) *Marktinformation China*. Beijing (German National Tourist Office).

Xu, Shengli (ed.) (2005) *Marktinformation China*. Beijing (German National Tourist Office).

Xu, Xin (2002) The Use and Abuse of 'Essentialist' Formulations in East-West and Sino-American Relations. In *Ritsumeikan Journal of Asia Pacific Studies*, 12: 23–30.

Xue, Muqiao (1981) *Almanac of China's Economy*. Hong Kong (Eurasia Press).

Yamashita, Shinji, Din, Kadir H. and Eades, James S. (eds) (1997) *Tourism and Cultural Development in Asia and Oceania*. Bangi (Penerbit Universiti Kebangsaan Malaysia).

Yatsko, Pamela and Tasker, Rodney (1998) Outward Bound: Just in Time, Ordinary Chinese Catch the Travel Bug. In *Far Eastern Economic Review*, 26 March.

Yau, Oliver H.M. (1988) Chinese Cultural Values: Their Dimensions and Marketing Implications. In *European Journal of Marketing*, 22: 5: 44–57.

Yi, Shaohua (2001) Protocols of WTO and Adjustment of China's Outbound Travel Policies. In *International Economics and Trade Research*, 5: 24–27 (in Chinese).

You, Xinran (2001) A Cross Cultural Comparison of Travel Push and Pull Factors: United Kingdom and Japan. In *International Journal of Hospitality and Tourism Administration*, 1: 3: 3–19.

Yu, Larry (1997) Travel Between Politically Divided China and Taiwan. In *Asia Pacific Journal*, www.hotel-online.com.

Yu, Xin and Weiler, Betty (2000) Chinese Pleasure Travellers in Australia: An Analysis of Perceived Importance and Levels of Satisfaction. In Ewen, M. (ed.) *Peak Performance in Tourism and Hospitality Research, 2000 Australian Tourism and Hospitality Research Conference*. 2–5 February, La Trobe University, Melbourne. 242–253.

Yu, Xin and Weiler, Betty (2001) Mainland Chinese Pleasure Travellers to Australia: A Leisure Behavior Analysis. In *Tourism, Culture & Communication*, 3: 81–91.

Yu, Xin, Weiler, Betty and Ham, Sam (2001) Intercultural Communication and Mediation: A Framework for Analysing the Intercultural Competence of Chinese Tour Guides. In *Journal of Vacation Marketing*, 8: 1: 75–87.

Yu, Y., Zhang, J. and Ren, L. (2003) A Study on the Tour Behaviour Decision – A Case of the Aged Tourism Market in Jiangxi Province. In *Tourism Tribune*, 3: 38–41 (in Chinese).

Yu, Yongding (2004) China's Macroeconomic Situation since 2003. In *China & World Economy*, 12: 4: 3–20.

Yuk, Pan Kwan (2005) Tourism Warning as China Relaxes Visa Rules. In *Financial Times*, 26 July.

Zhang, Guangrui (1985) China Ready for New Prospect for Tourism Development. In *Tourism Management*, 6: 2: 141–143.

Zhang, Guangrui (1989) Ten Years of Chinese Tourism: Profile and Assessment. In *Tourism Management*, 10: 1: 51–62.

Zhang, Guangrui (1993) Tourism crosses the Taiwan Straits. In *Tourism Management*, 14: 3: 228–231.

Zhang, Guangrui (2003a) China's Tourism since 1978: Policies, Experiences and Lessons learned. In Lew, Alan A. *et al.* (eds) *Tourism in China*. New York (Haworth). 13–34.

Zhang, Guangrui (2003b) Tourism Research in China. In Lew, Alan A. *et al.* (eds) *Tourism in China*. New York (Haworth). 67–82.

Zhang, Guangrui (2005) China's Outbound Tourism (2003–2005): Present Situation and Future Perspectives (in Chinese). In: *Green Book of China's Tourism 2003–2005 China's Tourism Development: Analysis and Forecast (2003–2005)*. Beijing (Social Sciences Academic Press). (*Green Book of China's Tourism No. 4*), 70–86 (in Chinese).

Zhang, Guangrui, Pine, Ray and Zhang Hanqin Qiu (2000) China's International Tourism Development: The Present and Future. In *International Journal of Contemporary Hospitality Management*, 12: 5: 282–290.

Zhang, Guangrui, Wei, Xiaoan and Liu, Deqian (eds) (2005) *Green Book of China's Tourism 2003–2005. China's Tourism Development: Analysis and Forecast (2003–2005)*. Beijing (Social Sciences Academic Press). (*Green Book of China's Tourism No. 4*).

Zhang, Hanqin Qiu (2004) Accession to the World Trade Organization: Challenges for China's Travel Service Industry. In *International Journal of Contemporary Hospitality Management*, 16: 6: 369–372.

Zhang, Hanqin Qiu and Chow, Ivy (2004) Application of Importance-Performance Model in Tour Guides' Performance: Evidence from Mainland Chinese Outbound Visitors in Hong Kong. In *Tourism Management*, 25: 81–91.

Zhang, Hanqin Qiu and Heung, Vincent C.S. (2001) The Emergence of the Mainland Chinese Outbound Travel Market and its Implications for Tourism Marketing. In *Journal of Vacation Marketing*, 8: 1: 7–12.

Zhang, Hanqin Qiu and Lam, Terry (1999) An Analysis of Mainland Chinese Visitors' Motivations to Visit Hong Kong. In *Tourism Management*, 20: 5: 587–594.

Zhang, Hanqin Qiu and Lam, Terry (2004) Human Resources Issues in the Development of Tourism in China: Evidence from Heilongjiang Province. In *International Journal of Contemporary Hospitality Management*, 16: 1: 45–51.

Zhang, Hanqin Qiu and Qu, Hailin (1996) The Trends of China Outbound Travel to Hong Kong and their Implications. In *Journal of Vacation Marketing*, 2: 4: 373–381.

Zhang, Hanqin Qiu and Wu, Ellen (2004) Human Resources Issues Facing the Hotel and Travel Industry in China. In *International Journal of Contemporary Hospitality Management*, 16: 7: 424–428.

Zhang, Hanqin Qiu, Chong, King and Ap, John (1999) An Analysis of Tourism Policy Development in Modern China. In *Tourism Management*, 20: 471–485.

Zhang, Haiqin Qiu, Chong, King and Jenkins, C.L. (2002) Tourism Policy Implementation in Mainland China: An Enterprise Perspective. In *International Journal of Contemporary Hospitality Management*, 14: 1: 38–42.

Zhang, Hanqin Qiu, Jenkins, Carson L. and Qu, Hailin (2003) Mainland Chinese Outbound Travel to Hong Kong and Its Implications. In Lew, Alan A. *et al.* (eds) *Tourism in China*. New York (Haworth). 277–293.

Zhang, Hanqin Qiu, Lam, Terry and Bauer, Thomas (2001) Analysis of Training and Education Needs of Mainland Chinese Tourism Academics in the Twenty-first Century. In *International Journal of Contemporary Hospitality Management*, 13: 6: 274–279.

Zhang, Hanqin Qiu, Pine, Ray and Lam, Terry (2005) *Tourism and Hotel Development in China*. New York (Haworth).

Zhang, Hanqin Qiu, Qu, Hailin and Tang, Venus Mo Yin (2004) A Case Study of Hong Kong Residents' Outbound Leisure Travel. In *Tourism Management*, 25: 267–273.

Zhang, Jianzhong (2003c) Researches on Current Status, Development Trend and Related Policies on the Outbound Tourism in China. In Wang, Xinjun (ed.) *Proceedings of 1st International Forum on Chinese Outbound Tourism and Marketing. Shenzhen 17–18 Nov. 2003, Shenzhen.* 75–79.

Zhang, Kevin H. (2004) The Evolution of China's Urban Transformation: 1949–2000. In Chen, Aimin, Liu, Gordon G. and Zhang, Kevin H. (eds) *Urban Transformation in China*. Aldershot (Ashgate Publishing). 25–39.

Zhang, Lili (1998) *An Economic Probe into the Chinese Tourism Development in Recent Times*. Tianjin (Tianjin Publishing House) (in Chinese).

Zhang, Liqing (2004) Coping with China's Balance of Payments Surplus: Why and How? In *China & World Economy*, 12: 4: 79–87.

Zhang, Wen (1997) China's Domestic Tourism: Impetus, Development and Trends. In *Tourism Management*, 18: 8: 565–571.

Zhang, Zeyu (1980) China's Growing Tourism Industry. In *Beijing Review*, 23: 27: 14–17.

Zhao, Jian (1989) Overprovision in Chinese Hotels. In *Annals of Tourism Research*, 10: 1: 63–65.

Zhao, Suisheng (ed.) (2000a) *China and Democracy. Reconsidering the Prospects for a Democratic China*. London (Routledge).

Zhao, Suisheng (2000b) Chinese Nationalism and Authoritarianism in the 1990s. In Zhao, Suisheng (ed.) *China and Democracy. Reconsidering the Prospects for a Democratic China*. London (Routledge). 253–270.

Zhao, Xinluo (1994) Barter Tourism Along the China-Russia Border. In *Annals of Tourism Research*, 21: 2: 401–403.

Zheng, Taylor Tie (2004) *China – The Future Travel Market*. Global Refund presentation, Schiphol (Global Refund).

Index